Green Energy and Technology

Climate change, environmental impact and the limited natural resources urge scientific research and novel technical solutions. The monograph series Green Energy and Technology serves as a publishing platform for scientific and technological approaches to "green"—i.e. environmentally friendly and sustainable—technologies. While a focus lies on energy and power supply, it also covers "green" solutions in industrial engineering and engineering design. Green Energy and Technology addresses researchers, advanced students, technical consultants as well as decision makers in industries and politics. Hence, the level of presentation spans from instructional to highly technical.

Indexed in Scopus.

Indexed in Ei Compendex.

Fatima Aderouane

Design, Construction, and Global Performance Analysis of Offshore Wind Turbines

Springer

Fatima Aderouane
Tangiers, Morocco

ISSN 1865-3529 ISSN 1865-3537 (electronic)
Green Energy and Technology
ISBN 978-3-031-86009-6 ISBN 978-3-031-86010-2 (eBook)
https://doi.org/10.1007/978-3-031-86010-2

This Springer imprint is published by the registered company Springer Nature Switzerland AG
The registered company address is: Gewerbestrasse 11, 6330 Cham, Switzerland

Bottom-founded and Floating Offshore Wind Turbines

Offshore wind foundation types. From left to right: monopile, jacket, twisted jacket, tension-leg floating platform, semi-submersible platform, and spar-buoy. Some of these foundations will be discussed below. *Source* wikimedia

Preface

This book is intended to provide genericised criteria for the design, construction, installation, and survey of Bottom-founded Offshore Wind Turbines, which possess permanently sited substructures and foundations of offshore wind turbines attached on and supported by the sea floor; and criteria for the design, construction, installation, and survey of permanently sited Floating Offshore Wind Turbines. Accordingly, this book addresses three principal areas: the permanent or floating substructure, the station keeping system and onboard machinery, equipment, and systems including applicable marine systems and associated equipment and machinery, safety systems and associated equipment, and lifesaving appliances and machinery. The requirements for ancillary offshore wind farm installations such as meteorological measuring towers, accommodation units, and transformer platforms are not addressed in this book.

Tangiers, Morocco
Fatima Aderouane

Acknowledgements The Class Rules referenced throughout this book are adopted on behalf of, and with the permission from, the American Bureau of Shipping. With grateful thanks to the editorial team at Springer Nature and Straive, and to my colleagues and friends for their time ensuring this book is technically correct and current.

Competing Interests The author has no competing interests to declare that are relevant to the content of this manuscript.

Preface

Contents

Part II Floating Offshore Wind Turbines

Abbreviations and Acronyms

ACI	American Concrete Institute
AF	Apparent Fixity Model
AISC	American Institute of Steel Construction
API	American Petroleum Institute
ASTM	American Society for Testing and Materials
AWS	American Welding Society
BEM	Blade Element Momentum Method
BOWT	Bottom-founded Offshore Wind Turbine
CFD	Computational Fluid Dynamics
DEL	Damage Equivalent Load
DLC	Design Load Case
DLL	Dynamic Link Library
DOF	Degree of Freedom
DS	Dynamic Stall
DW	Dynamic Wake
FE(M)	Finite Element (Method)
FEA	Finite Element Analysis
FOWT	Floating Offshore Wind Turbine
FVW	Free Vortex Wake
GBF	Gravity-Based Foundation
IEC	International Electrotechnical Commission
ILA	Integrated Load Analysis
ISO	International Organization for Standardization
LDD	Load Duration Distribution
NACE	National Association of Corrosion Engineers
NPD	Norwegian Petroleum Directorate
PISA	Pile Soil Analysis Method
PSI	Pile-Soil Interaction
QTF	Quadratic Transfer Function
RAO	Response Amplitude Operator
RFC	Rain Flow Counting

RNA	Rotor-Nacelle Assembly
SSI	Soil–Structure Interaction
TLP	Tension Leg Platform
VIM	Vortex-Induced Motion
VIV	Vortex-Induced Vibration
WTG	Wind Turbine Generator (RNA and Tower)

Nomenclature

1P	Rotor Frequency
2D	Two-dimensional
3D	Three-dimensional
3P	Blade-passing Frequency (for 3-blade wind turbine)
nP	Multiples of Rotor Frequency by n times (n=1,2,3…)

List of Figures

List of Tables

Chapter 12

Chapter 14

Chapter 15

Chapter 18

Chapter 19

Chapter 22

Chapter 23

Part I
Bottom-Founded Offshore Wind Turbines

Chapter 1
General

1 Introduction

This Part 1 provides criteria for the design, construction, installation, and survey of Bottom-Founded Offshore Wind Turbines, which possess permanently sited substructures and foundations of offshore wind turbines attached on and supported by the sea floor. The criteria applicable to floating offshore wind turbines are provided in Part 2. The requirements for ancillary offshore wind farm installations such as meteorological measuring towers, accommodation units, and transformer platforms are not addressed in this book. For the requirements for such bottom-founded installations, refer to the applicable Class Rules for the design, construction, and classification of Offshore Installations.

2 Scope and Conditions of Classification

The general requirements for the conditions of classification are typically contained in the Class Rules relating to the conditions for classification of offshore units and structures.

3 Classification Symbols and Notations

The following Class notations are designated for the Bottom-founded Offshore Wind Turbine. Additionally, at the request of the Owner, the Rotor-Nacelle Assembly (RNA) and tower may be included in the classification. An optional notation for the RNA and the tower will be added to Class Record upon the satisfactory compliance with the requirements given in Chap. 1, Sect. 4.1. Where the RNA and tower are not

F. Aderouane, *Design, Construction, and Global Performance Analysis of Offshore Wind Turbines*, Green Energy and Technology, https://doi.org/10.1007/978-3-031-86010-2_1

included in the classification, the RNA and tower installed on the Bottom-founded Offshore Wind Turbine are required to have a type certificate in accordance with IECRE OD-501 or other recognised standards. The type certificate will be reviewed by Class solely to confirm it is current and verify that the information of the installed RNA and tower is consistent with the design information, criteria and limitations considered in the classification of the Bottom-founded Offshore Wind Turbine. Class will not review or be responsible for the accuracy of the type certificate. Where the tower and its connection to the RNA are not covered by the wind turbine type certificate and the optional class notation is not requested, the tower and its connection to the RNA are to be included in the scope of classification. Where interface structures of the tower connecting to the substructure are not covered by the wind turbine type certificate and the optional class notation is not requested, these interface structures are to be included in the scope of classification. Use of the RNA which does not have a type certificate or the wind turbine having a non-horizontal-axis configuration are subject to special consideration by Class. Design documents are to be submitted for review. Scope of the design review is to be defined on a case-by-case basis. In this book, the Bottom-founded Offshore Wind Turbine is considered as an unmanned structure.

3.1 Offshore Wind Turbines Built Under Class Survey

Offshore wind turbines built and constructed to the satisfaction of the Class Surveyors and to the requirements of this book or to their equivalent, where approved by Class may be classed, and distinguished in the Class Record by an appropriate symbol. For example:

A1 Offshore Wind Turbine (bottom-founded)

4 Additional Class Notations

4.1 Notation for Rotor-Nacelle Assembly (RNA) and Tower

The RNA and the tower may be considered for the classification at the Owner's request. A Bottom-founded Offshore Wind Turbine may be classed and distinguished in the Record by the optional class notation RNA where:

(1) The RNA and the tower comply with the Class Type Approval requirements.
(2) The site-specific assessment demonstrates, as a minimum, which loads and deflections calculated for the Bottom-founded Offshore Wind Turbine under the site-specific conditions do not exceed those calculated for the RNA and the tower approved by the Class Type Approval.

4.2 Design Fatigue Life

4.2.1 New Construction

The design fatigue life value notation is assigned and published in the Class Record. The "design fatigue life" refers to the target value set by the owner or designer, not the value calculated in the analysis. The design fatigue life is to be verified to comply with the fatigue criteria set by Class, and as outlined in this book. The class notation FL (number of years), Year will be assigned to identify the design fatigue life in years and the year of maturation of fatigue life in the defined site. For example, FL (25), 2045 for a wind turbine installation built in 2020 if the minimum design fatigue life specified is 25 years. The class notation FL (number of years), Year is mandatory to be assigned and published in the Class Record.

4.2.2 Life Extension of Bottom-Founded Offshore Wind Turbines

When the operating life of a Bottom-founded Offshore Wind Turbine exceeds the design fatigue life specified in the FL (number of years), Year notation for which it was classed, an evaluation is to be made and appropriate actions are to be taken to extend the use of the Bottom-founded Offshore Wind Turbine up to the new operating life under the site-specific conditions. If the environmental criteria for the site or operating area have been revised since the original approval due to new environmental data or changing environmental conditions, the Coastal State may require the use of new environmental criteria for the life extension. For the life extension of a Bottom-founded Offshore Wind Turbine, the following actions will be required:

- Structural strength analysis.
- Design review and surveys related to any major modifications affecting class items.
- Verification from the original fatigue analysis that the actual fatigue values of all the structural elements are still higher than the proposed extended fatigue life.
- New fatigue analysis covering all the structural elements in accordance with FL requirements, as applicable.
- Identification of structural elements or details with a fatigue life below the new intended design fatigue life and proposed actions to increase the fatigue life of those elements or details.
- Design review and surveys of structural modifications proposed because of the fatigue analysis.
- Enhanced survey programme to monitor those structural elements or details with lower fatigue life which cannot be modified or renewed on site.
- Special survey, including underwater inspection, to determine the structural condition at the time of the life extension.

Survey requirements for extension of use are given in Chap. 7, Sect. 3.1.10. After the life extension is completed in accordance with this book, the existing FL notation with total number of fatigue life and year of maturation is to be updated accordingly.

4.3 Rules for Classification

4.3.1 Application

The requirements in this book are applicable to the Bottom-founded Offshore Wind Turbine. The design criteria specified in this book are intended for the Bottom-founded Offshore Wind Turbine to achieve the normal safety class as defined in IEC 61,400–3-1 (2019). This book is applicable to those features that are permanent in nature and can be verified by plan review, calculation, physical survey, or other appropriate means. Any statement in this document regarding other features is to be considered as guidance to the designer, Fabricator, and Owner.

4.3.2 References

References are made in this book to genericised Class Rules and other criteria issued by Class and other organisations. Unless otherwise noted, the applicable edition of a reference is the one officially issued and available on the date the Agreement for Classification is accepted by Class. Where an edition or date associated with a reference is given, it means that edition is relevant to the topic being presented in this book. Class may consider at its discretion, upon the request of the Owner, the application of other appropriate alternative methods and recognised codes of practice.

4.3.3 Alternatives

Any departure from the requirements of this book may be considered by Class based on suitable engineering analyses or risk assessments. In the case of such departures, classification is subject to Class's approval upon a demonstration of acceptable levels of safety in line with the principles of this book and recognised and accepted current offshore wind industry practice. Risk acceptance criteria are subject to approval by Class. Refer to the Class guidance on risk assessment application for the marine and offshore industries contains an overview of risk assessment techniques and additional information. Using a risk assessment approach to justify alternatives may be applicable either to the Bottom-founded Offshore Wind Turbine as a whole or to individual systems, subsystems, or components. As appropriate, remote hazards outside of the bounds of the system under consideration must be considered. Such account must include incidents relating to remote hazards directly affecting or being influenced by the system under consideration. Class will consider the application of risk-based techniques in the design of the Bottom-founded Offshore Wind Turbine as well as surveys during construction and surveys for maintenance of class. Portions of the Bottom-founded Offshore Wind Turbine not included in the risk assessment are to comply with the applicable parts of the Class Rules. The following are the responsibility of the Owner:

(1) Risk acceptance criteria.
(2) Hazard identification.
(3) Risk assessment.
(4) Risk management.
(5) Compliance with the applicable requirements of the Coastal State or other governmental authorities.

5 Design Documentation to Be Submitted to Class

The design documentation to be submitted is to describe the data, tools, procedures and methodologies of design and analysis which are employed to establish the design of the Bottom-founded Offshore Wind Turbine. The intended design life is also to be stated. Where model testing is used as a basis for a design, the applicability of the test results depends on the demonstration of the adequacy of the methods employed, including enumeration of sources of error, limits of applicability, and methods of extrapolation to full scale. Preferably, procedures are to be reviewed and agreed upon before model testing is done. As required in the subsequent paragraphs, calculations are to be submitted to demonstrate the sufficiency of the proposed design. Such calculations are to be presented in a logical and well-referenced fashion employing a consistent system of units. Where the calculations are in the form of computer analysis, the submission is to provide input and output data with computer generated plots for the structural model. A programme description (not code listings), user manuals, and the results of programme verification sample problems may be required to be submitted. The design documentation to be submitted is to include the reports, calculations, plans, specifications, and other documentation where applicable. The extensiveness of the submitted documentation is to reflect:

(1) The uniqueness of a specific design of the Bottom-founded Offshore Wind Turbine within an offshore wind farm.
(2) The level of experience with conditions in an area where the Bottom-founded Offshore Wind Turbine is to be located.

Design documentation should be submitted electronically to Class.

5.1 Reports

Reports by consultants and other specialists used as a basis for design are to be submitted for review. The contents of reports on offshore wind farm conditions, environmental considerations, foundation data, and materials are, in general, to comply with the recommended list of items given below.

5.1.1 Offshore Wind Farm Conditions

A report on offshore wind farm conditions is to present the configuration of an offshore wind farm and exact locations of all individual offshore wind turbines, transformer platform, service and accommodation units and any other supporting structures and facilities in the offshore wind farm where applicable. The report is also to contain information on wind turbine RNAs, particularly those properties that are used as the input or as the basis of the input for the load calculation and structural assessment for the Support Structure.

5.1.2 Environmental Considerations

Reports on environmental considerations, as described in 3/1, are to describe all environmental phenomena appropriate to the areas of construction, transportation, installation, maintenance, and repair. The types of environmental phenomena to be accounted for, as appropriate to the type and location of the Bottom-founded Offshore Wind Turbine, include wind, waves, current, temperature, tide, marine growth, chemical components and density of air and water, snow and ice, earthquake, and other pertinent phenomena. The establishment of environmental conditions is to be based on appropriate original data or, when permitted, data from analogous areas. Demonstrably valid statistical models are to be employed to perform the extrapolation to long-term values. Calculations required to establish the pertinent design environmental parameters are to be submitted for review. The report on environmental considerations is also to contain the calculations which quantify the effects or loadings on the Bottom-founded Offshore Wind Turbine if they are not provided in other documentation.

5.1.3 Foundation Data

Reports on foundation data are to present the findings of investigations or, when applicable, data from analogous areas on geophysical, geological, and geotechnical considerations existing at and near the installation site of an offshore wind turbine. The way such data is established and the specific items to be assessed are to comply with Chap.5, Sects. 4.1, 4.2 and 4.3. The report is to contain a listing of references to cover the investigation, sampling, testing, and interpretive techniques employed during and after the site investigation. The report is to include a description of the predicted soil-structure interaction, such as $p - y$ data, to be used in the design. As appropriate to the planned structure, the items which may be covered in the reports are axial and lateral pile capacities and response characteristics, the effects of cyclic loading on soil strength, scour, settlements and lateral displacements, dynamic interaction between soil and structure, the capacity of pile groups, slope stability, bearing and lateral stability, soil reactions on the structure, and penetration resistance. Recommendations relative to any special anticipated problem regarding installation are to be included in the report. Items such as the following are to be included, as appropriate: hammer sizes, soil erosion during installation, bottom preparation, and procedures to be followed in the case that pile installation procedures significantly deviate from those anticipated. The documentation for the foundation design is to be submitted in accordance with Chap.5, Sect. 4.

5.1.4 Materials and Welding

Reports on structural materials and welding may be required for metallic structures, concrete structures, or welding procedures where materials or procedures do not conform to those provided in Chap. 2 of this book. For metallic structures, where it is intended to employ new alloys not defined by a recognised specification, reports are to be submitted indicating the adequacy of the material's metallurgical properties, fracture toughness, yield and tensile strengths, and corrosion resistance, with respect to their intended application and service temperatures. For concrete gravity foundation structures, when it is not intended to test or define material properties in accordance with applicable standards of the American Society for Testing and Materials (ASTM) as listed in Chap.2, Sect. 1 of this book, a report is to be provided indicating the standards to be employed and their relative adequacy with respect to the corresponding ASTM standards.

5.2 Design Data and Calculations

Design and analysis calculations are to be submitted for items relating to loadings and structural responses for in-place and marine operations, structural dynamic properties, and foundation designs. Calculations are in general to include the items listed below. Calculations to be provided in association with environmental considerations and foundation data are described in Chap. 1, Sect. 5.

5.2.1 Loadings

Calculations for loadings are to be submitted in accordance with Chap. 4.

5.2.2 Structural Dynamic Properties

Calculations of natural periods of the Support Structure are to be submitted for review. A resonance diagram (Campbell diagram) depicting the relationship between the rotor speeds and the natural periods of turbine components, and the Support Structure is to be submitted. As applicable, the calculation of vibration amplitudes, velocities, and accelerations of the Support Structure may also be required.

5.2.3 Structural Responses

The stress and deflection calculations to be submitted are to include nominal element or member stresses and deflections. Calculations are also required for stresses in localised areas and structural joints, dynamic responses of the structure, and fatigue life of critical members and joints. For a pile-supported structure, calculations for stresses in each pile and the load capacity of the connection between the structure and the pile are to be submitted. For a gravity structure, calculations are to be submitted for the effects of soil's reaction on the foundation structure. For a Self-Elevating Unit to be used as the Substructure of an offshore wind turbine, the applicable calculations required in the Class Rules for the design, construction, and classification of mobile offshore units are to be submitted.

5.2.4 Marine Operations

Calculations are to be submitted in compliance with Chap.5, Sect. 4 of this book. For structural responses resulting from the marine operations described in Chap.5, Sect. 4.7, calculations are to demonstrate the adequacy of structural elements, members, or local structure, as deemed necessary. In addition, the calculations are to demonstrate, as applicable, that deflections resulting from applied loadings and overall structural displacement and settlement do not impair the structural performance of the Support Structure.

5.2.5 Other Calculations

For the Bottom-founded Offshore Wind Turbine carrying the optional RNA notation, a site-specific assessment is to be performed to calculate the loads and deflections of the type-approved RNA and tower under site-specific conditions. As required, additional calculations which demonstrate the adequacy of an overall design may be required. Such calculations are to include those performed in the design of corrosion protection system.

5.3 *Plans and Specifications*

Plans or specifications depicting or describing arrangements and details of major items of the Bottom-founded Offshore Wind Turbine are to be submitted for review or approval in a timely manner. Where deemed appropriate, and when requested by the Owner, a schedule for information submission and plan approval can be jointly established by the Owner and Class. This schedule, which affects the time required for review of submitted data and Class will adhere to as far as reasonably possible, is to reflect the construction schedule and the complexity of the Bottom-founded Offshore Wind Turbine. These plans are to include the following, where applicable.

(1) Arrangement plans, elevations, and plan views clearly showing in sufficient detail the overall configuration, dimensions and layout of the rotor, nacelle, tower, substructure, foundation, etc.
(2) Layout plans indicating the locations, dimension, and weights of main turbine components (e.g., blade, hub, shaft, nacelle, etc.) and the components (e.g., electrical, mechanical and control systems, etc.) in nacelle housing.
(3) Layout of secondary structures, fenders, ladders, access platform, boat landing, J-tube, etc.
(4) Structural plans indicating the complete structural arrangement, dimensions, member sizes, plating and framing, material properties, and details of connections and attachments.
(5) Tower scantlings and tower-RNA connection details, where applicable.
(6) Tower-hull connection details.
(7) Tower pedestal and foundation (transit piece supporting the tower) details.
(8) Pile plans indicating arrangements, nominal sizes, thicknesses, and penetration.
(9) Welding details and procedures, and schedule of non-destructive testing.
(10) Corrosion control systems.
(11) Various information in support of novel features utilised in the design where applicable.
(12) In-Service Inspection Plan.

For a Self-Elevating Unit to be used as the Substructure of an offshore wind turbine, the additional plans and design data required in the Class Rules for the design, construction, and classification of mobile offshore units are to be submitted.

5.4 *Information Memorandum*

An information memorandum is to be prepared and submitted to Class. Class will review the contents of the memorandum to establish consistency with other data submitted for the purpose of obtaining classification. Class will not review the contents of the memorandum for their accuracy, or the features described in the memorandum for their adequacy. An information memorandum is to contain, as appropriate to the installation, the following:

(1) Specifications of the RNA and the tower suitable to be installed for the site.
(2) Valid wind turbine type certificate of the RNA and the tower.
(3) Description of the wind turbine control system concept and the control system (structure of the control system, sequences of the start and stop procedures, behaviour of the turbine during normal operation and on detection of malfunctions, statement of trigger criteria).
(4) Description of turbine control system fault modes and frequency and duration of fault events.
(5) Description of the procedure for manual intervention after the activation of any wind turbine protection functions.
(6) Description of the wind turbine braking systems and their behaviour (structure of the braking systems, mode of operation, characteristic quantities, time constants).
(7) Functional description of the wind turbine locking devices.
(8) Description of the wind turbine control software used for load simulation (interfaces to the load analysis, such as programme modules for blade pitching, as well as identification of critical load cases).
(9) Description of the process that confirms that the control system (including software) on the wind turbine has the same behaviour as modelled in the load simulation.
(10) Computer software interface of the turbine control system linked to the load analysis programme if an independent analysis is requested.
(11) Site plan indicating general features at the installation site and the layout of the offshore wind farm.
(12) Environmental design criteria, including the recurrence interval used to assess environmental phenomena.
(13) Plans showing the general arrangement of the Support Structure and its connection to the RNA.
(14) Description of the safety and protective systems provided.
(15) Description of the modes of operation.
(16) Description of the export or inter-array power cable termination at the substructure.
(17) Listing of governmental authorities having cognisance over the offshore wind farm.
(18) Listing of any novel features.
(19) Brief description of any monitoring proposed, and security systems installed on the Support Structure.
(20) Description of transportation, installation, and maintenance procedures.

6 Operating Manual

The Operating Manual of the Bottom-founded Offshore Wind Turbine is to be submitted for review by Class solely to verify the operational procedures and conditions are consistent with the design information, criteria and limitations considered in the classification. Class is not responsible for the operation of offshore wind turbines. The Operating Manual is to be retained by the Owner or operator and made readily available to the Class Surveyor and to service personnel prior to conducting any maintenance or inspection. The Operating Manual required by this chapter does not need to be in addition to that required by the Coastal State or other governmental authorities.

Chapter 2
Materials and Welding

1 Structural Steels

1.1 General

1.1.1 Scope

This chapter covers the general specifications for materials used for the construction of offshore steel structures. When material other than steel is used as a structural material, documentation is to indicate the mechanical properties, toughness, fatigue, and corrosion characteristics of the proposed material. Where such materials are used in combination with steel, galvanic effects are to be accounted for, as applicable. This chapter is not intended for metals used in reinforced or prestressed concrete, which is addressed in Chap. 2, Sect. 5. All materials are to be suitable for intended service conditions; they are to be of good quality, defined by a recognised specification and free of injurious imperfections. Note: the following requirements are adapted from the ABS Classification Rules for Offshore Wind Turbines.

1.1.2 Material Selection

Materials used are required to exhibit satisfactory formability and weldability characteristics. As required, documentation is to be submitted to substantiate the applicability of proposed steel. Reference can be made to Tables 2 and 3 of this book for ASTM and API steel grades and to the Class Rules for the design, construction, and classification of offshore installations, for guidance on selection of acceptable grades of steel.

F. Aderouane, *Design, Construction, and Global Performance Analysis of Offshore Wind Turbines*, Green Energy and Technology, https://doi.org/10.1007/978-3-031-86010-2_2

1.1.3 Corrosion Control

Details of corrosion control systems (such as coatings, sacrificial anodes or impressed current systems) are to be submitted with adequate supporting data to show their suitability. Such information is to indicate the extent to which the possible existence of stress corrosion, corrosion fatigue, and galvanic corrosion due to dissimilar metals are to be considered. Where the intended sea environment contains unusual contaminants, any special corrosive effects of such contaminants are also to be considered. Appropriate coatings may be used to achieve satisfactory corrosion protection for miscellaneous parts such as bolts and nuts.

1.1.4 Toughness

Materials are to exhibit fracture toughness which is satisfactory for the intended application as supported by previous satisfactory service experience or appropriate toughness tests. Where the presence of ice is judged as a significant environmental factor, material selection may require special consideration.

1.1.5 Through Thickness Stress

In cases where principal loads, from either service or weld residual stresses, are imposed perpendicular to the surface of a structural member, the use of steel with improved through thickness (Z-direction) properties may be required.

1.2 Steel Properties

1.2.1 General

Material specifications are to be submitted for review or approval. Due regard is to be given to established practices in the country in which material is produced and the purpose for which the material is intended.

1.2.2 Tensile Properties

In Table 1, the designation Group I, II or III is used to categorise tensile properties.

Table 1 Steel tensile properties

Group	Yield strength fy	
	ksi	MPa
I	$fy < 40$	$fy < 275$
II	$40 < fy \leq 60$	$275 \leq fy \leq 415$
III	$60 \leq fy \leq 100$	$415 \leq fy \leq 690$

1.2.3 Toughness

Appropriate supporting information or test data are to indicate that the toughness of the steels will be adequate for their intended application and minimum service temperature. Criteria indicative of adequate toughness are contained in Chap. 2, Sect. 1.3.

1.2.4 Bolts and Nuts

Bolts and nuts are to have mechanical and corrosion characteristics comparable to the structural elements being joined and are to be manufactured and tested in accordance with recognised material standards.

1.3 Toughness Criteria for Steel Selection

1.3.1 General

When members are subjected to significant tensile stress, fracture toughness is to be considered in the selection of materials.

1.3.2 Steel Classification

Steels are to be classified as Groups I, II or III according to their tensile properties as listed in Table 1. It should be noted that the yield strengths given in Table 1 are provided only as a means of categorising steels.

Some of the typical ASTM and API steels belonging to the groups of Table 1 are shown in Tables 2 and 3. Steels other than those mentioned therein may be used, provided that their chemical composition, mechanical properties, and weldability are like those listed.

Table 2 Structural steel plates and shapes

Group	Specification and grade	Yield strength		Tensile strength	
		ksi	MPa	ksi	MPa
I	ASTM A36	36	250	58–80	400–550
	ASTM A131 Grade A	34	235	58–75	400–515
	ASTM A285 Grade C	30	205	55–75	380–515
	ASTM A131 Grades B, D	34	235	58–75	400–515
	ASTM A516 Grade 65	35	240	65–85	450–585
	ASTM A573 Grade 65	35	240	65–77	450–530
	ASTM A709 Grade 36T2	36	250	58–80	400–550
	ASTM A131 Grade E	34	235	58–75	400–515
II	ASTM A572 Grade 42	42	290	60 min	415 min
	ASTM A572 Grade 50	50	345	65 min	450 min
	ASTM A588 (to 2 in. thick)	50	345	70 min	485 min
	ASTM A709 Grades 50T2, 50T3	50	345	65 min	450 min
	ASTM A131 Grade AH32	46	315	68–85	470–585
	ASTM A131 Grade AH36	51	350	71–90	490–620
	API Spec 2H-Grade 42	42	290	62–80	425–550
	API Spec 2H-Grade 50 (to 21/2 in thick)	50	345	70–90	485–620
	(over 21/2 in thick)	47	325	70–90	485–620
	API Spec 2W-Grade 42 (to 1 in thick)	42–67	290–460	62 min	425 min
	(over 1 in thick)	42–62	290–430	62 min	425 min
	Grade 50 (to 1 in thick)	50–75	345–515	65 min	450 min
	(over 1 in thick)	50–70	345–485	65 min	450 min
	Grade 50T (to 1 in thick)	50–80	345–550	70 min	485 min
	(over 1 in thick)	50–75	345–515	70 min	485 min
	API Spec 2Y-Grade 42 (to 1 in thick)	42–67	290–460	62 min	425 min
	(over 1 in thick)	42–62	290–430	62 min	425 min
	Grade 50 (to 1 in thick)	50–75	345–515	65 min	450 min
	(over 1 in thick)	50–70	345–485	65 min	450 min
	Grade 50T (to 1 in thick)	50–80	345–550	70 min	485 min
	(over 1 in thick)	50–75	345–515	70 min	485 min
	ASTM A131 Grades DH32, EH32	46	315	68–85	470–585
	Grades DH36, EH36	51	350	71–90	490–620
	ASTM A537 Class 1 (to 21/2 in thick)	50	345	70–90	485–620
	ASTM A633 Grade A	42	290	63–83	435–570
	Grades C, D	50	345	70–90	480–620
	ASTM A678 (80) Grade A	50	345	70–90	485–620
	ASTM A992	50–65	345–450	65 min	450 min

(continued)

Table 2 (continued)

Group	Specification and grade	Yield strength		Tensile strength	
		ksi	MPa	ksi	MPa
	ASTM A131 Grades DH32, EH32	46	315	68–85	470–585
	Grades DH36, EH36	51	350	71–90	490–620
III	ASTM A537 Class 2	60	415	80–100	550–690
	ASTM A633 Grade E	60	415	80–100	550–690
	ASTM A678 (80) Grade B	60	415	80–100	550–690
	API Spec 2W-Grade 60 (to 1 in thick)	60–90	415–620	75 min	515 min
	(over 1 in thick)	60–85	415–585	75 min	515 min
	API Spec 2Y-Grade 60 (to 1 in thick)	60–90	415–620	75 min	515 min
	(over 1 in thick)	60–85	415–585	75 min	515 min
	ASTM A710-Grade A Class 3 (to 2 in thick)	75	515	85 min	585 min
	(2 in. to 4 in thick)	65	450	75 min	515 min
	(over 4 in thick)	60	415	70 min	485 min

Table 3 Structural steel pipes

Group	Specification and grade	Yield strength		Tensile strength	
		ksi	MPa	ksi	MPa
I	API 5L-Grade B	35	240	60 min	415 min
	ASTM A53 Grade B	35	240	60 min	415 min
	ASTM A135 Grade B	35	240	60 min	415 min
	ASTM A139 Grade B	35	240	60 min	415 min
	ASTM A381 Grade Y	35 35	240	60 min	415 min
	ASTM A500 Grade A	33–39	230–270	45 min	310 min
	ASTM A501	36	250	58 min	400 min
	ASTM A106 Grade B	35	240	60 min	415 min
	ASTM A524 (strength varies with thickness)	30–35	205–240	55–85	380–585
II	API 5L95 Grade X42 (2% max. cold expansion)	42	290	60 min	415 min
	API 5L95 Grade X52 (2% max. cold expansion)	52	360	66 min	455 min
	ASTM A500 Grade B	42–46	290–320	58 min	400 min
	ASTM A618 Grade Ia, Ib and II (to 3/4 in thick)	50	345	70 min	485 min
	API 5L95 Grade X52 (with SR5, SR6, or SR8)	52	360	66 min	455 min

1.3.3 Toughness Characteristics

Satisfactory toughness characteristics can be demonstrated by any one of the following:

(1) Demonstration of past successful application, under comparable conditions, of a steel produced to a recognised standard (such as those of the ASTM, API, or other recognised standard).
(2) Demonstration that a steel manufactured by a particular producer using a specific manufacturing process has minimum toughness levels representative of those listed herein.
(3) Charpy impact testing in accordance with Table 4.

1.4 Minimum Service Temperature

Minimum service temperature refers to the temperature of the material and is to be established in accordance with Chap. 2, Sects. 1.4.1 to 1.4.3 below. This temperature is to be based on meteorological data taken over a period of not less than 10 years for the lowest average daily temperature.

1.4.1 Material Below the Splash Zone

For material below the splash zone (refer to Chap. 5, Sect. 2.1.6), the service temperature is defined as 0°C (32°F). A higher service temperature may be used if adequate supporting data can be presented relative to the lowest average daily water temperature applicable to the depths involved.

1.4.2 Material Within or Above the Splash Zone

For material within or above the splash zone, the service temperature is the same as the lowest average daily atmospheric temperature. A higher service temperature may be used if the material above the waterline is warmed by adjacent sea water temperature or by auxiliary heating.

1.4.3 Special Conditions

In all cases where material temperature is reduced by localised cryogenic storage or other cooling conditions, such factors are to take account of establishing the minimum service temperature.

1.4.4 General

This chapter covers specifications for materials for concrete used in the construction of the Support Structure of offshore wind turbines. It includes the metals used in

Table 4 Charpy toughness specification for steels

Group	Section size	Energy absorption (Longitudinal)	
		ft-lb	Joules
I	6 mm < t < 19 mm (0.25 in < t < 0.75 in)	15	20
	t > 19 mm (0.75 in)	20	27
II, III	t > 6 mm (0.25 in)	25	34
Group	Test temperature	Minimum service temperature (as determined by 2/1.4)	
I, II	30 °C (54 °F) below Minimum Service Temperature		
III	−40 °C (−40°F)	−10 °C (32 °F)	
	−50 °C (−58 °F)	−10 °C (14 °F)	
	−50 °C (−58 °F)	−20 °C (−4 °F)	
	−60 °C (−76 °F)	−30 °C (−22 °F)	
Group	Test Temperature	Minimum Service Temperature (as determined by 2/1.4)	
I, II	At Minimum service temperature		
III	Same as *Note 1b*		

Notes: Test Temperatures—The following applies for service temperatures down to −30 °C (−22 °F). For lower service temperatures, test requirements are to be specially considered. For structural members and joints whose performance is vital to the overall integrity of the structure and which experience an unusually severe combination of stress concentration, rapid loading, cold working, and restraint, the impact test guidelines of Table 4 are to be met at test temperatures as given below: For structural members and joints which sustain significant tensile stress and whose fracture may pose a threat to the survival of the structure, the impact test guidelines of Table 4 are to be met at test temperatures as given below: For primary structural members subject to significant tensile stresses and whose usage warrants impact toughness testing, the impact test guidelines of Table 4 are to be met at the following test temperatures. For structural members which have sufficient structural redundancy so that their fracture would not pose a threat to the survivability of the structure, the toughness criteria specified for c above may be relaxed provided the materials used in such cases are appropriate for the loading conditions, loading rates, and temperatures encountered in service. Impact tests are not necessary for section sizes below 6 mm (0.25 in) in thickness. Energy values in Table 4 are minimum average values for full-size longitudinal specimens. Alternative toughness criteria which may be applied are: Under-sized longitudinal specimens: proportional reduction in Table 4 energy values in accordance with ASTM A20 or equivalent. Transverse specimens: 2/3 of the energy values shown in Table 4 but in no case less than 20 J (15 ft-lb). Longitudinal or transverse specimens: lateral expansion is not to be less than 0.5 mm (0.02 in), or 0.38 mm (0.015 in), respectively. Nil-ductility temperature (NDT) as determined by drop weight tests is to be 5 °C (10 °F) below the test temperature indicated in note 1 above. Other fracture toughness tests as appropriate. The minimum number of specimens to be tested per heat is to be three; however, this number is to be increased in accordance with usage of the material (refer to ASTM A673 or equivalent). 1.5 Materials for Concrete Construction reinforced or prestressed concrete. All materials are to be suitable for intended service conditions and are to be of good quality, defined by recognised specifications and free of injurious defects. Materials used in the construction of concrete structures are to be selected with due attention given to their strength and durability in the marine environment. Materials which do not conform to the requirements of this chapter may be considered for approval upon presentation of sufficient evidence of satisfactory performance.

1.4.5 Zones

Particular attention is to be given in each of the following zones (refer to Chap. 5, Sect. 2.1.5):

(1) **Submerged zone**: chemical deterioration of the concrete, corrosion of the reinforcement and hardware, and abrasion of the concrete.
(2) **Splash zone**: freeze–thaw durability, corrosion of the reinforcement and hardware, chemical deterioration of the concrete, and fire hazards.
(3) **Ice zone**: freeze–thaw durability, corrosion of the reinforcement and hardware, chemical deterioration of the concrete, fire hazards, and abrasion of the concrete.
(4) **Atmospheric zone**: freeze–thaw durability, corrosion of reinforcement and hardware, and fire hazards.

1.4.6 Cement

Type

Cement is to be equivalent to Types I or II Portland cement as specified by ASTM C150 or Portland-pozzolan cement as specified by ASTM C595. ASTM C150 Type III Portland cement may be specially approved for applications.

Tricalcium Aluminate

The tricalcium aluminate content of the cement is to be in the 5–10% range.

Oil Storage

For environments which contain detrimental sulphur bearing materials (such as where oil storage is planned and the oil is expected to contain sulphur compounds which are detrimental to concrete durability), the maximum content of tricalcium aluminate is to be at the lower end of the 5–10% range. Alternatively, pozzolans, or pozzolans and fly ash, may be added or a suitable coating employed to protect the concrete.

1.4.7 Water

Cleanliness

Water used in mixing concrete is to be clean and free from injurious amounts of oils, acids, alkalis, salts, organic materials, or other substances that may be deleterious to concrete or steel.

Non-potable Water

If non-potable water is proposed for use, the selection of proportions of materials in the concrete is to be based on test concrete mixes using water from the same source. The strength of mortar test cylinders made with non-potable water is not to be less than 90% of the strength of similar cylinders made with potable water. Strength test comparisons are to include 7–28-day strength data on mortars prepared and tested in accordance with recognised standards such as ASTM C109.

1.4.8 Chloride or Sulphide Content

Water for structural concrete or grout is to not contain more than 0.07% chlorides as Cl by weight of cement, nor more than 0.09% sulphates as SO^4 when tested by ASTM D512. Chlorides in mix water for prestressed concrete or grout are to be limited to 0.04% by weight of cement. Total chloride content, as Cl, of the concrete prior to exposure is not to exceed 0.10% by weight of the cement for normal reinforced concrete and 0.06% by weight of cement for prestressed concrete.

1.4.9 Aggregates

General

Aggregates are to conform to the requirements of ASTM C33 or equivalent. Other aggregates may be used if there is supporting evidence that they produce concrete of satisfactory quality. When specially approved, lightweight aggregates like those specified in ASTM C330 may be used for conditions that do not pose durability problems.

Washing

Marine aggregates are to be washed with fresh water before use to remove chlorides and sulphates so that the total chloride and sulphate content of the concrete mix does not exceed the limits defined Chap. 2, Sect. 1.5.5.

Size

The maximum size of the aggregate is to be such that the concrete can be placed without voids. It is recommended that the maximum size of the aggregate is to not be larger than the smallest of the following: one-fifth of the narrowest dimension between sides of forms; one-third of the depth of slabs; three-fourths of the minimum clear spacing between individual reinforcing bars, bundles of bars, prestressing tendons, or post-tensioning ducts.

1.4.10 Admixtures

General

The admixture is to be shown capable of maintaining the same composition and performance throughout the work as the product used in establishing concrete proportions. Admixtures containing chloride ions are not to be used if their use will produce a deleterious concentration of chloride ions in the mixing water.

Recognised Standards

Admixtures are to be in accordance with applicable recognised standards such as ASTM C260, ASTM C494, ASTM C618 or equivalents.

Pozzolan Content

Pozzolan or pozzolan and fly ash content is not to exceed 15% by weight of cement unless specially approved.

1.4.11 Steel Reinforcement

Steel reinforcement used in offshore concrete structures is to be suitable for its intended service and in accordance with recognised standards.

Reinforcement for Non-prestressed Concrete

Non-prestressed reinforcement is to be in accordance with one of the following specifications or its equivalents:

(1) Deformed reinforcing bars and plain bars: ASTM A615.
(2) Bar and rod mats: ASTM A184.

(3) Plain wire for spiral reinforcement: ASTM A82, ASTM A704.
(4) Welded plain wire fabric: ASTM A185.
(5) Deformed wire: ASTM A496.
(6) Welded deformed wire fabric: ASTM A497.

Welded Reinforcement

Reinforcement which is to be welded is to have the properties needed to produce satisfactory welded connections. Welding is to be in accordance with recognised specifications such as AWS D1.1 or is to be proven to produce connections of satisfactory quality.

Steel Reinforcement for Prestressed Concrete

Steel reinforcement for prestressed concrete is to be in accordance with one of the following specifications or equivalent:

(1) Seven-wire strand: ASTM A416.
(2) Wire: ASTM A421.

Other Materials

Other prestressing tendons may be approved upon presentation of evidence of satisfactory properties.

1.4.12 Concrete

The concrete is to be designed to assure sufficient strength and durability. A satisfactory method for quality control of concrete is to be used which is equivalent to ACI 318. Mixing, placing, and curing of concrete are to conform to recognised standards.

1.4.13 Water-Cement Ratios

Unless otherwise approved, water-cement ratios and 28-day compressive strengths of concrete for the three exposure zones are to be in accordance with Table 5.

Table 5 Water-cement ratios and compressive strengths

Zone	Maximum w/c ratio	Minimum 28-day cylinder compressive strength
Submerged	0.45	35 MPa (5000 psi)
Splash and atmospheric	0.40–0.45*	35 MPa (5000 psi)

* Depending upon severity of exposure

1.4.14 Other Durability Requirements

Cement Content

Minimum cement content is to verify an adequate amount of paste for reinforcement protection and be not less than 355 kg/m^3 (600 lb/yd^3).

Freeze–Thaw Durability

When freeze–thaw durability is required, the concrete is to contain entrained air in accordance with a recognised standard such as ACI 211.1. Attention is to be paid to the appropriate pore distribution of the entrained air and the spacing between pores in the hardened concrete. The calculated spacing factors are not to exceed 0.25 mm (0.01 in).

Scouring

When severe scouring action is expected, the coarse aggregate is to be as hard as the material causing the abrasion. Also, the sand content of the concrete mix is to be kept as low as possible, and air entrainment is to be limited to the minimum appropriate to the application.

1.4.15 Grout for Bonded Tendons

General

Grout for bonded tendons is to conform to ACI 318 or equivalent.

Chlorides and Sulphates

Grout is not to contain chlorides or sulphates in amounts which are detrimental to the structure. Limitations are included in 2/1.5.5.

Contents

Grout is to consist of Portland cement and potable water, or Portland cement, sand, and potable water. Admixtures may be used only after sufficient testing to indicate that their use is beneficial and that they are free of harmful quantities of chlorides, nitrates, sulphides, sulphates, or any other material which has been shown to be detrimental to the steel or grout.

Sand

Sand, if used, is to conform to ASTM C144 or equivalent, except that gradation may be modified as necessary to obtain increased workability.

Preparation

Proportions of grouting materials are to be based on results of tests on fresh and hardened grout prior to beginning work. The water content is to be the minimum necessary for proper placement but in no case more than 50% of the content of cement by weight. Grout is to be properly mixed and screened.

Temperature

Temperature of members at the time of grouting is to be above 10°C (50°F) and is to be maintained at this temperature for at least 48 h.

2 Welding and Fabrication

2.1 Introduction

Welding for steel structures is to comply with the pertinent requirements of a recognised code, such as the AWS D1.1 Structural Welding Code—Steel. While the requirements of this chapter are to be addressed using a recognised reference code, the reference code may not provide coverage of all necessary items. Therefore, this chapter provides additional requirements which, as the need arises, extend the scope of the code to make it suitable for classification purposes. Also, because of the possible wide variation of requirements which may exist among selected welding codes, Chap. 2, Sects. 2.5 through 2.7, provide some specific requirements which are intended to verify a basic degree of uniformity in the welding performed for structures classed accordingly.

2.2 *Overview*

2.2.1 Plans and Specifications

Submitted plans or specifications are to be in accordance with Chap. 1, Sect. 5, and they are to clearly indicate the extent of welding for the main parts of the structure. The plans or specifications are to indicate the extent of non-destructive inspection of the weld. The welding process, filler metal and joint design are to be indicated on plans or in separate specifications submitted for approval, which are to distinguish between manual and automatic welding. The Surveyor is to be informed of the planned sequences and procedures to be followed in the erection and welding of the main structural members. In all instances, welding procedures and filler metals are to be applied which will produce sound welds that have strength and toughness comparable to that of the base material.

2.2.2 Workmanship and Supervision

It is to be demonstrated that all welders and welding operators to be employed in the construction of structures to be classed are properly qualified and are experienced in the type of work proposed and in the proper use of the welding processes and procedures to be followed. Enough skilled supervisors are to be employed to verify thorough supervision and control of all welding operations. Inspection of welds employing methods outlined in Chap. 2, Sect. 2.4.9, is to be carried out to the satisfaction of the Surveyor.

2.2.3 Welding Procedures

Procedures for the welding of all joints, including types of electrodes, edge preparations, welding techniques and proposed positions, are to be established before construction begins. Details of proposed welding procedures and sequences may be required to be submitted for review, depending on the intended application. Special precautions are to be taken for welding thick sections regarding joint preparation, preheat, welding sequence, heat input and interpass temperature. Ultrasonic inspection may be called for to verify the absence of injurious laminations for material used where through thickness (Z-direction) properties are important.

2.3 *Preparation for Welding*

2.3.1 Edge Preparation and Fitting

Edge preparations are to be accurate and uniform and the parts to be welded are to be fitted in accordance with the approved joint detail. All means employed for correcting improper fitting are to be to the satisfaction of the Surveyor. Where excessive root openings are encountered for butt weld connections, weld buildup of the edges may be approved by the Surveyor, depending upon the location of the joint and the welding procedures employed. Unless specially approved, such buildup of each edge, where permitted, is not to exceed $t/2$ or 12.5 mm (1/2 in), whichever is less, where t is the thickness of the thinner member being welded. Where sections to be butt welded differ in thickness and have an offset on any side of more than 3 mm (1/8 in), a suitable transition taper is to be provided. In general, the transition taper length is to be not less than three times the offset. The transition may be formed by tapering the thicker member or by specifying a weld joint design which will provide the required transition.

2.3.2 Alignment

Means are to be provided for maintaining the members to be welded in correct position and alignment during the welding operation. In general, strongbacks or other appliances used for this purpose are to be so arranged as to allow for expansion and contraction during production welding. The removal of such items is to be carried out to the satisfaction of the Surveyor.

2.3.3 Cleanliness

All surfaces to be welded are to be free from moisture, grease, loose mill scale, excessive rust, and paint. Primer coatings of ordinary thicknesses, thin coatings of linseed oil or equivalent coatings may be used, provided it is demonstrated that their use has no adverse effect on the production of satisfactory welds. Slag and scale are to be removed not only from the edges to be welded but also from each pass or layer before the deposition of subsequent passes or layers. Weld joints prepared by arc-air gouging may require additional preparation by grinding or chipping and wire brushing prior to welding, to minimise the possibility of excessive carbon on the surfaces. Compliance with these cleanliness requirements is of prime importance in the welding of higher-strength steels (refer to Chap. 2, Sect. 1.3, especially those which are quenched and tempered.

2.3.4 Tack Welds

Tack welds of consistent good quality, made with the same grade of filler metal as intended for production welding and deposited in such a manner as not to interfere with the completion of the final weld, need not be removed, provided they are found upon examination to be thoroughly clean and free from cracks or other defects. Preheat may be necessary prior to tack welding when the materials to be joined are highly restrained. Special consideration is to be given to using the same preheat as specified in the welding procedure when tack welding higher-strength steels, particularly those materials which are quenched and tempered. These same precautions are to be followed when making any permanent welded markings.

2.3.5 Run-On and Run-Off Tabs

When used, run-on and run-off tabs are to be designed to minimise the possibility of high stress concentrations and base-metal and weld-metal cracking.

2.4 Production Welding

2.4.1 Environment

Proper precautions are to be taken to verify that all welding is done under conditions where the welding site is protected against the deleterious effects of moisture, wind, and severe cold.

2.4.2 Sequence

Welding is to be planned to progress symmetrically so that shrinkage on both sides of the structure will be equalised. The ends of frames and stiffeners are to be left unattached to the plating at the subassembly stage until connecting welds are made in the intersecting systems of plating, framing and stiffeners at the erection stage. Welds are not to be carried across an unwelded joint or beyond an unwelded joint which terminates at the joint being welded, unless specially approved.

2.4.3 Preheat and Postweld Heat Treatment

The use of preheat is to be considered when welding higher-strength steels, materials of thick cross-section, materials subject to high restraint, and when welding is performed under high humidity conditions or when the temperature of the steel is below 0°C (32°F). The control of interpass temperature is to be specially considered when welding quenched and tempered higher-strength steels. When preheat is used, the temperature is to be in accordance with the accepted welding procedure. Postweld heat treatment, when specified, is to be carried out using an approved method.

2.4.4 Low-Hydrogen Electrodes or Welding Processes

Unless otherwise approved, the use of low-hydrogen electrodes or welding processes is required for welding all higher-strength steels and may also be considered for ordinary-strength steel weldments subjected to high restraint. When using low-hydrogen electrodes or processes, proper precautions are to be taken to verify that the electrodes, fluxes, and gases used for welding are clean and dry.

2.4.5 Back Gouging

Chipping, grinding, arc-air gouging or other suitable methods are to be employed at the root or underside of the weld to obtain sound metal before applying subsequent beads for all full-penetration welds. When arc-air gouging is employed, the selected technique is to minimise carbon buildup and burning of the weld or base metal. Quenched and tempered steels are not to be flame gouged using oxy-fuel gas.

2.4.6 Peening

The use of peening is not recommended for single-pass welds and the root or cover passes on multipass welds. Peening, when used to correct distortion or to reduce residual stresses, is to be effected immediately after depositing and cleaning each weld pass.

2.4.7 Fairing and Flame Shrinking

Fairing by heating or flame shrinking, and other methods of correcting distortion or defective workmanship in fabrication of main strength members and other members which may be subject to high stresses, are to be carried out only with the expressed approval of the Surveyor. These corrective measures are to be kept to an absolute minimum when higher-strength quenched and tempered steels are involved, due to high local stresses and the possible degradation of the mechanical properties of the base material.

2.4.8 Weld Soundness and Surface Appearance

Production welds are to be sound, crack-free, and free from lack of fusion or penetration, slag inclusions and porosity. The surfaces of welds are to be visually inspected and are to be regular and uniform with a minimum amount of reinforcement and free from undercut and overlap and free from injurious arc strikes. Contour grinding when required by an approved plan or specification or where deemed necessary by the Surveyor is to be carried out to the satisfaction of the Surveyor.

2.4.9 Inspection of Welds

Inspection of welded joints in important locations is to be carried out preferably by established non-destructive test methods such as radiographic, ultrasonic, magnetic-particle or dye-penetrant inspection. Approved acceptance criteria, or the Class guide for performing non-destructive inspections, are to be used in evaluating radiographs and ultrasonic indications (refer to also Chap. 2, Sect. 2.4.11). Radiographic or ultrasonic inspection, or both, is to be used when the overall soundness of the weld cross-section is to be evaluated. Magnetic-particle or dye-penetrant inspection may be used when investigating the outer surface of welds, as a check of intermediate weld passes such as root passes, and to check back chipped, ground, or gouged joints prior to depositing subsequent passes. Surface inspection of important tee or corner joints in critical locations, using an approved magnetic-particle or dye-penetrant method, is to be conducted to the satisfaction of the Surveyor. Some steels, especially higher-strength steels, may be susceptible to delayed cracking. When welding these materials, the final non-destructive testing is to be delayed for a suitable period to permit detection of such defects. Weld run-on or run-off tabs may be used where practicable and these may be sectioned for examination. The practice of taking weld plugs or samples by machining or cutting from the welded structure is not recommended and is to be used only in the absence of other suitable inspection methods. When such weld plugs or samples are removed from the welded structure, the holes or cavities thus formed are to be properly prepared and welded, using a suitable welding procedure as established for the original joint.

2.4.10 Extent of Inspection of Welds

General

The minimum extent of non-destructive testing to be conducted is indicated in Chap. 2, Sect. 2.4.10 below. The distribution of inspected welds is to be based on the classification of application of the welds, as mentioned in Chap. 2, Sect. 2.4.10, and the variety of weld sizes used in the structure. Non-destructive testing is to be carried out forming and postweld heat treatment, and procedures are to be adequate to detect delayed cracking. Welds which are inaccessible or difficult to inspect in service are to be subject to increased levels of non-destructive inspection. Non-destructive examination of full penetration butt welds is to be carried out by radiographic or ultrasonic methods. Where a method (such as radiography or ultrasonics) is selected as the primary non-destructive method of inspection, the acceptance standards of such a method govern. Where inspection by any method indicates the presence of defects that could jeopardise the integrity of the structure, removal and repair of such defects are to be carried out to the satisfaction of the attending Surveyor. Should the ultrasonic method be used as the primary inspection method, such testing is to be supplemented by a reasonable amount of radiographic inspection to confirm adequate quality control is being achieved. To assess the extent of surface imperfections in welds made in Group III steels used in critical structural locations, representative inspection by the magnetic-particle or dye-penetrant method is to also be accomplished.

Plans

A plan for non-destructive testing of the structure is to be submitted. This plan is to include but is not restricted to visual inspection of all welds, representative magnetic-particle or dye penetrant inspection of tee and fillet welds not subject to ultrasonic inspection, and the inspection of all field welds by appropriate means. The extent and method of inspection are to be indicated on the plan, and the extent of inspection is to be based on the function of the structure and the accessibility of the welds after the structure is in service. For automated welds for which quality assurance techniques indicate consistent satisfactory performance a lesser degree of inspection may be permitted.

Classification of Application

Welds are to be designated as being special, primary, or secondary depending on the function and severity of service of the structure in which the welds are located. Special welds are those occurring in structural locations of critical importance to the integrity of the structure or its safe operation. Secondary welds are those occurring in locations of least importance to the overall integrity of the structure. Primary welds are those occurring in locations whose importance is intermediate between the special and secondary classifications.

Extent of Non-destructive Inspection—Steel Jacket Type Structures

In general, the number of penetration type welds (i.e., butt, T, K and Y joints) to be inspected in each classification is to be based on the percentages stated below. Alternatively, the extent of radiographic and ultrasonic inspection may be based on other methods, provided the alternative will not result in a lesser degree of inspection.

Where the extent of welds to be inspected is stated as a percentage, such as 20% of primary welds, this means that complete inspection of 20% of the total number of welds considered to be primary is required. All welds considered special are to be inspected 100% by the ultrasonic or radiographic method. Twenty percent of all welds considered primary are to be inspected by the ultrasonic or radiographic method. Welds considered to be secondary are to be inspected on a random basis using an appropriate method. In locations where ultrasonic test results are not considered dependable, the use of magnetic-particle or dye-penetrant inspection as a supplement to ultrasonic inspection is to be conducted. For T, K, or Y joints, approval may be given to substituting magnetic-particle or dye-penetrant inspection for ultrasonic inspection when this will achieve a sufficient inspection quality. Magnetic-particle or dye-penetrant inspection of fillet welds is to be accomplished for all permanent fillet welds used in jacket construction, all load-bearing connections, and all fillet welds in Special Application areas of the deck structure. The random inspection of other deck fillet welds is to be carried out at the discretion of the Surveyor.

Additional Inspection—Special Conditions

Additional inspection may be required depending on the type and use of the structure, the material and welding procedures involved, and the quality control procedures employed.

Additional Inspection—Production Experience

If the proportion of unacceptable welds becomes abnormally high, the frequency of inspection is to be increased.

High Through Thickness (Z-Direction) Stresses

At important intersections, welds which impose high stresses perpendicular to the member thicknesses (Z-direction loading) are to be ultrasonically inspected to assure freedom from lamellar tearing after welding.

2.4.11 Acceptance Criteria

As stated in Chap. 2, Sect. 2.4.9, recognised acceptance criteria such as those issued by the AWS are to be employed. When employing Class guide for performing non-destructive inspections, Class A and Class B criteria are to be applied as follows.

(1) Class A acceptance criteria are to be used for Special Application Structure and critical locations within Primary Application Structure such as circumferential

welds of cylindrical and built-up columns or legs, weld intersections of external plating in the Support Structure, etc.

(2) Class B acceptance criteria are to be used for Primary Application Structure where Class A acceptance does not apply.

(3) Twice Class B acceptance criteria are to be used for Secondary Application Structure.

When radiographic or ultrasonic inspection is specified for other types of connections, such as partial penetration and groove type tee or corner welds, modified procedures and acceptance criteria are to be specified which adequately reflect the application.

2.4.12 Repair Welding

Defective welds and other injurious defects, as determined by visual inspection, non-destructive test methods, or leakage under hydrostatic tests, are to be excavated in way of the defects to sound metal and corrected by rewelding, using a suitable repair welding procedure to be consistent with the material being welded. Removal by grinding of minor surface imperfections such as scars, tack welds and arc strikes may be permitted. Special precautions, such as the use of both preheat and low-hydrogen electrodes, are to be considered when repairing welds in higher-strength steel, materials of thick cross-section or materials subjected to high restraint.

2.5 Butt Welds

2.5.1 Manual Welding Using Stick Electrodes

Manual welding using stick electrodes may be employed for butt welds in members not exceeding 6.5 mm (1/4 in) in thickness without bevelling the abutting edges. Members exceeding 6.5 mm (1/4 in) are to be prepared for welding using an appropriate edge preparation, root opening and root face (land) to provide for welding from one or both sides. For welds made from both sides, the root of the first side welded is to be removed to sound metal by an approved method before applying subsequent weld passes on the reverse side. When welding is to be deposited from one side only, using ordinary welding techniques, appropriate backing (either permanent or temporary) is to be provided. The backing is to be fitted so that spacing between the backing and the members to be joined is in accordance with established procedures. Unless specially approved otherwise, splices in permanent backing strips are to be welded with full penetration welds prior to making the primary weld.

2.5.2 Submerged-Arc Welding

Submerged-arc welding, using wire-flux combinations for butt welds in members not exceeding 16 mm (5/8 in) in thickness, may be employed without bevelling the abutting edges. Members exceeding 16 mm (5/8 in) are normally to be prepared for welding using an appropriate edge preparation, root opening and root face (land) to provide for welding from one or both sides. When it is determined that sound welds can be made without gouging, the provisions of Chap. 2, Sect. 2.4.5 are not applicable. Where the metal is to be deposited from one side only, using ordinary welding techniques, backing (either permanent or temporary) is to be provided and the members are to be bevelled and fitted in accordance with established procedures.

2.5.3 Gas Metal-Arc and Flux Cored-Arc Welding

Manual semi-automatic or machine automatic gas metal-arc welding, and flux cored-arc welding using wire-gas combinations and associated processes, may be ordinarily employed utilising the conditions specified in Chap. 2, Sect. 2.5.1 except that specific joint designs may differ between processes.

2.5.4 Electroslag and Electrogas Welding

The use of electroslag and electrogas welding processes will be subject to special consideration, depending upon the specific application and the mechanical properties of the resulting welds and heat-affected zones.

2.5.5 Special Welding Techniques

Special welding techniques employing any of the basic welding processes mentioned in Chap. 2, Sect. 2.5 will be specially considered, depending upon the extent of the variation from the accepted technique. Such special techniques include one side welding, narrow-gap welding, tandem-arc welding, open-arc welding, and consumable-nozzle electroslag welding. The use of gas tungsten-arc welding will also be subject to special consideration, depending upon the application and whether the process is used manually or automatically.

2.6 *Fillet Welds*

2.6.1 General

The sizes of fillet welds are to be indicated on detail plans or on a separate welding schedule and are subject to approval. The weld throat size is not to be less than 0.7 times the weld leg size. Fillet welds may be made by an approved manual or automatic technique. Where the gap between the faying surfaces of members exceeds 2 mm (1/16 in) and is not greater than 5 mm (3/16 in), the weld leg size is to be increased by the amount of the opening. Where the gap between members is greater than 5 mm (3/16 in), fillet weld sizes and weld procedures are to be specially approved by the Surveyor. Completed welds are to be to the Surveyor's satisfaction. Special precautions such as the use of preheat or low-hydrogen electrodes or low hydrogen welding processes may be required where small fillets are used to attach heavy members or sections. When heavy sections are attached to light members, the weld size may be required to be modified.

2.6.2 Tee Connections

Except where otherwise indicated under Chap. 2, Sect. 2.6, the fillet weld requirement for tee connections is to be determined by the lesser thickness member being joined. Where only the webs of girders, beams or stiffeners are to be attached, it is recommended that the unattached face plates or flanges be cut back. Except for girders of thickness greater than 25 mm (1 in), reduction in fillet weld sizes may be specially approved in accordance with either (1) or (2) specified below. However, in no case is the reduced leg size to be less than 5 mm (3/16 in).

(1) Where quality control facilitates working to a gap between members being attached of 1 mm (0.04 in) or less, a reduction in fillet weld leg size of 0.5 mm (0.02 in) may be permitted provided that the reduced leg size is not less than 8 mm (5/16 in).

(2) Where automatic double continuous fillet welding is used and quality control facilitates working to a gap between members being attached of 1 mm (0.04 in) or less, a reduction in fillet weld leg size of 1.5 mm (1/16 in) may be permitted provided that the penetration at the root is at least 1.5 mm (1/16 in) into the members being attached and the reduced leg size is not less than 5 mm (3/16 in).

2.6.3 Lapped Joints

Lapped joints are not to have overlaps of less width than twice the thinner plate thickness plus 25 mm (1 in). Both edges of an overlapped joint are to have continuous fillet welds in accordance with Chap. 2, Sect. 2.6.1 or Chap. 2, Sect. 2.6.4.

2.6.4 Overlapped End Connections

Overlapped end connections of structural members which are effective in the overall strength of the unit are to have continuous fillet welds on both edges equal in leg size to the thickness of the thinner of the two members joined. All other overlapped end connections are to have continuous fillet welds on each edge of leg sizes such that the sum of the two is not less than 1.5 times the thickness of the thinner member.

2.6.5 Overlapped Seams

Unless specially approved, overlapped seams are to have continuous welds on both edges of the sizes required by the approved plans and are to be in accordance with the applicable provisions of Chap. 2, Sect. 2.6

2.6.6 Plug Welds or Slot Welds

Upon consideration by Class, plug welds or slot welds may be specially approved for specific applications. Where used in the body of doublers and similar locations, such welds may be spaced about 300 mm (12 in) between centres in both directions. Slot welds are not to be filled with weld metal. For plate thicknesses up to 13 mm (1/4 in), fillet sizes are to be equal to plate thickness but not greater than 9.5 mm (3/8 in); for thicknesses over 13 mm (1/4 in) to 25 mm (1 in) fillet sizes are to be 16 mm (5/8 in) maximum.

2.7 Full Penetration Corner or Tee Joints

Measures taken to achieve full penetration corner or tee joints, where specified, are to be to the satisfaction of the attending Surveyor. Ultrasonic inspection of the member in way of the connection may be called for to assure the absence of injurious laminations prior to fabrication which could interfere with the attainment of a satisfactory welded joint.

Chapter 3
Environmental Conditions

1 General

Environmental conditions to which the Bottom-founded Offshore Wind Turbine is expected to be exposed during its design life are to be determined using adequate data for the areas in which the Support Structure is to be transported and installed. For the Support Structure requiring substantial near-shore construction (e.g., concrete gravity installations), environmental studies are to be appropriate with the duration of construction operations and the relative severity of expected conditions. Environmental phenomena that influence the transport, installation, maintenance, and operation of the Bottom-founded Offshore Wind Turbine are to be described in terms of relevant characteristic parameters. Statistical data and realistic statistical and mathematical models which describe the range of expected variations of environmental conditions are to be employed. All data used are to be fully documented with the sources and estimated reliability. Methods employed in developing available data into design criteria are to be described and submitted in accordance with Chap. 1, Sect. 5.1. Probabilistic methods for short-term, long-term, and extreme-value prediction are to employ statistical distributions appropriate to environmental phenomena under consideration, as evidenced by relevant statistical tests, confidence limits and other measures of statistical significance. Hindcasting methods and models are to be fully documented if they are used to derive the environmental data. Suitable data and analyses supplied by consultants will be accepted as the basis for design. For installations in areas where published design standards and data exist, such standards and data may be cited in the design documentation.

F. Aderouane, *Design, Construction, and Global Performance Analysis of Offshore Wind Turbines*, Green Energy and Technology, https://doi.org/10.1007/978-3-031-86010-2_3

2 Environmental Factors to Be Considered

In general, the design of the Support Structure of an offshore wind turbine requires investigation of the following environmental factors.

(1) Wind.
(2) Waves.
(3) Currents.
(4) Tides, storm surges, and water levels.
(5) Air and sea temperatures.
(6) Air density.
(7) Ice and snow accumulation.
(8) Marine growth.
(9) Seismicity.
(10) Sea ice or lake ice.

Other phenomena, such as tsunamis, submarine slides, seiche, abnormal composition of air and water, air humidity, salinity, ice drift, icebergs, ice scouring, etc., may require investigation depending upon site specific conditions and a specific design of the Support Structure. The required investigation of seabed and soil conditions is described in Chap. 5, Sect. 4.3.

3 Wind

3.1 General

Statistical wind data are normally to include information on the frequency of occurrence, duration, and direction of various wind speeds at the location where offshore wind turbines are to be installed. If onsite measurements are taken, the duration of individual measurements and the height above sea-level of measuring devices are to be stated. A wind speed value is only meaningful when qualified by its elevation and time-averaging duration. In the absence of site data, published data and data from nearby land and sea stations may be used upon the agreement with Class.

3.2 Wind Properties

3.2.1 Wind Speed and Turbulence

A wind condition is typically represented by a mean wind speed and a standard deviation of wind speed. The turbulence intensity, which measures the variation of wind speed relative to the mean wind speed, is defined as the ratio of the wind

Fig. 1 Vector components of turbulent wind velocity

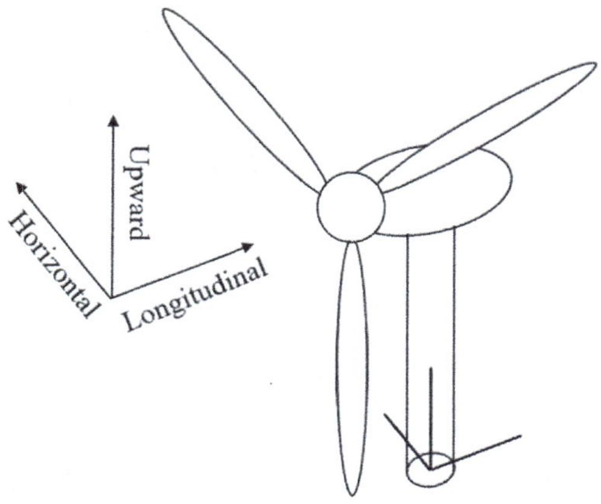

speed standard deviation to the mean wind speed (i.e. coefficient of variance of wind speed). In this book, the mean wind speed, denoted as V_{hub}, , at turbine hub height with 10-min averaging duration is employed to define the turbulent wind conditions in the Design Load Cases (DLCs) in Chap. 4, Sect. 3.2. The steady wind conditions as referred in the DLCs in Chap. 4, Sect. 3.2 are defined by the mean wind speed with 1-min or 3-s averaging duration. These turbulent and steady wind conditions are used to calculate the rotor generated aerodynamic loads as described in Chap. 4, Sect. 4. Mean wind speeds with 1-min averaging time duration are used in the calculation of steady wind forces exerted on exposed components of the Support Structure, as described in Chap. 4, Sect. 42. Conversion of a mean wind speed from one averaging duration to another is to be based on site-specific wind conditions. Use of conversion factors in recognised industry standards appropriate to the installation site may be acceptable and is subject to approval by Class. For wind speeds given in terms of the "fastest mile of wind", V_f, the corresponding time-averaging period t in seconds is given by $t = 3600V_f$, where V_f is the fastest mile of wind at a reference height of 10 m (32.8 ft), in miles per hour. The turbulence of wind over 10 min is considered stationary and can be modelled by a power spectral density function and a coherence function. The turbulence model is to include the effects of varying wind speed, shears and directions and allow rotational sampling through varying shears. The three vector components of turbulent wind velocity, as depicted in Fig. 1 are defined as:

(1) Longitudinal—Along the direction of the mean wind speed.
(2) Lateral—Horizontal and normal to the longitudinal direction.
(3) Upward—Normal to both the longitudinal and lateral directions and pointing upward.

3.2.2 Wind Profile

The mean wind speed profile (vertical wind shear) is to be defined by the power law:

$$V(z) = V_{\text{hub}}(z/z_{\text{hub}})^{\alpha}$$

where:

$V(z)$ wind profile of the 10-min mean wind speed as a function of height, z, , above the SWL, in m/s (ft/s).

V_{hub} 10-min mean wind speed at hub height, in m/s (ft/s).

α power law exponent, whose value is given in Chap. 3, Sects. 3.5 and 3.6, (dimensionless).

z height above the SWL, in m (ft).

z_{hub} hub height above the SWL, in m (ft).

The power law wind profile can also be applied to the mean wind speeds with 1-min or 1-h averaging duration, except that the exponent is to be determined based on site-specific conditions or using published data. For extratropical storm wind conditions, the mean wind speed profile may be represented by the following logarithmic wind shear law. It is expressed using the 1-h mean wind speed at 10 m (32.8 ft) above the SWL.

$$V(z,t) = V(z,t_0)\left[1 - 0.41 I_u(z) ln\left(\frac{t}{t_0}\right)\right] \text{ for } t < t_0$$

where:

$V(z,t)$ mean wind speed at height z and corresponding to an averaging time, t, , in m/s (ft/s).

z height above the SWL, in m (ft).

t averaging time shorter than $t_0 = 3{,}600$ s, in seconds.

$V(z,t_0)$ 1-h mean wind speed at height z, , in m/s (ft/s)

$$V_o\left[1 + C ln\left(\frac{z}{10\phi}\right)\right]$$

t_0 reference averaging time (1 h), in seconds = 3600 s

$C0.0573\sqrt{1 + \frac{0.15V_0}{\phi}}$ (dimensionless)

V_0 1-h mean wind speed at 10 m (32.8 ft) above the SWL, in m/s (ft/s).

$I_u(z)$ turbulence intensity (i.e., the ratio of the wind speed standard deviation to the mean wind speed) at height z, , (dimensionless)

$$= 0.06\left[1 + \frac{0.043V_0}{\phi}\right]\left(\frac{z}{10\phi}\right)^{-0.22}$$

ϕ unit conversion factor, (dimensionless).

1 when using SI units (m, m/s).

3.28 when using US Customary units (ft, ft/s).

For tropical cyclones (also termed as hurricanes or typhoons), the wind profile provided in Appendix 2 is to be applied. For squalls, the wind profile recommended in API RP 2MET may be applied. Other wind profile models may also be used if they can be justified by site-specific data.

3.2.3 Wind Spectrum and Spatial Coherence

Site-specific spectral density of wind speed and spatial coherence are to be determined based on measured wind data. Unless site conditions indicate otherwise, the Kaimal spectrum and the exponential coherence model, as defined in Appendix 1, Sect. 1, or the Mann uniform shear turbulence model, as recommended in Annex C of IEC 61,400–1 (2019), is to be applied. For the extratropical storm wind, API RP 2MET (2019) recommends using the NPD (Frøya) wind spectrum in conjunction with the two-point coherence function. Refer to Appendix 1, Sect. 3 for the definitions. For the tropical cyclone storm wind, the wind spectrum in Appendix 1, Sect. 5 may be applied.

3.3 Long-Term and Extreme-Value Predictions

Long-term and extreme-value predictions for sustained and gust winds are to be based on recognised techniques and clearly described in the design documentation. Preferably, the statistical data used for the long-term distributions of wind speed are to be based on the same averaging periods of wind speeds as are used for the determination of loads.

3.4 Wind Conditions

A wind condition for the design of the Bottom-founded Offshore Wind Turbine is represented by a steady mean flow and an associated turbulence, as well as a mean wind direction and a change of the wind direction. The design wind conditions are further categorised into (a) the normal wind conditions, which occur more frequently than once per year; and (b) The extreme wind conditions representing rare wind conditions with a given return period. The normal and extreme wind conditions for the design of the Bottom-founded Offshore Wind Turbine are specified in Chap. 3, Sects. 3.5 and 3.6. The design load conditions defined in Chap. 4, Sect. 3 indicate how the wind condition is to be applied in combination with other environmental conditions.

3.5 *Normal Wind Conditions*

3.5.1 Normal Wind Profile Model (NWP)

The normal mean wind speed profile (vertical wind shear) is to be defined by the power law specified in Chap. 3, Sect. 3.3.2, where the power law exponent $\alpha = 0.14$. The 10-min mean wind speed at hub height, V_{hub}, is defined in Tables 1 and 2 of chap. 4.

3.5.2 Normal Turbulence Model (NTM)

The standard deviation of turbulence of the normal turbu1lence model, denoted as σ_{NTM}, , is defined as the 90% quantile in the probability distribution of wind speed standard deviation conditioned upon a given 10-min mean wind speed at hub height (V_{hub}). The value of the turbulence standard deviation is to be determined using appropriate statistical techniques applied to measured and preferably de-trended data. Where the site assessment is not available, the recommended approach provided in paragraph 6.4.3 of IEC 61,400–3-1 (2019) may be used to estimate the standard deviation as appropriate to the installation site. The Normal Turbulence Model (NTM) is to be applied together with the Normal Wind Profile Model (NWP) as defined in Chap. 3, Sect. 3.5.1.

3.6 *Extreme Wind Conditions*

The extreme wind conditions are represented by extreme wind shear events, peak wind speeds due to storms, extreme operating gust and turbulence, and rapid changes in wind speed and direction.

3.6.1 Extreme Wind Speed Model (EWM)

The Extreme Wind Speed Model (EWM) is defined as a turbulent wind model with a given return period as specified in Tables 1 and 2 of Chap. 4 for the definition of Design Load Cases (DLCs). For the EWM, when site data are not available, the wind profile of 10-min mean wind speed is to be represented by the power law model as follows:

$$V(z) = V_{\text{hub}}(z/z_{\text{hub}})^{0.11}$$

where:

V_{hub} 10-min mean wind speed at hub height, in m/s (ft/s)

$$= V_{10\text{min},n-\text{yr}}$$

$V_{10\text{min},n-\text{yr}}$ 10-min mean wind speed at hub height with a return period of n years, where n is specified in Table 1 of Chap. 1, in m/s (ft/s).

z height above the SWL, in m (ft).

z_{hub} hub height above the SWL, in m (ft).

The standard deviation of longitudinal turbulent wind speed of extreme wind condition, σ_1, at hub height is to be calculated as:

$$\sigma_1 = 0.11 \times V_{\text{hub}}$$

For the extratropical storm wind, the logarithmic wind shear law given in Chap. 3, Sect. 3.2.2 may be used to calculate the wind profile and standard deviation of extreme wind conditions. For the tropical cyclone storm wind, the wind profile, the standard deviation, the turbulence intensity, and the gust factor provided in Appendix 2, Sect. 1 are to be applied. Where applicable, extreme squall events with 1-year and 50-year return periods in the time domain are to be considered as additional extreme wind conditions.

3.6.2 Extreme Operating Gust (EOG)

The Extreme Operating Gust (EOG) is represented by the hub height gust magnitude, V_{gust}, as defined in the following equation:

$$V_{\text{gust}} = \min\left\{1.35\left(V_{3\text{sec},1-\text{yr}} - V_{\text{hub}}\right);\ 3.3\left(\frac{\sigma_{\text{NTM}}}{1 + \frac{0.1D}{\Lambda_1}}\right)\right\}$$

where:

$V_{3sec,1-yr}$ 3-s mean wind speed at hub height with a return period of 1 year, in m/s (ft/s).

V_{hub} 10-min mean wind speed at hub height, as defined in Table 1 of Chap. 4, in m/s (ft/s).

σ_{NTM} longitudinal turbulence standard deviation defined in Chap. 3, Sect. 3.5.2, in m/s (ft/s).

Λ_1 longitudinal turbulence length scale, in m (ft).

$0.7z$ when $z \leq 60$ m (196.8 ft).

42 m (137.8 ft) when $z \geq 60$ m (196.8 ft).
D rotor diameter, in m (ft).
The time history of transient wind speed at height z is to be defined by:

$$V(z,t) = \begin{cases} V(z) - 0.37V_{gust}\sin\left(\frac{3\pi}{T}\right)\left[1 - \cos\left(\frac{2\pi}{T}\right)\right] & 0 \leq t \leq T \\ V(z) & \text{otherwise} \end{cases}$$

where:

$V(t)$ normal wind profile defined in Chap. 3, Sect. 3.5.1, in m/s (ft/s).

z height above the SWL, in m (ft).

T 10.5 s.

3.6.3 Extreme Turbulence Model (ETM)

The Extreme Turbulence Model (ETM) is to be represented by the Normal Wind Profile Model (NWP) model specified in Chap. 3, Sect. 3.5.1 and the turbulence whose longitudinal component of the standard deviation is given by:

$$\sigma_1 = cI_{ref}\left[0.072\left(\frac{V_{ave}}{c} + 3\right)\left(\frac{V_{hub}}{c} - 4\right) + 10\right]$$

where:

c 2 m/s (6.56 ft/s).

V_{hub} 10-min mean wind speed at hub height, as defined in Table 1 of Chap. 4, in m/s (ft/s).

V_{ave} site-specific annual mean wind speed at hub height, in m/s (ft/s).

I_{ref} expected value of turbulence intensity at hub height when $V_{hub} = 15$ m/s (49.2 ft/s).

3.6.4 Extreme Direction Change (EDC)

The magnitude, θ, , of the Extreme Direction Change (EDC) is to be calculated by:

$$\theta_e = \pm 4\arctan\left(\frac{\sigma_{NTM}}{V_{hub\left(1+\frac{0.1D}{\wedge_1}\right)}}\right) - 180^\circ \leq \theta_e \leq 180^\circ$$

where:

σ_{NTM} longitudinal turbulence standard deviation defined in Chap. 3, Sect. 3.5.2, in m/s (ft/s).

V_{hub} 10-min mean wind speed at hub height, as defined in Table 1 of Chap. 4, in m/s (ft/s).

$\wedge_1$ longitudinal turbulence length scale, defined in Chap. 3, Sect. 3.6.2, in m (ft).

D rotor diameter, in m (ft).

The time history of transient extreme direction change, $\theta(t)$, is defined by:

$$\theta(t) = \begin{cases} 0^\circ & t < 0 \\ \pm 0.5\theta_e\left[1 - \cos\left(\frac{\pi t}{T}\right)\right] & 0 \leq t \leq T \\ \theta_e & t > Y \end{cases}$$

where $T = 6$ s is the duration of the extreme direction change. The sign in the equation is to be chosen such that the most unfavourable transient loading occurs. At the end of the time history of direction change, the direction is assumed to remain a constant value (θ_e). The wind speed is to follow the normal wind profile (NWP) model in Chap. 3, Sect. 3.5.1.

3.6.5 Extreme Coherent Gust with Direction Change (ECD)

The Extreme Coherent Gust with Direction Change (ECD) is to have a magnitude of:

$V_{cg} = 15$ m/s or 49.21 ft/s.

The time history of transient wind speed at height z, is defined by:

$$V(z,t) = \begin{cases} V(z) & t < 0 \\ (V(z) + 0.5V_{\mathrm{cg}}[1 - \cos(\frac{\pi t}{T})] & 0 \le t \le T \\ V(z) + V_{\mathrm{cg}} & t > Y \end{cases}$$

where:

$V(z)$ normal wind profile defined in Chap. 3, Sect. 3.5.1, in m/s (ft/s).

T rise time of gust wind, in seconds 10 s.

z height above the SWL, in m (ft).

The rise in wind speed is assumed to occur simultaneously with the time history of direction change:

$$\theta(t) = \begin{cases} 0^\circ & t < 0 \\ \pm 0.5\theta_{\mathrm{cg}}(V(z) + 0.5V_{\mathrm{cg}}[1 - \cos(\frac{\pi t}{T})] & 0 \le t \le T \\ +\theta_{\mathrm{cg}} & t > Y \end{cases}$$

where:

θ_{cg} magnitude of direction change, in degree

$$\begin{cases} 180^\circ & V_{hub} \le 4m/s(13.12ft/s) \\ 720^\circ(V_{hub}/\phi)^{-1} & V_{hub} \ge 4m/s(13.12ft/s) \end{cases}$$

V_{hub} 10-min mean wind speed at hub height, in m/s (ft/s).

Trise time of gust wind, in seconds 10 s.
Φunit conversion factor, (dimensionless).
1 when using SI units (m, m/s).
3.28 when using US Customary units (ft, ft/s).

3.6.6 Extreme Wind Shear (EWS)

The Extreme Wind Shear (EWS) is to be applied in both vertical and horizontal directions. The two extreme wind shears are considered independent events and therefore not to be applied simultaneously. The time history of transient positive and negative vertical shear is given by:

$$V(z,t) = \begin{cases} V_{\mathrm{hub}}\left(\frac{z}{z_{\mathrm{hub}}}\right)^{\alpha} \pm \left(\frac{z - z_{\mathrm{hub}}}{D}\right)\left(2.5\phi + 0.2\beta\sigma_{\mathrm{NTM}}\left(\frac{D}{\Lambda_1}\right)^{\frac{1}{4}}\right)[1 - \cos(\frac{2\pi t}{T})] & 0 \le t \le T \\ V_{\mathrm{hub}\left(\frac{z}{z_{\mathrm{hub}}}\right)^{\alpha}} & \text{otherwise} \end{cases}$$

The time history of transient horizontal shear is given by:

$$V(z,t)=\begin{cases} V_{hub}\left(\frac{z}{z_{hub}}\right)^{\alpha}\pm\left(\frac{y}{D}\right)\left(2.5\phi+0.2\beta\sigma_{NTM}\left(\frac{D}{\Lambda_1}\right)^{\frac{1}{4}}\right)\left[1-\cos\left(\frac{2\pi t}{T}\right)\right] & 0\le t\le T \\ V_{hub\left(\frac{z}{z_{hub}}\right)^{\alpha}} & \text{otherwise} \end{cases}$$

where:

V_{hub} 10-min mean wind speed at hub height, as defined in Table 1 of Chap. 4, in m/s (ft/s).

$\beta\sigma_{NTM}$ longitudinal turbulence standard deviation defined in Chap. 3, Sect. 3.5.2, in m/s (ft/s).

Λ_1 longitudinal turbulence length scale, defined in Chap. 3, Sect. 3.6.2, in m (ft).

z height above the SWL, in m (ft).

z_{hub} hub height above the SWL, in m (ft).

y horizontal distance from hub in the crosswind direction, in m (ft).

D rotor diameter, in m (ft).

α 0.14, (dimensionless).

β 6.4, (dimensionless).

T 12 s.

ϕ unit conversion factor, (dimensionless).

1 when using SI units (m, m/s).
3.28 when using US Customary units (ft, ft/s).

The sign for the transient wind shear is to be determined such that the most unfavourable transient loading occurs.

3.7 *Survival Wind Conditions*

The survival wind conditions are described by the Survival Wind Speed Model (SurWM) for the robustness assessment for the Bottom-Funded Offshore Wind Turbines in tropical cyclone-prone areas.

3.7.1 Survival Wind Speed Model (SurWM)

The Survival Wind Speed Model (SurWM) is like the Extreme Wind Speed Model (EWM) defined in Chap. 3, Sect. 3.6.1, but with a return period longer than 50 years, as specified in Chap. 4, Sect. 3.2. When site data are not available, the wind profile and turbulence spectrum for the SurWM are to be defined according to Chap. 3, Sects. 3.2.2 and 3.2.3.

4 Waves

4.1 General

The development of wave data is to reflect conditions at the installation site and the type of the Support Structure. Statistical wave data from which design parameters are determined are normally to include the frequency of occurrence of various wave height groups, associated wave periods and directions. Published data and previously established design criteria for specific areas may be used where such exist. Hindcasting techniques that adequately account for the shoaling and fetch limited effects on wave conditions at the site may be used to augment available data. Analytical wave spectra employed to augment available data are to reflect the shape and width of the data and are to be appropriate to the general site conditions. As applicable, wave data are to be developed to determine the following:

(1) Provision for air gap.
(2) Maximum mud line shear force and overturning moment.
(3) Dynamic response of the Support Structure.
(4) Maximum stress.
(5) Fatigue.
(6) Wave impact on the local structure.

All long-term and extreme value predictions employed for the determination of design wave conditions are to be fully described and based on recognised techniques. Because the wave-induced responses may be increased due to the change of wave period and direction, consideration is to be given to waves of less than the maximum height but with a different period and/or direction. Waves that cause the most unfavourable effects on the overall structure may also differ from waves having the most severe effects on individual structural components. In addition to the most severe wave conditions, frequent waves of smaller heights are to be investigated to assess their effect on the fatigue and dynamic responses. The wave conditions for the design of the Bottom-founded Offshore Wind Turbine are described in Chap. 3, Sect. 4.2 through Sect. 4.6. The design load conditions defined in Chap. 4, Sect. 3 specify how these wave conditions are to be applied in combination with other design environmental conditions.

4.2 Normal Sea State (NSS)

The Normal Sea State (NSS) is a stochastic design sea state represented by a significant wave height, a peak spectral period, and a wave direction. It is to be determined based on the site-specific long-term joint probability distribution of metocean parameters conditioned upon a given 10-min mean wind speed at hub height, V_{hub}. The Normal Sea State (NSS) is used in Chap. 4, Sect. 3 to define the Design Load Cases (DLCs) requiring either strength analysis or fatigue analysis. For strength analyses, the Normal Sea State (NSS) can be characterised by the expected value of significant wave height, $H_{S,\text{NSS}}$, conditioned upon a given value of V_{hub} (i.e., $H_{S,\text{NSS}} = E[H_S|V_{\text{hub}}]$). A range of peak period, T_p, , associated with each significant wave height is to be determined for load calculations. The resultant most unfavourable responses are to be used in the design of the Support Structure. For fatigue analyses, the number and resolution of the Normal Sea State (NSS) are to be determined in such a manner that the fatigue damage associated with the full long-term distribution of metocean parameters can be sufficiently accounted for.

4.3 Severe Sea State (SSS)

The Severe Sea State (SSS) is a stochastic design sea state to be applied in combination with the normal wind conditions as specified in Chap. 3, Sect. 3.5 for the strength analysis of the Support Structure when the turbine is in the power production mode. The Severe Sea State (SSS) is represented by a significant wave height, $H_{S,\text{SSS}}(n - yr))$, a peak spectral period and a wave direction. It is to be determined by extrapolation of site-specific long term joint probability distribution of metocean parameters to the extent that the joint occurrence of $H_{S,\text{SSS}}(n - yr))$ and a given value of 10-min mean wind speed, V_{hub}, at hub height has a specified return period of n years, where n is defined in Table 1 of Chap. 4. A series of V_{hub} is to be selected within the range of mean wind speed corresponding to power production. As a conservative estimation, the 100-year return significant wave height independent of wind speed may be used to approximate $H_{S,\text{SSS}}(n - \text{yr}))$. A range of peak period associated with each significant wave height is to be determined for the load calculation. The resultant most unfavourable responses are to be used in the design of the Support Structure.

4.4 Extreme Sea State (ESS)

The Extreme Sea State (ESS) is a stochastic extreme design sea state with a given return period as defined in Table 1 of Chap. 4 for specific design load cases. The Extreme Sea State (ESS) is represented by a significant wave height, $H_{s,n-\text{yr}}$, with a return period of n years, a peak spectral period, and a wave direction. The values of $H_{s,n-\mathrm{yr}}$ are to be determined from onsite measurements, hindcast data, or both for the installation site. A range of peak spectral periods appropriate to site-specific metocean conditions is to be determined for load calculations, and the resultant most unfavourable responses are to be used in the design of the Support Structure.

4.5 Survival Sea State (SurSS)

The Survival Sea State (SurSS) condition is like the Extreme Sea State (ESS) defined in Chap. 3, Sect. 4.4, but with a return period longer than 50 years, as specified in Chap. 4, Sect. 3.3.

4.6 Breaking Waves

Where breaking waves are likely to occur at the installation site, the loads exerted by those breaking waves are to be assessed in the design. Breaking wave criteria are to be appropriate to the installation site and based on recognised methods. In shallow water the empirical limit of the wave height is approximately 0.78 times the local water depth. In deep water, a theoretical limit of wave steepness prior to breaking is Chap. 1, Sect. 4.3. Further guidance on breaking wave hydrodynamics can be found in IEC 61,400–3-1 (2019), Annex B.

5 Currents

5.1 Currents

Data for currents are to include information on current speed, direction, and variation with depth. The extent of information needed is to be commensurate with the expected severity of current conditions at the site in relation to other load-causing phenomena, experience in adjacent or similar areas, and the type of structure and foundation to be installed. Onsite data collection may be required for previously unstudied areas or areas expected to have unusual or severe conditions. Consideration is to be given to the following types of current, as appropriate to the installation site:

- Wind-generated current.
- Tide, density, circulation, and river-outflow generated sub-surface current.
- Near shore, breaking wave induced surface currents running parallel to the coast.
- The direction of wind generated surface current velocity is assumed to be aligned with the wind direction.

For near shore, the breaking wave induced surface current runs in parallel to the coastline, current speed may be estimated by the following equation:

$$U_b = 2\zeta\sqrt{gH_b}$$

where:

U_b speed of the near shore, breaking wave induced surface current running parallel to the coastline, in m/s (ft/s).

ζ sea floor slope, in rad.

g acceleration of gravity, in m/s^2 (ft/s^2).

H_b breaking wave height, in m (ft).

Current velocity profiles with depth are to be based on site-specific data or recognised empirical relationships. Unusual profiles due to bottom currents and stratified effects due to river out-flow currents are to be accounted for. For the design of offshore wind turbines to be installed on the US Outer Continental Shelf, the current profile is to be determined in accordance with Annex H, Annex I and Annex J of API RP 2MET (2019). The current models for the design of the Bottom-founded Offshore Wind Turbine are described in Chap. 3, Sects. 5.2 to 6.4. The design load conditions defined in Chap. 4, Sect. 3 specify how the current models are to be applied in combination with other design environmental conditions.

5.2 *Normal Current Model (NCM)*

The Normal Current Model (NCM) is to be determined based on the long-term joint probability distribution of metocean parameters at the installation site. For strength analyses, the Normal Current Model (NCM) is to represent the site-specific wind generated current conditioned upon a given 10-min mean wind speed at hub height (i.e., V_{hub}). Tide and storm generated sub-surface currents are not included. For fatigue analyses, the Normal Current Model (NCM) is to be determined in such a manner that the fatigue damage associated with the full long-term distribution of metocean parameters can be sufficiently accounted for.

5.3 *Extreme Current Model (ECM)*

The extreme current model is defined as site-specific currents with a given return period as specified in Table 1 of Chap. 4. The Extreme Current Model (ECM) is to be applied in combination with the Extreme Wind Model (EWM) defined in Chap. 3, Sect. 3.6, the Extreme Sea State (ESS) defined in Chap. 3, Sect. 4.4, and the Extreme Water Level Range Model (EWLR) defined in Chap. 3, Sect. 6.3, with consideration of their joint occurrence probabilities as required in Chap. 4, Sect. 3.2(10).

5.4 Survival Current Model (SurCM)

The Survival Current Model (SurCM) is like the Extreme Current Model (ECM) defined in Chap. 3, Sect. 6.3, but with a return period longer than 50 years, as specified in Chap. 4, Sect. 3.

6 Tides, Storm Surges, and Water Levels

6.1 General

Tides can be classified as lunar or astronomical tides, wind tides, and pressure differential tides. The combination of the latter two is commonly called the storm surge. The water depth at any location consists of the mean depth, defined as the vertical distance between the sea floor and an appropriate near-surface datum, and a fluctuating component due to astronomical tides and storm surges. Astronomical tide variations are bounded by the highest astronomical tide (HAT) and the lowest astronomical tide (LAT). Storm surge is to be estimated from available statistics or by mathematical storm surge modelling. The still water level (SWL) referred in the definition of environmental conditions for load calculation is to be taken as the highest still water level (HSWL), which is defined as the sum of the highest astronomical level and the positive storm surge. Definitions of various water levels referred in this book are illustrated in Fig. 1.

Variations in the elevation of the daily tide may be used in determining the elevations of boat landings, barge fenders and the top of the splash zone for corrosion protection of structure. Water depths assumed for various types of analysis are to be clearly stated. The water level ranges for the design of the Bottom-founded Offshore Wind Turbine are defined in Chap. 3, Sect. 6. The design load conditions defined in Chap. 4, Sect. 3, specify how the water level ranges are to be applied in combination with other design environmental conditions (Fig. 2).

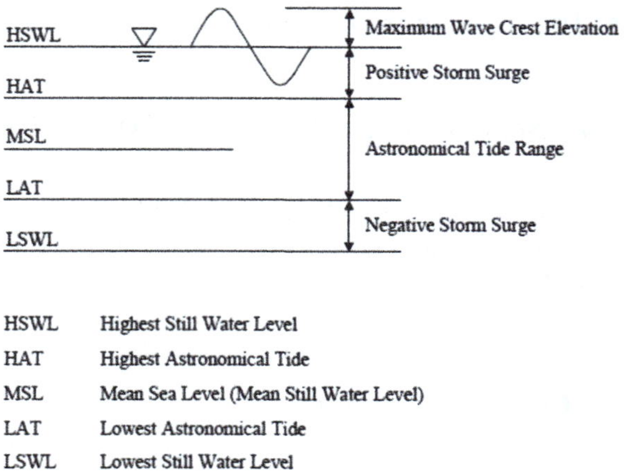

Fig. 2 Definitions of water levels

6.2 *Normal Water Level Range (NWLR)*

The Normal Water Level Range (NWLR) is defined as the variation in water level with a return period of one year. In the absence of site-specific long-term probability distribution of water levels, the normal water level range may be approximated by the variation between highest astronomical tide (HAT) and lowest astronomical tide (LAT). Load calculations for the strength analysis are to be performed based on the water level within the Normal Water Level Range (NWLR) to determine the most unfavourable responses. The influence of water level variation on fatigue loads is also to be considered, if deemed necessary.

6.3 *Extreme Water Level Range (EWLR)*

The Extreme Water Level Range (EWLR) is to be assumed for load cases associated with extreme wave conditions with a given return period as specified in Table 1 of Chap. 4. Load calculations for the strength analysis are to be performed based on the water level within the Extreme Water Level Range (EWLR) to determine the most unfavourable responses of the Support Structure. In the absence of the long-term joint probability distribution of metocean parameters including water level, the following water levels are to be considered as a minimum:

- The highest still water level (HSWL), defined as a combination of highest astronomical tide (HAT) and positive storm surge, with a given return period as specified in Table 1 of Chap. 4.
- The lowest still water level (LSWL), defined as a combination of lowest astronomical tide (LAT) and negative storm surge, with a given return period as specified in Table 1 of Chap. 4.

- The water level associated with the highest breaking wave load.

6.4 Survival Water Level Range (SurWLR)

The Survival Water Level Range (SurWLR) is defined in a manner like the Extreme Water Level Range (EWLR) in Chap. 3, Sect. 6.3, but with a return period longer than 50 years, as specified in Chap. 4, Sect. 3.

7 Other Conditions

7.1 Temperature

Extreme values of air, sea and seabed temperatures are to be expressed in terms of return periods and associated highest and lowest values. Wind speed data are typically presented with respect to a specific reference temperature. Temperature data is also to be used to evaluate the selection of air density, structural materials, ambient ranges and conditions for machinery and equipment design, and for determination of thermal stresses, as relevant to the Bottom-founded Offshore Wind Turbine.

7.2 Air Density

The air density is to be measured in conjunction with the wind conditions at the installation site. Where there are no site data for the air density, the value of air density is to be determined according to ISO 2533 and corrected as appropriate for annual average temperature at the installation site.

7.3 Ice and Snow Accumulation

For an installation site where ice and snow may accumulate, estimates are to be made of the extent to which ice and snow may accumulate. Data are to be derived from actual field measurements, laboratory data or data from analogous areas.

7.4 *Marine Growth*

Marine growth is to be considered in the design of the Bottom-founded Offshore Wind Turbine. Estimates of the rate and extent of marine growth may be based on experience and available field data. Particular attention is to be paid to increases in hydrodynamic loading due to increased diameters and surface roughness of members caused by marine fouling as well as to the added weight and increased inertial mass of submerged structural members. The types of fouling likely to occur and their effects on corrosion protection coatings are to be considered.

7.5 *Seismicity and Earthquake Related Phenomena*

The effects of earthquakes on the Bottom-founded Offshore Wind Turbine located in areas known to be seismically active are to be considered.

7.5.1 Levels of Earthquake Conditions

The magnitudes of the parameters characterising the earthquakes with return periods appropriate to the design life and the intended safety level of the Bottom-founded Offshore Wind Turbine are to be determined. Two levels of earthquake conditions are to be considered to address the risk of damage and structure collapse, respectively:

(1) ***Strength Level.*** Ground motion which has a reasonable likelihood of not being exceeded at the site during the design life of the Bottom-founded Offshore Wind Turbine.
(2) ***Ductility Level.*** Ground motion for a rare, intense earthquake to be applied to evaluate the risk of structural collapse.

7.5.2 Regional and Site-Specific Data

The anticipated seismicity of an area is, to the extent practicable, to be established based on suitable regional and site-specific data including the following:

(1) Magnitudes and recurrence intervals of seismic events.
(2) Proximity to active faults.
(3) Type of faulting.
(4) Attenuation of ground motion between the faults and the site.
(5) Subsurface soil conditions.
(6) Records from past seismic events at the site where available, or from analogous sites.

7.5.3 Other Earthquake Related Phenomena

Seismic data are to be used to establish quantitative Strength Level and Ductility Level earthquake criteria describing the earthquake induced ground motion expected during the design life of the Bottom-founded Offshore Wind Turbine. In addition to ground motion, and as applicable to the site in question, the following earthquake related phenomena are to be considered.

(1) Liquefaction of subsurface soils submarine slides.
(2) Tsunamis.
(3) Acoustic overpressure shock waves.

7.6 Sea Ice or Lake Ice

For an installation site where ice hazards may occur, the effects of sea ice or lake ice on the Support Structure of an offshore wind turbine are to be considered in the design. Depending on the ice conditions at the site, the Support Structure may encounter moving ice and fast ice cover. Statistical ice data of the site are to be used as the base for deriving the parameters such as ice thickness, ice crushing strength and pack ice concentration, etc., which are required for determining the ice loads. Impact, both centric and eccentric, is to be considered where moving ice may impact the Support Structure. Impact of smaller ice masses, which are accelerated by storm waves, and of large masses (multi-year floes and icebergs) moving under the action of current, wind, and Coriolis effect is to be considered in the design. The interaction between ice and the Support Structure produces responses both in the ice and the structure soil system, and this compliance is to be considered as applicable.

7.7 Lightning

Lightning protection is to be designed in accordance with IEC 61,400–24. It is not necessary for protective measures to extend to all parts of the Bottom-founded Offshore Wind Turbine if safety is not compromised.

7.8 Electric Network Conditions

The assessment of electric network conditions is to be performed in accordance with IEC 61,400–3-1 (2019).

Chapter 4
Loads

1 General

This chapter pertains to the identification, definition, and determination of the loads which the Bottom-founded Offshore Wind Turbine may experience in its pre-service (load-out, transportation, installation, and commissioning) and in-service (operation, maintenance, and repair) phases. As appropriate to the planned Bottom-founded Offshore Wind Turbine, the types of loads described in Chap. 4, Sect. 2 are to be considered in the design.

2 Types of Loads to Be Considered

Loads applied to the Bottom-founded Offshore Wind Turbine are, for the purposes of this book, categorised as dead loads, live loads, deformation loads and environmental loads.

2.1 Dead Loads

Dead loads are loads which do not change during the mode of operation under consideration. Dead loads include, but are not limited to, the following:

(1) Weight of rotor components (blades, hub, shaft, etc.) and equipment inside the nacelle (control and protection components, gearbox, drive train components, electrical generation components, cables, etc.)
(2) Weight of nacelle housing structure, tower, substructure, foundation, access platform, fenders, ladders, corrosion protection system, and other permanent structures.

F. Aderouane, *Design, Construction, and Global Performance Analysis of Offshore Wind Turbines*, Green Energy and Technology, https://doi.org/10.1007/978-3-031-86010-2_4

(3) External hydrostatic pressure and buoyancy calculated based on the SWL Static earth pressure

2.2 *Live Loads*

Live loads associated with the normal operation of a wind turbine are loads which may change during the mode of operation considered. Live loads acting after construction and installation may include, but are not limited to, the following.

(1) Forces exerted on the wind turbine due to lifting equipment during installation and maintenance of the wind turbine.
(2) Forces exerted by vessels moored to the structure or routine impact loads from a typical supply vessel that would normally service the offshore wind turbine
(3) Loads associated with helicopter operations, where relevant.
(4) Actuation loads generated by wind turbine operations and controls including torque control from a generator or inverter, yaw and pitch actuator loads, and mechanical braking loads. The range of actuator forces is to be considered as appropriate in the calculation of response and loading. In particular, the range of friction, spring force or pressure for mechanical brakes is influenced by temperature and ageing, which are to be considered when calculating the response and loading during any brake event.

Where applicable, the dynamic effects of the live loads on the Support Structure are to be considered. Live loads encountered during load-out, transportation and installation are to be determined for each specific operation involved. The dynamic effects of such loads, as necessary, are to be accounted for (refer to Chap. 5, Sect. 4. 7).

2.3 *Deformation Loads*

Deformation loads are loads due to deformations imposed on the Support Structure. The deformation loads include those due to temperature variations leading to thermal stress in the structure and, where necessary, loads due to soil displacements (e.g., differential settlement or lateral displacement) or due to deformations of adjacent structures. For concrete structures, deformation loads due to prestress, creep, shrinkage, and expansion are to be considered.

2.4 *Environmental Loads*

Environmental loads are loads due to the action of wind, wave, current, ice, snow, earthquake, marine growth, and other environmental phenomena as described in

Chap. 3. The characteristic parameters defining environmental loads are to be appropriate to the installation site and in accordance with the requirements specified in Chap. 3 of this book. The combination and severity of environmental conditions for the design of the Support Structure are specified in Chap. 4, Sect. 3. Calculations of environmental loads are to be in accordance with Chap. 4, Sect. 4. Environmental loads are to be applied from directions producing the most unfavourable effects on the Support Structure unless site-specific studies provide evidence in support of a less stringent requirement. Directionality may be considered in applying the environmental criteria.

3 Design Load Conditions

3.1 General

Design load conditions of the Bottom-founded Offshore Wind Turbine are to be represented by a set of Design Load Cases (DLCs), which are defined by various combinations of turbine operational modes, site specific environmental conditions, electrical network conditions and other applicable design conditions, such as specific transportation, assembly, maintenance, or repair conditions. All relevant DLCs with a reasonable probability of occurrence and covering the most significant conditions that an offshore wind turbine may experience are to be considered in the design. For the Bottom-founded Offshore Wind Turbine installed in a tropical cyclone region, Survival Load Cases (SLCs) may be assessed to verify its structural robustness. Assessment of the tower for the design load conditions defined in this chapter is required for the following situations (refer to Chap. 1, Sect. 3):

- The optional RNA notation is requested by the Owner.
- The optional RNA notation is not requested by the Owner, however the tower and its connection to the RNA and/or to the substructure are not included in the wind turbine type certificate.

3.2 Definition of Design Load Cases (DLCs)

As a minimum, the DLCs defined in Table 1 of Chaps. 1 and 2 are to be suitably assessed in the design of the Bottom-founded Offshore Wind Turbine. Other design load cases are to be considered, wherever they are deemed relevant to the structural integrity of a specific design of the Support Structure. If correlation exists between an extreme environmental condition and a fault condition of wind turbine, a realistic combination of the two is to be considered as a design load case. For each DLC defined in Tables 1 and 2 of Chaps. 1 and 2, the 'Type of Analysis' is denoted 'S' for the strength analysis or 'F' for the fatigue analysis. Results of the strength analysis are used in the structural assessment against acceptance criteria pertaining to the yielding and buckling. Results of the fatigue analysis are used in the structural assessment against criteria pertaining to fatigue performance. Those DLCs indicated with 'S,' are further classified as:

(1) 'N' representing normal design conditions, which are expected to occur frequently during the design life of the Bottom-founded Offshore Wind Turbine. The corresponding operational mode of the turbine is in a normal state or with minor faults or abnormalities.
(2) 'A' representing abnormal design conditions, which are less likely to occur than normal design conditions. They usually correspond to design conditions with severe faults that result in activation of system protection functions.
(3) 'T' representing the design conditions relevant to temporary operations including load-out, transportation, installation, maintenance, and repair.

The type of design conditions, 'N,' 'A,' or 'T' determines the safety factors in the strength design criteria to be applied to the Support Structure. These safety factors are defined in Chap. 5. The DLCs specified in Table 1 are in agreement with those required by "Table 2—Design load cases" of IEC 61,400–3-1 (2019), with various modifications to:

(1) Address the design requirements for the Support Structure to be installed in an offshore site where the extreme environmental condition is governed by tropical cyclones, (also known as hurricanes or typhoons).
(2) Provide further interpretations of the DLCs in terms of site-specific design requirements for the Support Structure.

The following statements pertain to the required DLCs in Table 1. Further guidance on the analysis requirements can be found in paragraph 7.4 and paragraph 7.5.4 of IEC 61,400–3-1 (2019):

(1) The DLCs defined in Table 1 are for the design of the Bottom-founded Offshore Wind Turbine, which represents the site-specific Support Structure of an offshore wind turbine. As such, DLC 1.1 required by IEC 61,400–3-1 (2019) for the calculation of ultimate loads acting on the RNA is not included in Table 1.
(2) The safety factors indicated in the column entitled 'Safety Factor' in Table 1 are specified in Chap. 5 of this book.
(3) The design environmental conditions referred in Table 1 for wind, waves, sea currents, and water level ranges are in accordance with the definitions specified in Chap. 3 of this book. Detailed references are listed in the table notes.
(4) Site-specific extreme wind speeds with various combinations of return periods and averaging time durations are used to define the environmental conditions in DLC's with serial number 6.1, 6.2, 6.3, 6.4, 7.1, 7.2, 8.2 and 8.3. This differs from IEC 61,400–3-1 (2019) where the extreme wind speeds in these DLCs are defined using the RNA's Reference Wind Speed (V_{ref}) in combination with various prescribed conversion factors for different averaging time durations or return periods.

Table 1 Design load cases

Design condition	DLC	Wind condition	Waves	Wind and wave directionality	Sea currents	Water level	Other conditions	Type of analysis	Safety factor
(1) Power production	1.2	NTM $V_{in} \leq V_{hub} \leq V_{out}$	NSS Joint prob. distribution of H_s, T_p, V_{hub}	MIS, MUL	NCM	NWLR or $\geq$ MSL		F	FDF
	1.3	ETM $V_{in} \leq V_{hub} \leq V_{out}$	NSS $H_s = E[Hs\|V_{hub}]$	COD, UNI	NCM	MSL		S	N
	1.4	ECD $V_{in} \leq V_{hub} \leq V_{out}$	NSS $H_s = E[H_s\|V_{hub}]$	MIS, wind direction change	NCM	MSL		S	N
	1.5	EWS $V_{in} \leq V_{hub} \leq V_{out}$	NSS $H_s = E[H_s\|V_{hub}]$	COD, UNI	NCM	MSL		S	N
	1.6	NTM $V_{in} \leq V_{hub} \leq V_{out}$	NSS $H_s = H_s, SSS(n - yr)$	MIS, MUL	NCM	NWLR	$n = 50$ in general; $n = 100$ for tropical cyclone-prone sites (refer to Note 1 for further guidance)	S	N

(continued)

Table 1 (continued)

Design condition	DLC	Wind condition	Waves	Wind and wave directionality	Sea currents	Water level	Other conditions	Type of analysis	Safety factor
(2) Power production	2.1	NTM $V_{in} \leq V_{hub} \leq V_{out}$	NSS $H_s = E[Hs\|V_{hub}]$	COD, UNI	NCM	MSL	Normal control system fault or loss of electrical network or primary layer control function fault	S	N
	2.2	NTM $V_{in} \leq V_{hub} \leq V_{out}$	NSS $H_s = E[Hs\|V_{hub}]$	COD, UNI	NCM	MSL	Abnormal control system fault or secondary layer protection function related fault	S	A
	2.3	EOG $V_{hub} = V_r \pm 2m/s(6.6ft/s) and V_{out}$	NSS $H_s = E[Hs\|V_{hub}]$	COD, UNI	NCM	MSL	External or internal electrical fault including loss of electrical network	S	A
	2.4	NTM $V_{in} \leq V_{hub} \leq V_{out}$	NSS $H_s = E[Hs\|V_{hub}]$	COD, UNI	NCM	MSL	Control system fault, electrical fault or loss of electrical network	S	FDF

(continued)

Table 1 (continued)

Design condition	DLC	Wind condition	Waves	Wind and wave directionality	Sea currents	Water level	Other conditions	Type of analysis	Safety factor
	2.5	NTM $V_{in} \leq V_{hub} \leq V_{out}$	NSS $H_s =$ $E[Hs\|V_{hub}]$	COD, UNI	NCM	NWLR or $\geq$ MSL	Low voltage ride through	S	N
(3) Start Up	3.1	NWP $V_{in} \leq V_{hub} \leq V_{out}$	NSS $H_s =$ $E[Hs\|V_{hub}]$	COD, UNI	NCM	NWLR or $\geq$ MSL		F	FDF
	3.2	EOG $V_{hub} =$ $V_r \pm 2m/s(6.6ft/s)andV_{out}$	NSS $H_s =$ $E[Hs\|V_{hub}]$	COD, UNI	NCM	MSL		S	N
	3.3	EOG $V_{hub} =$ $V_r \pm 2m/s(6.6ft/s)andV_{out}$	NSS $H_s =$ $E[Hs\|V_{hub}]$	MIS, wind direction change	NCM	MSL		S	N
(4) Normal Shut Down	4.1	NWP $V_{in} \leq V_{hub} \leq V_{out}$	NSS $H_s =$ $E[Hs\|V_{hub}]$	COD, UNI	NCM	NWLR or $\geq$ MSL		F	FDF
	4.2	EOG $V_{hub} =$ $V_r \pm 2m/s(6.6ft/s)andV_{out}$	NSS $H_s =$ $E[Hs\|V_{hub}]$	COD, UNI	NCM	MSL		S	N
(5) Emergency Stop	5.1	NTM $V_{hub} =$ $V_r \pm 2m/s(6.6ft/s)andV_{out}$	NSS $H_s =$ $E[Hs\|V_{hub}]$	COD, UNI	NCM	MSL		S	N

(continued)

Table 1 (continued)

Design condition	DLC	Wind condition	Waves	Wind and wave directionality	Sea currents	Water level	Other conditions	Type of analysis	Safety factor
(6) Parked (standing still or idling)	6.1	EWM Turbulent Wind Model $V_{hub} = V_{10min,n-yr}$	ESS $H_s = H_{s,n-yr}$	MIS, MUL	ECM *n*-yr Current	EWLR *n*-yr Water Level	$n = 50$ in general; $n = 100$ for tropical cyclone-prone sites (refer to Note 1 for further guidance)	S	N
	6.2	EWM Turbulent Wind Model $V_{hub} = V_{10min,n-yr}$	ESS $H_s = H_{s,n-yr}$	MIS, MUL	ECM *n*-yr Current	EWLR *n*-yr Water Level	Loss of electrical network $n = 50$ in general; $n = 100$ for tropical cyclone-prone sites (refer to Note 1 for further guidance)	S	A
	6.3	EWM Turbulent Wind Model $V_{hub} = V_{10min,n-yr}$	ESS $H_s = H_{s,n-yr}$	MIS, MUL	ECM *n*-yr Current	NWLR	Extreme yaw misalignment $n = 1$	S	N
	6.4	NTM $V_{hub} = V_{10min,n-yr}$	NSS Joint prob. distribution of H_s, T_p, V_{hub}	MIS, MUL	NCM	NWLR or $\geq$ MSL	$n = 1$	F	FDF

(continued)

Table 1 (continued)

Design condition	DLC	Wind condition	Waves	Wind and wave directionality	Sea currents	Water level	Other conditions	Type of analysis	Safety factor
(7) Parked and fault conditions	7.1	EWM Turbulent Wind Model $V_{hub} = V_{10min,n-yr}$	ESS $H_s = H_{s,n-yr}$	MIS, MUL	ECM n-yr Current	NWLR	$n = 1$	S	A
	7.2	NTM $V_{hub} = V_{10min,n-yr}$	NSS Joint prob. distribution of H_s, T_p, V_{hub}	MIS, MUL	No current	NWLR or $\geq$ MSL	$n = 1$	F	FDF
(8) Transport, assembly, maintenance and repair	8.1	To be defined by the Fabricator and/or Owner						S	T
	8.2	EWM Turbulent Wind Model $V_{hub} = V_{10min,n-yr}$	ESS $H_s = H_{s,n-yr}$	MIS, MUL	ECM n-yr Current	NWLR	$n = 1$	S	A
	8.3	NTM $V_{hub} = V_{10min,n-yr}$	NSS Joint prob. distribution of H_s, T_p, V_{hub}	MIS, MUL	NCM	NWLR or $\geq$ MSL	No grid during installation period $n = 1$	F	FDF
	8.4	To be defined by the Fabricator and/or Owner						F	FDF

Notes. The reduction of the return period of 100 years to 50 years in tropical cyclone prone areas may be acceptable provided additional safety factors are applied or robustness assessments for the Survival Load Cases are performed to justify such reduction. The symbols and abbreviations used in the table are summarised as follows: *A—abnormal; COD—co-directional (aligned) wind and wave direction; DLC—design load case; ECD—extreme coherent gust with direction change; ECM—extreme current model; EDC—extreme direction change; EOG—extreme operating gust; ESS—extreme sea state; ETM—extreme turbulence model; EWLR—extreme water level range; EWM—extreme wind speed model; EWS—extreme wind shear; F—fatigue; FDF—fatigue design factor;* $H_{n\text{-}yr}$*—extreme wave height with a return period of n years;* H_s*—significant wave height;* $H_{s,\,n\text{-}yr}$*—significant wave height of the extreme sea state with a return period of n years;* $H_{s,SSS}$*—significant wave height of the severe sea state with a return period of n years; MIS—misaligned wind and wave directions; MSL—mean sea level; MUL—multi-directional wind and wave; N—normal; n—return period of an extreme environmental condition; NCM—normal current model; NSS—normal sea state; NTM—normal turbulence model; NWLR—normal water level range; NWP—normal wind profile model; S—strength; SSS—severe sea state; T—transport, assembly, maintenance and repair; Tp—peak period of wave spectrum; UNI—uni-directional wind and wave directions;* $V_{10min,n\text{-}yr}$*—10-min mean wind speed at hub height with a return period of n years;* $V_{3s,n\text{-}yr}$*—3-s mean wind speed at hub height with a return period of n years;* $V_{(zhub)}$*—steady wind speed at hub height,* z_{hub}*;* $V_{1min,n\text{-}yr}$*—1-min mean wind speed at hub height with a return period of n years;* V_{hub}*—10- minute mean wind speed at hub height;* V_{in}*—cut-in wind speed;* V_{out}*—cut-out wind speed;* V_r*—rated wind speed;* $V_{r\,\pm\,2\,m/s}$*—(66 ft/s) sensitivity to the wind speeds in the given range*

Table 2 Design load cases for ice conditions

Design Condition	DLC	Ice Condition	Wind Condition	Water Level	Type of Analysis	Safety Factor
Power Production	D1	Horizontal load from temperature fluctuations	NTM $V_{hub} = V_{10min,n-yr}$ Wind Speed resulting in maximum thrust	NWLR	S	N
	D2	Horizontal load from water fluctuations or arch effect	NTM $V_{hub} = V_r \pm 2m/s(6.6ft/s) and V_{out}$ Wind Speed resulting in maximum thrust	NWLR	S	N
	D3	Horizontal load from moving ice floe at relevant velocities $h_{ice} = h_{ice,\ 50\text{-}yr}$ in open sea $h_{ice} = h_{ice,\ m}$ for landlocked waters	NTM $V_{hub} = V_r \pm 2m/s(6.6ft/s) and V_{out}$ Wind Speed resulting in maximum thrust	NWLR	S	N
	D4	Horizontal load from moving ice floe at relevant velocities $h_{ice} = h_{ice,\ 50\text{-}yr}$ in open sea $h_{ice} = h_{ice,\ m}$ for landlocked	NWP $V_{in} \le V_{hub} \le V_{out}$	NWLR	F	FDF
	D5	Vertical force from fast ice covers due to water level fluctuations	No wind load applied	NWLR	S	N

(continued)

Table 2 (continued)

Design Condition	DLC	Ice Condition	Wind Condition	Water Level	Type of Analysis	Safety Factor
Parked	D6	Pressure from hummocked ice and ice ridges	EWM $V_{hub} = V_{10min,n-yr}$	NWLR	S	N
	D7	Horizontal load from moving ice floe at relevant velocities $h_{ice} = h_{ice,\ 50\text{-}yr}$ in open sea $h_{ice} = h_{ice,\ m}$ for landlocked waters	NTM $V_{hub} = V_{10min,n-yr}$	NWLR	F	FDF

Notes The symbols and abbreviations used in the table are described in the *Notes* of Table 1 in addition to those summarised as follows. h_{ice}—ice thickness; $h_{ice,\ 50\text{-}yr}$—extreme ice thickness with a return period of 50 years; $h_{ice,\ m}$—long-term mean value of the annual maximum ice thickness

(5) The return period n of each extreme or severe environmental condition is specified in the column entitled 'Other Conditions' in Table 1. Any reduction to the return period is subject to special consideration by Class. The return period is not to be less than 50 years for the extreme environmental conditions of DLC 6.1 and DLC 6.2 and the severe wave conditions of DLC 1.6. Such return period is to be increased to 100 years for designing the Support Structure of an offshore wind turbine at an installation site where the extreme storm condition is expected to be determined by tropical cyclones (i.e., tropical cyclone-prone site, as indicated in Table 1). The increase in the return period is to address higher variations of the severity of extreme storm in tropical cyclone-prone offshore regions such as the Gulf of Mexico and the East Atlantic Coast in the US, where the extreme wind speed versus the return period curve typically shows a steep slope. The reduction of the return period of 100 years to 50 years in tropical cyclone prone areas may be acceptable provided additional safety factors are applied or robustness assessments for the Survival Load Cases (refer to Chap. 4, Sect. 3.3) are performed to justify such reduction.

(6) DLC 6.2 assume a loss of connection to electrical power network at an early stage of the storm containing the extreme wind condition. A nacelle yaw misalignment ranging between -180° and + 180° is required to be considered. Load calculations are to be based on the misalignment angle that results in the most unfavourable responses in the Bottom-founded Offshore Wind Turbine. The range of yaw misalignment may be reduced to account for the contribution

from an active or passive yaw control system, provided the designer can justify that:

- Such a system can achieve the assumed reduction of yaw misalignment under site specific conditions.
- An appropriate monitoring and maintenance programme is implemented to maintain the effectiveness of yawing control during the design life of the Bottom-founded Offshore Wind Turbine.

(7) For a site where the turbulent wind models defined in Chap. 3, Sect. 3 are considered insufficient to describe site-specific wind conditions, the DLC's in Table 1 involving the combined steady wind model and deterministic design wave height are to be evaluated for the strength analysis.

(8) For those load cases, including DLC 1.6, DLC 6.1, 6.2, 6.3, 7.1, and 8.2, which in general require time domain dynamic simulations for the combined turbulent wind and stochastic extreme waves, the simulation time duration may differ from the reference periods of wind speed and significant wave height. The extreme responses determined for the wave condition used in the simulation are to be equivalent to those obtained from 3-h simulations. IEC 61,400–3-1 (2019), paragraph 7.5.6 recommends at least six 1-h constrained wave simulations be performed. Other approaches may be taken if the estimated extreme response is shown not less than that obtained with the approach recommended by IEC 61,400–3-1 (2019), paragraph 7.5.6.

(9) Where a wind speed range is indicated in Table 1, wind speeds leading to the most unfavourable responses are to be considered for the structural and foundation design. When the range of wind speeds is represented by a set of discrete values, the interval between two adjacent discrete wind speeds is not to be greater than 2 m/s (66 ft/s), In addition, the turbine Rated Wind Speed (V_r), where applicable, is to be included as one of the discrete wind speeds to be used in the load calculation.

(10) Extreme metocean conditions in DLC 6.1, 6.2, 6.3, 7.1 and 8.2 are formed by combining the turbulent extreme wind (EWM), the stochastic extreme wave (ESS), the extreme current (ECM), and the extreme water level range (EWLR). The probability of joint occurrence of these environmental parameters is to be considered when establishing extreme metocean conditions. Consideration is to be given to the peak wind, peak wave, and peak current condition (refer to e.g., API RP 2MET, 2019), as appropriate to site conditions and a specific design of the Support Structure. Combining all individual extremes with the same return period is normally a conservative approach.

(11) For those DLCs denoted by 'S' in the 'Type of Analysis' column in Table 1 for the structural strength design, the effect of environmental loads is to be suitably combined with the effect of dead loads, live loads, and deformation loads (refer to Chap. 4, Sect. 2). Combinations of the load effects that produce the most severe local and/or global effects on the Support Structure are to be considered.

(12) If site-specific directional data are not available, the direction of applied environmental conditions is to be determined to produce the most unfavourable effect on the Support Structure. For DLCs with the serial number 6.1, 6.2, 6.3, 7.1 and 8.3, the misalignment between wind and wave directions is to be considered up to 90° for extreme environmental conditions governed by tropic cyclones.

(13) The effect of currents is required to be considered in those DLCs denoted by 'F' (i.e., fatigue analysis) in the 'Type of Analysis' column in Table 1.

(14) A new DLC 8.4 is added to account for the fatigue damage occurred during transport, assembly, maintenance, and repair.

(15) Wind and wave directionality for DLC 1.2, 1.6, 6.4, 7.2 and 8.3 is to consider the misalignment of wind and wave directions. If site-specific directional data are not available, the direction of applied environmental conditions is to be determined to produce the most unfavourable effect on the Support Structure.

For the Bottom-founded Offshore Wind Turbine to be installed at a site where ice is expected to occur, the DLCs specified in Table 2 for ice conditions are to be considered. The DLCs specified in Table 2 for ice conditions are in agreement with those required by "Table 3—Design load cases for sea ice" in IEC 61,400–3-1 (2019). The descriptions and analysis requirements for DLCs defined in Table 2 can be referred to Annex D of IEC 61,400–3-1 (2019), in conjunction with the amendments described as follows:

(1) The safety factors referred in Table 2 are defined in Chap. 5 of this book.

(2) The design environmental conditions referred in Table 2 for ice, wind and water level ranges are in accordance with the definitions in Chap. 3 of this book.

(3) Site-specific extreme wind speeds with a given return period and averaging time duration are used to define the extreme wind conditions.

(4) The return period chosen for the extreme ice conditions of DLC D3, D4 and D7 is not to be less than 50 years. Any reduction to the return period of environmental conditions is subject to special consideration by Class.

(5) Reference is also made to API RP 2N and ISO 19906 for the alternative methods of determining static and dynamic ice loads.

3.3 Definition of Survival Load Cases (SLCs)

Where the robustness assessment of the Support Structure in a tropical cyclone region is chosen to demonstrate the ultimate strength capacity of the Support Structure, the Survival Load Cases (SLCs) defined in this chapter are to be applied. As a minimum, the SLCs specified in Table 3 are to be assessed. The probability of joint occurrence of environmental parameters are to be considered when establishing survival metocean conditions, with the consideration of the peak wind, peak wave, and peak current condition (refer to API RP 2MET). The effect of environmental loads is to be combined with the effect of dead loads, live loads, and deformation loads. Combinations of the load effects that produce the most unfavourable effects on the Support Structure are to be used to assess the design adequacy. The safety factors applicable for the survival load case are described in Chap. 5. The differentiation of 'N,' 'A' and 'T' turbine operating conditions is not applicable to the survival load cases.

Table 3 Survival load cases—tropical cyclone regions

Design condition	Wind Condition	Waves	Wind and wave directionality	Sea currents	Water level	Other conditions
Parked (standing still or idling)	SurWM $V_{hub} = V_{10min,n-yr}$	SurSS $H_s = H_{s,n-yr}$	MIS, MUL	SurCM *n*-yr Currents	SurWLR *n*-yr Water Level	Intact blade(s)
	SurWM $V_{hub} = V_{10min,500-yr}$	SurSS $H_s = H_{s,500-yr}$	MIS, MUL	SurCM500-yr Currents	SurWLR 500-yr Water Level	Damaged blade(s), where relevant (1)
	SurWM $V_{hub} = V_{10min,m-yr}$	SurSS $H_s = H_{s,m-yr}$	MIS, MUL	SurCM*m*-yr Currents	SurWLR *m*-yr Water Level	Loss of electrical network (2)

Notes. "Damaged blade(s)" case is to be assessed if one turbine blade or multiple turbine blades cannot remain intact under the storm wind condition with a return period of 500 years (i.e., n < 500 years). For the 'Loss of electrical network' case, the return period 'm' for extreme environmental conditions is selected such that the joint event of the loss of yaw power and controls during the extreme environmental conditions has the probability of 1/500, or equivalently a return period of 500 years. The value of 'm' should be selected and justified by the designer. For sites expected to experience a loss of network power, or in the absence of information defining the network reliability, a return period of 500 years for environmental conditions may be used as a conservative value, unless back-up power is to be assessed. *The symbols and abbreviations used in the table are summarised as follows: $n-yr$: maximum return period (n years) of the storm wind condition that turbine blades can sustain and remain intact or 500 years, whichever is less; $m-yr$: : maximum return period (m years) of the storm wind condition as defined above in item (2).; $H_{s,n-yr}$: significant wave height with a return period of n years; $H_{s,500-yr}$: significant wave height with a return period of 500 years; $V_{10min,n-yr}$: 10-min mean wind speed at hub height with a return period of n years; $V_{10min,500-yr}$: : 10-min mean wind speed at hub height with a return period of 500 years; SurWM: survival wind model (Chap. 3,* Sect. 3.7*); SurSS: survival sea state (Chap. 3,* Sect. 4.5*); SurCM: survival current model (Chap. 3,* Sect. 5.4*); SurWLR: survival water level range (Chap. 3,* Sect. 6.4*); Other symbols and abbreviations used in the table are defined in the Notes of Table 1*

4 Determination of Environmental Loads

4.1 *General*

Environmental loads are to be determined using analytical methods compatible with the environmental condition models established in compliance with Chap. 3. Any recognised load calculation method may be employed provided it has been proven sufficiently accurate in practice, and it is shown to be appropriate to the structure's characteristics and site conditions. Model or onsite test data can also be used to establish environmental loads.

4.2 *Wind Loads*

Wind loads and local wind pressures are to be determined based on analytical methods or wind tunnel tests using a representative model of the offshore wind turbine. Static and dynamic wind load effects generated directly by inflowing wind and indirectly by wind generated responses of the RNA are to be considered.

4.2.1 Aerodynamic Loads Generated by the Rotor

Aerodynamic loads induced by airflow passing through the rotor are determined by the mean wind speed and turbulence across the rotor plane, rotor rotational speed, air density, and aerodynamic shapes of wind turbine components, as well as interactive effects such as aero-elasticity and rotational sampling. Aerodynamic loads due to these effects are to be calculated using recognised methods and computer programmes. The wind models in 3/2 and the DLCs in 4/3 specify the design load conditions for which aerodynamic loads generated by the rotor are to be calculated. For offshore wind turbines installed in a wind farm, the potential shadow effect and wake effect on the loads are to be considered for both the strength and fatigue analyses. For large wind farms, an increase in the turbulence intensity or terrain roughness is to be considered. The mutual influence of offshore wind turbines through the wake interaction behind the rotor is to be considered up to 10 times of the rotor diameter. Reference is made to IEC 61,400–1 for the guidance on the wake effect from neighbouring offshore wind turbines.

4.2.2 Wind Forces on Exposed Structural Components

For wind pressure normal to flat surfaces, such as nacelle and boat landing, or normal to the axis of members not having flat surfaces, such as tower and exposed above-water tubular members of the Support Structure, the wind loading can be considered as a constant (steady) force determined as follows:

$$F_w = \left(\frac{\rho}{2g}\right) C_s A V^2$$

where:

F_w wind drag load, in N (lb).
ρ weight density of air, in N/m^3 (lb/ft^3).
G acceleration of gravitation, in m/s^2 (ft/s^2)
C_s shape coefficient.
A projected area of member on a plane normal to the direction of the considered force, in m^2 (ft^2).
V wind speed at a given elevation above the SWL, in m/s (ft/s).

In the absence of experimental data, values in Table 4 for the shape coefficient (C_s) are to be applied. For any direction of wind approaching to the structure, the wind force on flat surfaces is to be assumed to act normal to the surface. The wind force on cylindrical objects is to be assumed to act in the direction of the wind. The conversion of mean wind speeds with different averaging time durations and the applicable wind profile are to follow 3/3.2.1.

The area of open truss works commonly used for derricks and crane booms may be approximated by taking 30% of the projected area of both the windward and leeward sides with the shape coefficient taken in accordance with Table 4. Where one structural member shields another from direct exposure to the wind, shielding may be considered. Two structural components are to be separated by not more than seven times the width of the windward component for a reduction to be taken in the wind load on the leeward member. Where applicable, cyclic loads due to vortex induced vibration (VIV) of structural members exposed to wind loads are to be investigated. Both drag and lift components of load due to vortex induced vibration are to be considered. The effects of wind loading on structural members or components that are not normally exposed to wind loads after installation are to be considered, especially during fabrication or transportation phases.

Table 4 Values of C_s

Shape	Values of C_s
Sphere	0.40
Cylindrical shape	0.50
Major flat surfaces and overall projected area of the Support Structure	1.00
Isolated structural shapes (cranes, angles, beams, channels, etc.)	1.50
Under deck areas (smooth)	1.00
Under-deck areas (exposed beams and girders)	1.30
Derricks or truss cranes (each face)	1.25
Sides of buildings	1.50

4.2.3 Wave Loads

A sufficient range of realistic wave periods and wave crest positions relative to the structure are to be investigated to determinate the maximum wave loads on the Support Structure. Consideration is also to be given to other wave induced effects such as wave impact loads, dynamic amplification, and fatigue of structural members. The need for analysis of these effects is to be assessed based on the configuration and behavioural characteristics of the Support Structure, the wave conditions and experience. For structures composed of members having diameters which are less than 20% of the wave lengths being considered, semi-empirical formulations such as Morison's equation are an acceptable basis for determining wave loads. For structures composed of members whose diameters are greater than 20% of the wavelengths being considered, or for structural configurations that alter the incident flow field, diffraction forces and the hydrodynamic interaction of structural members are to be accounted for in the design. The hydrodynamic force acting on a cylindrical member, as given by Morison's equation, is expressed as the sum of the force vectors indicated in the following equation:

$$F = F_D + F_I$$

where:

F Hydrodynamic force vector per unit length along the member, acting normal to the axis of the member, in N/m (lb/ft).
F_D Drag force vector per unit length, in N/m (lb/ft).
F_I Inertia force vector per unit length, in N/m (lb/lf).

The drag force vector for a stationary, rigid member is given by:

$$F_D = \left(\frac{\rho}{2g}\right) D C_d u_n |u_n|$$

where:

ρ Weight density of water, in N/m^3 (lb/ft^3).
g Acceleration of gravitation, in m/s^2 (ft/s^2).
D Projected width of the member in the direction of the cross-flow component of velocity (in the case of a circular cylinder, D denotes the diameter), in m (ft).
C_D Drag coefficient.
u_n Component of the fluid velocity vector normal to the axis of the member, in m/ s (ft/s).
$|u_n|$ Absolute value of u_n, in m/s (ft/s).

The inertia force vector for a stationary, rigid member is given by:

$$F_I = \left(\frac{\rho}{g}\right)\left(\frac{\pi D^2}{4}\right) C_M a_n$$

where:

C_M Inertia coefficient based on the displaced mass of fluid per unit length.
a_n Component of the fluid acceleration vector normal to the axis of the member, in m/s^2 (ft/s^2).

For the compliant turbine Support Structures which exhibit substantial rigid body oscillations due to the wave action, the modified form of Morison's equation given below may be used to determine the hydrodynamic force:

$$F_D = \left(\frac{\rho}{2g}\right) D C_D (u_n - \overline{u}_n)|u_n - \overline{u}_n|$$

$$F_I = \left(\frac{\rho}{2g}\right)\left(\frac{\pi D^2}{4}\right) a_n + \left(\frac{\rho}{g}\right)\left(\frac{\pi D^2}{4}\right) C_m (a_n - \overline{a}_n)$$

where:

$\overline{u}_n$ component of the velocity vector of the structural member normal to its axis, in m/s (ft/s).
$\overline{a}_n$ component of the acceleration vector of the structural member normal to its axis, in m/s^2 (ft/s^2).
C_m added mass coefficient

$$C_m^{-1}$$

For shapes of structural member other than circular cylinders, the term $\frac{\pi D^2}{4}$ in the above equations is to be replaced by the actual cross-sectional area of the shape. Values of u_n and a_n in Morison's equation are to be determined using a recognised wave theory appropriate to the wave heights, wave periods, and water depth at the installation site. Values for the coefficients of drag and inertia in Morison's equation are to be determined based on model tests, full-scale measurements, or previous studies which are appropriate to the structural configuration, surface roughness, and pertinent flow parameters (e.g., Reynolds number). Reference is made to API RP 2A-WSD for the recommended values. For structural configurations which alter the incident wave field, diffraction theories of wave loading are to be employed to account for both the incident wave force (i.e., Froude-Kylov force) and the force resulting from the diffraction of the incident wave due to the presence of the structure. The hydrodynamic interaction of structural members is to be considered. For installation sites where the ratio of water depth to wavelength is less than 0.25, nonlinear effects of wave action are to be considered. This may be fulfilled by modifying linear diffraction theory to account for nonlinear effects or by performing model tests. Wave force calculations are to consider shallow water effects which may increase the current due to blockage effects and alter wave kinematics. Where the effect of breaking waves is to be considered in the design, guidance for the breaking wave hydrodynamics and the loads exerted by a breaking wave on a structure may be referred to IEC 61,400–3-1 (2019), Annex B and C.

5 Current Loads

Current induced loads on immersed structural members are to be determined based on analytical methods, model test data or full-scale measurements. When currents and waves are superimposed, the current velocity is to be added vectorially to the wave induced particle velocity prior to computation of the total force. Current profiles used in the design are to be representative of the expected conditions at the installation site. Where appropriate, flutter and dynamic amplification due to vortex shedding are to be considered. For calculation of current loads in the absence of waves, the lift force normal to flow direction, and the drag force may be determined as follows:

$$F_L = \left(\frac{\rho}{2g}\right) A C_L U^2$$

$$F_D = \left(\frac{\rho}{2g}\right) A C_D U^2$$

where:

F_L total lift force per unit length, in N/m (lb/ft).
F_D total drag force per unit length, in N/m (lb/ft).
ρ weight density of water, in N/m^3 (lb/ft^3).
g acceleration of gravitation, in m/s^2 (ft/s^2).
A projected area per unit length in a plane normal to the direction of the force, in m^2/m (ft^2/ft).
$C_L L$ lift coefficient.
C_D drag coefficient (refer to 4/4.2.3).
U local current velocity, in m/s (ft/s).

For long cylindrical members with large length-diameter ratios, lift force may become significant and is to be considered in the design. The source of C_L values employed is to be documented. Vortex induced vibration (VIV) is a resonant response caused by vortex shedding at resonant frequencies in current. Effects of VIV on fatigue and increased drag loads are to be suitably assessed for structural members. Dynamic amplification due to vortex shedding is to be considered if deemed necessary.

5.1 Ice and Snow Accumulation Induced Loads

At locations where an offshore wind turbine is subjected to ice and snow accumulation, increased weight and change in effective area of structural members due to accumulated ice and snow are to be considered. Particular attention is to be paid to increases in aerodynamic and hydrodynamic loading due to the change in size and surface roughness of both non-rotating and rotating parts of an offshore wind turbine caused by ice and snow accumulation.

5.2 Earthquake Loads

For the Bottom-founded Offshore Wind Turbine located in seismically active areas, the Strength Level and Ductility Level earthquake induced ground motions (refer to 3/7.5) are to be determined based on seismic data applicable to the installation site. Earthquake ground motions are to be described by either applicable ground motion records or response spectra consistent with the return period appropriate to the design life of the Bottom-founded Offshore Wind Turbine. Available standardised spectra applicable to the region of the installation site are acceptable provided such spectra reflect site-specific conditions affecting frequency content, energy distribution, and duration. These conditions include:

- The type of active faults in the region.
- The proximity of the site to the potential source faults.
- The attenuation or amplification of ground motion between the faults and the site.
- The soil conditions at the site.

The ground motion description used in the design is to consist of three components corresponding to two orthogonal horizontal directions and the vertical direction. All three components are to be applied to the structure simultaneously. When a standardised response spectrum, such as the one given in the API RP 2A-WSD, is used for structural analysis, input values of ground motion (spectral acceleration representation) to be used are not to be less severe than the following.

- 100% in both orthogonal horizontal directions.
- 50% in the vertical direction.

When three-dimensional, site-specific ground motion spectra are developed, the actual directional accelerations are to be used. If single site-specific spectra are developed, accelerations for the remaining two orthogonal directions are to be applied in accordance with the factors given above. If time history method is used for structural analysis, at least three sets of ground motion time histories are to be employed. The time histories are used to account for the potential sensitivity of the structure's response to variations in the phasing of the ground motion records. Structural appurtenances and turbine equipment are to be designed to resist earthquake induced accelerations at their foundations. As appropriate, effects of soil liquefaction, shear failure of soft mud and loads due to acceleration of the hydrodynamic added mass by the earthquake, submarine slide, tsunamis and earthquake generated acoustic shock waves are to be considered.

5.3 Marine Growth

The following effects of anticipated marine growth are to be accounted for in the design.

- Increase in hydrodynamic diameter.
- Increase in surface roughness used in the determination of hydrodynamic coefficients (e.g., lift, drag and inertia coefficients).
- Increase in dead load and inertial mass.

The amount of accumulation assumed for a specific design is to reflect the extent of and interval between cleaning of submerged structural parts.

5.4 Ice Loads

Ice loads acting on the Support Structure may include both static and dynamic loads. Static loads can be generated by temperature fluctuations, ice features lodged against the structure, or changes in water level in a fast ice cover. Dynamic loads are normally caused by moving ice interactions with the Support Structure. The methods to determine ice loading on the Support Structure of an offshore wind turbine are described in Annex D of IEC 61,400–3-1 (2019). Further reference is made to API RP 2N and ISO 19906. The global forces exerted by ice on the Support Structure as whole and local concentrated loads on structural elements are to be considered. The effects of rubble piles on the development of larger areas and their forces on the Support Structure need to be considered. Ice jamming between legs is to be accounted for where the Support Structure is designed to consist of multiple legs. Where relevant, liquefaction of the underlying soil due to repetitive compressive failures of the ice against the Support Structure is to be considered.

Chapter 5
Structure and Foundation Design

1 General

This chapter provides general requirements to be applied to the structural and foundation design of the Bottom-founded Offshore Wind Turbine. The criteria in 5/2 through 5/5 of this chapter which deal specifically with the design of steel structures, concrete structures, and foundation, are to be applied along with the requirements of this chapter. Wherever references in Chapter 5 are made to the API RP 2A, other applicable recognised industry standards such as ISO standards may also be used. Design documentation of structures and foundations to be submitted for review is specified in 1/5. The design assessment of design life extension and reuse of the existing Support Structure of an offshore wind turbine is to be in accordance with the Class Rules for the design, construction, and classification of offshore installations, with the exception that the load, strength, and fatigue calculations are to follow Chaps. 4 and 5 of this book. The design criteria in this chapter 5 are to be applied in the design of the substructure and the foundation of the Bottom-founded Offshore Wind Turbine. The effect of the RNA and tower on the substructure and the foundation design is to be considered. Compliance of the tower with the design criteria in this chapter for the substructure is required for the following situations (refer to 1/3):

- The optional RNA notation is requested by the Owner.
- The optional RNA notation is not requested by the Owner, however the tower and its connection to the RNA and/or to the substructure are not included in the wind turbine type certificate.

F. Aderouane, *Design, Construction, and Global Performance Analysis of Offshore Wind Turbines*, Green Energy and Technology, https://doi.org/10.1007/978-3-031-86010-2_5

2 Format of Design Approach

The design requirements of this book are specified in terms of the Working Stress Design (WSD) approach for steel components of the Support Structures and the Load and Resistance Factor Design (LRFD) approach for concrete components of the Support Structures. For steel components of the Support Structures, the use of alternative design criteria based on the LRFD approach is to be in accordance with 5/2.6.4. In addition, consideration is to be given to the serviceability of the structure relative to excessive deflection, vibration and, in the case of concrete, cracking.

2.1 General Design Considerations

2.1.1 Design Load Conditions

The steel Support Structure of an offshore wind turbine is to be designed and analysed for the loads to which it is likely to be exposed during construction, transportation, installation, and in-service operations. Loadings to be investigated for in-service operations are to include at least those relating to realistic operating and environmental conditions combined with dead loads, live loads and deformation loads that are appropriate to the function and operation of an offshore wind turbine. Load combinations are to reflect the design load conditions as specified in Chap. 4. With reference to Chaps. 3 and 4, both stochastic and deterministic formats are used to establish the design load conditions. Where relevant, consideration is to be given to the effect of dynamic amplification. The influence of less severe environmental loads in combination with operational loads is to be investigated for their potential to produce maximum peak stresses in the Support Structure. For the Support Structures located in seismically active areas, earthquake loads (refer to 3/7.5 and 4/5.2) are to be combined with dead loads, live loads and deformation loads appropriate to the offshore wind turbine operation and function which may be occurring at the onset of an earthquake.

2.1.2 Design Life

The design life of the Bottom-founded Offshore Wind Turbine is not to be less than 20 years. Upon the request by the Owner, a shorter design life may be accepted by Class for the purpose of demonstrating viability of a new design concept or conducting pilot operations. Continuance of classification beyond the design life is subject to the survey requirements in 7/3.1.11 and additional engineering analyses.

2.1.3 Air Gap

A minimum air gap of 1.5 m (5 ft) is to be provided between the 50-year return maximum wave crest elevation above the highest still water level (HSWL, refer to Fig. 2 of Chap. 3) and the lowest edge of the Support Structure for which wave forces are not included in the design. Consideration is to be given to the effect of initial and expected long-term settlements of the Support Structure, wave run-up, tilting of the Support Structure and, where appropriate, tsunamis. Local wave crest elevation is to be considered as appropriate. Appurtenances and localised components of the Support Structure are to be designed as appropriate to the local increases in wave pressure due to irregularity of waves and proximity to the Support Structure. The local design wave pressures on appurtenances are not to be less than those used for global structure design at the same elevation.

2.1.4 Structural Dynamic Properties

The ratio of the natural frequencies of the Support Structure, including the tower, substructure, and foundation, to the excitation frequencies of the various sources is to be determined. Excitation frequencies are to include rotor rotation frequencies, blade passing frequencies, natural frequencies of blades and other RNA components, and design wave periods. The overall damping contributed from various sources such as aerodynamic damping, hydrodynamic damping, structural damping and soil and foundation damping is to be determined and appropriately considered when deriving structural dynamic properties. Consideration is also to be given to the oscillatory loading on the Support Structure generated by vortex induced vibrations transverse to the wind or current direction. Variations of soil properties and the possible occurrence of scour, corrosion, marine growth, and sand movement during the design life may change the natural frequencies of the Support Structure and are to be suitably considered in the design.

2.1.5 Long-Term and Secondary Effects

Consideration is to be given to the following effects, as appropriate to the planned Support Structure of an offshore wind turbine:

(1) Local vibration due to machinery, equipment, and vortex shedding.
(2) Stress concentrations at critical joints.
(3) Secondary stresses induced by large deflection.
(4) Cumulative fatigue.
(5) Corrosion.
(6) Abrasion due to ice.
(7) Freeze–thaw action on concrete and coatings.

2.1.6 Zones of Exposure

Measures taken to mitigate the effects of corrosion are to be specified and described by the following definitions for corrosion protection zones.

(1) ***Submerged Zone***. That part of the Support Structure below the Splash Zone.
(2) ***Splash Zone***. That part of the Support Structure. Characteristically, the Splash Zone is not easily accessible for field painting, nor protected by cathodic protection.
(3) ***Atmospheric Zone***. That part of the Support Structure above the Splash Zone.

Additionally, for offshore wind turbines located in areas subject to floating or submerged ice, the portion of the Support Structure expected to encounter floating or submerged ice is to be designed with consideration for such contact.

2.2 Considerations for Particular Types of Substructures

2.2.1 General

Specific design considerations listed in this chapter are to be considered for particular types of substructures of an offshore wind turbine as additional factors that affect the safety and performance of the structure. These design considerations are not intended to supplant or modify other criteria specified in this book.

2.2.2 Pile-Supported Steel Structures

Consideration is to be given to the following effects, as appropriate to the substructure:

(1) The soil-pile interaction and the loads imposed on the substructure during towing and launching are to be considered.
(2) Carefully controlled installation procedures are to be developed so that the bearing loads of the structure on the soil are kept within acceptable limits until the piles are driven.
(3) As applicable, special procedures are to be used to handle long, heavy piles until they are self-supporting in the soil. Pile driving delays are to be minimised to avoid setup of the pile sections.
(4) Natural periods of the integrated RNA, tower, substructure, and foundation are to be checked to verify that the Bottom-founded Offshore Wind Turbine is not in resonance with excitation loads having significant energy contents.
(5) Instability of structural members due to submersion is to be considered, with due account for second-order effects produced by factors such as geometrical imperfections.

(6) Connections other than welded joints such as clamps, connectors, and bolts, joining diagonal braces to the column or piles to the substructure, the strength and fatigue resistance are to be assessed by analytical methods or testing.

2.2.3 Concrete or Steel Gravity Structures

Considerations are to be given to the following effects, as appropriate to the substructure:

(1) The procedure for transporting and positioning the structure and the accuracy of measuring devices used during these procedures are to be documented.
(2) Effects of repeated loadings on soil properties, such as pore pressure, water content, shear strength and stress strain behaviour, are to be investigated.
(3) Soil reactions against the base of the structure during installation are to be evaluated. Consideration is to be given to the occurrence of point loading caused by sea floor irregularities. Suitable grouting between base slab and sea floor may be employed to reduce concentration of loads.
(4) The strength and durability of construction materials are to be maintained. Where sulphate attack is anticipated, mitigation measures are to be implemented by choosing appropriate cements, incorporating pozzolans in the mix, or applying suitable coatings on the surfaces.
(5) Instability of structural members due to submersion is to be considered, with due account for second-order effects produced by factors such as geometrical imperfections.
(6) Where necessary, protection against horizontal sliding along the sea floor is to be provided by means of skirts, shear keys or equivalent means.
(7) The long-term resistance to abrasion, cavitation, freeze–thaw durability, and strength retention of the concrete are to be considered. Means are to be provided to minimise reinforcing steel corrosion.
(8) Provision is to be made to always maintain adequate negative buoyancy to resist the uplift forces from wind, waves, currents, and overturning moments.

2.2.4 Concrete-Steel Hybrid Foundation Structures

Where necessary, the underside of the concrete base is to be provided with skirts or shear keys to resist horizontal loading. Steel or concrete keys or equivalent means may be used in the design. The steel portions of a steel–concrete hybrid structure are to be designed in accordance with 5/2; the concrete portions are to be designed in accordance with 5/3. Any effects of the hybrid structure interacting on itself in areas such as corrosion protection are to be considered. Special attention is to be paid to the design of the connections between steel and concrete components. Applicable design considerations in 5/2.2.3 for concrete bases are also to be considered.

2.2.5 Self-Elevating Unit to Be Used as the Substructure

A self-elevating mobile offshore unit converted to the site dependent substructure of an offshore wind turbines is to be designed in accordance with this book in conjunction with the Class Rules for the design, construction, and classification of mobile offshore units, wherever applicable. When selecting a unit for a particular site, consideration is to be given to soil conditions at the installation site. The bearing capacity and sliding resistance of the foundation are to be investigated. The foundation design is to be in accordance with 5/4.4. As applicable, the footprints left by a self-elevating unit and scour are to be considered in the foundation design. In the structural analysis, the leg to hull connections and soil/structure interaction are to be carefully considered. The upper and lower guide flexibility, stiffness of the elevating/holding system, and any special details regarding its interaction with the leg are to be taken into consideration. For units with spud cans, the legs may be assumed pinned at the reaction point. For mat supported units, the soil structure interaction may be modelled using discrete elastic elements (springs). While used as the site dependent substructure, the calculated loads are to demonstrate that the maximum holding capacity of the jacking system will not be exceeded. Units with spudcans are to be pre-loaded during installation to minimise the possibility of significant settlement under severe storm conditions.

2.3 Steel Structures

2.3.1 General

The requirements of this chapter are to be applied in the design and analysis of the steel substructure of an offshore wind turbine. 5/1 is to be consulted regarding general design requirements. Items to be considered in the design of welded connections are specified in 2/2.

2.3.2 Materials

The requirements of this chapter are specified for the substructure constructed of steel, which is manufactured and has properties as specified in 2/1. Where it is intended to use steel or other materials having properties differing from those specified in 2/1, their applicability will be considered by Class upon reviewing the specifications of alternative materials and the proposed methods of fabrication.

2.3.3 Corrosion Protection

Materials are to be protected from corrosion by using a corrosion protection system including coatings. The system is to be effective from the time the substructure is initially placed on site. Where the sea environment contains unusual contaminants, any special corrosive effects of such contaminants are to be considered. For the design of protection systems, reference is to be made to the publications from NACE International: SP0176 and SP0108, or other recognised standards.

2.3.4 Access for Inspection

In the design of the substructure, consideration is to be given to providing access for inspection during construction and, to the extent practicable, for survey after construction. Any openings on the substructure for the purpose of providing access to an offshore wind turbine are to be evaluated to verify there is no adverse effect on the integrity of the structure.

2.4 Design Criteria

The steel substructure of an offshore wind turbine is to be designed and analysed for the loads as specified in 5/2.5. The resultant structural responses of the substructure are to satisfy the acceptance criteria in 5/2.6.9 through 5/2.13.

2.5 Loading Conditions

Loading conditions for the structural analysis and design of the steel substructure are to be determined in accordance with 5/3. The Design Load Cases (DLCs) in Table 4 of chap. 4 are to be assessed as a minimum requirement of design load conditions. DLCs for ice conditions are to be considered in accordance with 4/3.2 for the installation site where ice is expected to occur. Where the robustness assessment of the substructure and the foundation in a tropical cyclone region is performed, the Survival Load Cases (SLCs) in Table 1 of Chap. 4 are to be assessed as a minimum.

2.6 Structural Analysis

The following is general guidance on choosing appropriate approaches to perform structural analyses for the substructure of an offshore wind turbine. The designer is to verify that the structural analysis method is suitable for specific structural behaviours and can lead to accurate analysis results.

(1) The nature of loads and loading combinations as well as the local environmental conditions are to be taken into consideration in the selection of design methods. Methods of analysis and their associated assumptions are to be compatible with the overall design principles. When assessing structural instability as a mode of failure, the effects of initial stress and geometric imperfections are to be considered. Construction tolerances are to be consistent with those used in the structural stability assessment.
(2) Dynamic analyses are to be considered to assess the effects of environmental or other types of loads where dynamic amplification is expected. When performing fatigue analyses, a long-term distribution of the stress range, with proper consideration of dynamic effects, is to be obtained for relevant loadings anticipated during the design life of the substructure.
(3) The interaction among environmental loadings and between external loads and structural responses are to be assessed by applying environmental loads simultaneously to the substructure. Other rational methods for calculating loadings are acceptable provided it is demonstrable that the adopted method leads to an equivalent design.
(4) Soil-pile interaction can affect overall dynamic behaviour of the substructure. The simulation of foundation soil-pile interaction in the structural analysis is to be properly carried out based on site specific soil conditions.
(5) For static loads, plastic methods of design and analysis can be employed only when the properties of the steel and the connections are such that they exclude the possibility of brittle fracture, allow for formation of plastic hinges with sufficient plastic rotational capability, and provide adequate fatigue resistance.
(6) Under dynamic loads, when plastic strains may occur, the considerations specified in v) are to be satisfied and any buckling and destabilising nonlinear effects are to be considered.
(7) When considering the effects of earthquakes in seismically active areas (refer to 3/7.5 and 4/5.2), dynamic analyses are to be performed using the recognised methods, such as those recommended in API RP 2A.
(8) Where the Survival Load Cases in accordance with 4/3.3 are assessed, the ultimate strength analysis method provided in API RP 2SIM is to be applied.

3 Strength Design Criteria

3.1 General

The design of the steel substructure of an offshore wind turbine can be based on either the Working Stress Design (WSD) approach or the Load and Resistance Factor Design (LRFD) approach as specified in 5/2.6.3 and 5/2.6.4, respectively. However, it is not permitted to mix elements of these two approaches in the design for the same structural component in the substructure. The strength design criteria specified in this paragraph are to be applied to those design load cases relevant to the strength assessment. A minimum set of Design Load Cases (DLCs) for the strength assessment is defined in Table 1 of Chap. 4 and, if applicable, Tables 2 and 3 of chap. 4 and 4, where the 'Type of Analysis' column is denoted 'S.'

3.1.1 Working Stress Design (WSD) Approach

When the strength design of the steel substructure is based on the WSD approach, the design acceptance criteria are to be expressed in terms of appropriate basic allowable stresses in accordance with the requirements specified in this chapter. Linear, elastic methods can be employed in the analysis of the substructure provided proper measures are taken to prevent general and local buckling failure, and the interaction between soil and structure is adequately considered. The loading conditions for the strength analysis are to be in accordance with 5/2.3. A factor of 1.0 applies to all categories of load. When a live load is considered a favourable load that relieves total load responses, the minimum value of this live load, as appropriate to the operation mode being considered, is to be used in the load combination. For the Design Load Cases, the factors of safety specified in this paragraph are to be applied in conjunction with the normal (N) and abnormal (A) design conditions as well as the design conditions (T) related to transport, installation, maintenance, and repair operations, as defined in 4/3.2. Where the Survival Load Cases in accordance with 4/3.3 are assessed, structural members may reach their ultimate capacities or beyond, provided the global strength of the substructure is not compromised.

Individual Stresses in Structural Members

Individual stress components or direct combinations of such stresses in a structural member are not to exceed the allowable stress as obtained from the following equation:

$$F_{allowable} = \frac{F_y}{F.S}$$

where:

$F_{allowable}$ allowable stress.
F_y specified minimum yield strength.
$F.S$ factor of safety.

- For the normal design conditions (designated 'N' in the column entitled as 'Safety Factor' in Tabled 1 and 2 of chap. 4 and 4)

1.5 for axial or bending stress
2.5 for shear stress

- For the abnormal design conditions (designated 'A' in the column entitled as 'Safety Factor' in Table 1 and 2 of chap. 4 and 4)

1.25 for axial or bending stress
2.0 for shear stress

- For the design load conditions involving the transport, assembly on site, maintenance, and repair operations (DLC 8.1 in Table 1 of Chap. 4)

1.67 for axial or bending stress
2.75 for shear stress

Buckling Strength of Structural Members Subject to a Single Action

Buckling is to be considered for a structural element subject to compressive axial load or bending moment. The computed compressive or bending stress is not to exceed the allowable stress as obtained from the following equation:

$$F_{allowable} = \frac{F_{cr}}{F.S}$$

where

$F_{allowable}$ allowable stress.
F_y critical buckling strength of a structural member under axial compression or critical bending strength of a structural member under bending moment, as defined in the applicable Class guide for buckling and ultimate strength assessment for offshore structures.
$F.S$ factor of safety.

- For the normal design conditions (designated 'N' in the column entitled 'Safety Factor' in Tables 1 and 2 of chap. 4 and 4)

$$\frac{1.5}{\psi}$$

- For the abnormal design conditions (designated 'A' in the column entitled 'Safety Factor' in Tables 1 and 2 of chap. 4 and 4)

$$\frac{1.25}{\psi}$$

- For the design load conditions involving the transport, assembly on site, maintenance, and repair operations (DLC 8.1 in Table 1 of Chap. 4)

$$\frac{1.67}{\psi}$$

ψ adjustment factor, as defined in the Class guide for buckling and ultimate strength assessment for offshore structures.

Structural Members Subject to Combined Axial Load and Bending

Structural members under axial tension or compression in combination with bending are to be designed according to the applicable Class guide for buckling and ultimate strength assessment for offshore structures in conjunction with the utilisation factors, which are the reciprocals of the corresponding safety factors specified in 5/2.6.3.

Allowable Stress of Plated Structures

For plated structures where the equivalent stress is determined using the von Mises equivalent stress criterion, the equivalent stress is not to exceed the allowable stress as obtained from the following equation:

$$F_{allowable} = \frac{F_y}{F.S}$$

where:

$F_{allowable}$ allowable stress.
F_y specified minimum yield strength.
$F.S$ factor of safety.

- For the normal design conditions (designated 'N' in the column entitled 'Safety Factor' in Tables 1 and 2 of chap. 4 and 4)

  1.33

- For the abnormal design conditions (designated 'A' in the column entitled 'Safety Factor' in Tables 1 and 2 of chap. 4 and 4)

  1.11

- For the design load conditions involving the transport, assembly on site, maintenance, and repair operations (DLC 8.1 in Table 1 of Chap. 4)

  1.5

Buckling Strength of Plated Structures

The buckling strength of plated structures is to be designed according to the applicable Class guide for buckling and ultimate strength assessment for offshore structures in conjunction with the utilisation factors, which are the reciprocals of the corresponding safety factors specified in 5/2.6.3.

3.1.2 Load and Resistance Factor Design (LRFD) Approach

In lieu of the WSD approach described in 5/2.6.3, the design of the steel substructure can also be based on the LRFD approach. The load combination is to be in accordance with 5/2.5 in general. For the Design Load Cases specified in 5/2.4, the partial safety factors (γf) specified in Table 1 are to be applied to environmental loads. A partial safety factor of 1.0 is to be applied to dead loads, live loads, and deformation loads, where they are combined with the design environmental loads. Where a dead load, a live load, or a deformation load is considered as a favourable load that relieves total load responses, a partial safety factor of 0.9 is to be applied. Where a live load is considered a favourable load, the minimum value of this live load is to be used in the load combination.

Table 1 Partial safety factors (γf) for environmental loads

Normal (N)	Abnormal (A)	Transport, assembly, maintenance and repair (T)
1.35	1.1	1.5

Note The design conditions represented by N, A and T are defined in 4/3.2. Alternatively, the partial load factors specified in IEC 61,400–3-1 can be applied to the load effects determined by the dynamic analysis of the Bottom-founded Offshore Wind Turbine using unfactored loads. For design of tubular members and their connections (joints), resistance factors, member strength capacity equations, and strength criteria are to be determined in accordance with API RP 2A-LRFD or ISO 19902. For structural members with other types of sections within the scope of the AISC Steel Construction Manual (LRFD part), resistance factors and member strength capacity equations are to be determined based on the relevant design requirements specified in that standard. For the ultimate capacity of flat plate and the cylindrical shell structure, reference is made to the applicable Class guide for buckling and ultimate strength assessment for offshore structures or other recognised industry standards. A minimum resistance factor of $\frac{1.05}{\psi}$ where the adjustment factor ψ is defined in 5/2.6.3, is to be applied. Where the resistance concerns bolted connections and fillet and partial penetration welds, the minimum resistance factor of 1.30 is to be applied. Where the Survival Load Cases in accordance with 4/3.3 are assessed, all partial safety factors are to be taken as 1.0, and structural members can reach their ultimate capacities or beyond, provided the global strength of the substructure is not compromised

3.2 Structural Response to Earthquake Loads

Structures located in seismically active areas are to be designed to possess adequate strength and stiffness to withstand the effects of the Strength Level earthquake, as well as sufficient ductility to remain stable during rare motions of greater severity associated with the Ductility Level earthquake. Refer to 3/7.5 and 4/5.2 for the definitions of the Strength Level and Ductility Level earthquakes as well as earthquake loads. The sufficiency of the structural strength and ductility is to be demonstrated by strength and as required, ductility analyses. For the Strength Level earthquake, the strength analysis is to demonstrate that the structure is adequately sized for strength and stiffness to maintain all nominal stresses within their yield or buckling limits. In the ductility analysis, it is to be demonstrated that the structure has the capability of absorbing the energy associated with the Ductility Level earthquake without reaching a state of incremental collapse. The safety factors for the strength design of structural members, as those defined in 5/2.6.3 for the normal design conditions, are to be reduced by a factor of 1.5, except that for the plated structures designed to 5/2.6.3, the allowable stress is to be taken as the specified minimum yield strength (F_y). In the case that the design is based on the LRFD approach, the load factors for all load types considered in the load combinations are to be taken as 1.0. Pile-soil performance and pile design requirements are to be determined based on case-by-case studies.

3.3 Fatigue Assessment

The fatigue assessment is to be performed to verify adequate capacity against fatigue failure within the design life of the Bottom-founded Offshore Wind Turbine. The fatigue assessment is to be performed for structural members and joints where fatigue is a probable mode of failure, or for which experience is insufficient to justify safety from possible cumulative fatigue damage. Emphasis is to be given to structural members and connections that are difficult to inspect and repair once the Bottom-founded Offshore Wind Turbine is in service and to those areas susceptible to corrosion-accelerated fatigue. For structural members and connections that require a detailed assessment of cumulative fatigue damage, the calculated fatigue life is not to be less than the design life of the Bottom-founded Offshore Wind Turbine times the safety factors for fatigue life [i.e., fatigue design factors (FDFs)] as specified in Table 2. Fatigue resistance of structural details is to be evaluated in accordance with the applicable Class guidance for performing fatigue assessments of offshore structures. Suitable S–N curves are to be selected, with special attention being given to the application and limitation of those curves. The loading conditions for the fatigue assessment are to be in accordance with 5/2.5. A minimum set of Design Load Cases (DLCs) for the fatigue assessment is specified in Table 1 of Chap. 4 and, if applicable, Table 2 of chap. 4, where 'F' in the column titled 'Type of Analysis' designates the

Table 2 Safety factors for fatigue life of structures (fatigue design factors [FDF])

Importance	Inspectable and repairable	
	Yes	No
Non-critical	1	3
Critical	3	5

Notes "*Critical*" *indicates that failure of these structural items would result in the rapid loss of structural integrity and produce an event of unacceptable consequence. A Fatigue Design Factor of 1.0 is applicable to Inspectable and repairable non-critical structural members above the splash zone. Diver or ROV inspectable and repairable redundant framing. For critical structure members and the turbine tower structure installed above the splash zone, a Fatigue Design Factor of 2.0 may be applied provided that these structures can be inspected during periodic survey or when structural damage is suspected such that critical crack development can be detected and repaired*

fatigue assessment. Loading history of the substructure during transport and installation phases is to be accounted for. Fatigue analyses are to be carried out using an appropriate loading spectrum or time series in accordance with the accepted theories in calculating accumulated damage. In the case that the design is based on the LRFD approach, the load factors for all load categories are to be taken as 1.0 in the fatigue assessment.

3.4 *Stresses in Connections*

Connections of structural members are to be developed to verify effective load transmission between joined members, to minimise stress concentration and to prevent excessive punching shear. Connection details are also to be designed to minimise undue constraints against overall ductile behaviour and to minimise the effects of post-weld shrinkage. Undue concentration of welding is to be avoided. The design of tubular joints may be in accordance with the relevant requirements in API RP 2A.

3.4.1 Structure-Pile Connections

The connection of the substructure of an offshore wind turbine to the foundation and to the turbine tower is to be accomplished by positive, controlled means such as welding, grouting, or other mechanical connectors. Such connections are to be capable of withstanding anticipated static, dynamic, and long-term cyclic loadings. Details of mechanical connectors are to be submitted for review. In the design of a grouted connection, consideration is to be given to the use of mechanical shear connectors as their presence increases the strength of the connection and alleviates the effect of long-term grout shrinkage. Adequate clearance in the annulus is to be provided for proper placement of the grout. Reliable means for the introduction of the grout to the annulus are to be provided to achieve complete filling of the annulus and to minimise the possibility of dilution of the grout and the creation of voids in the grout. Wipers or similar devices are to be used to minimise intrusion of mud into the annulus between the pile and the substructure leg during installation. The particulars of grouting mixtures used in the grouted pile connections are to be submitted for review. For grouted connections undertaking axial loads, the design guidance can be found in API RP 2A. Special attention is to be paid to the limitation of geometric configurations imposed by API RP 2A. For those grouted pile connections expected to undertake bending moments, their strengths are to be assessed by suitable analysis methods or by tests. The assessment results are to be submitted for review. For grouted connections whose geometries are not covered by the existing design criteria, special consideration is to be given to the effects of reduced confinement on allowable bond stress, and suitable analyses or tests are to be submitted for review by Class. For bolted flange connections, special care is to be taken to verify evenness of contact surface to avoid overstressing of bolts. The design and installation are to be in accordance with recognised standards such as the AISC Steel Construction Manual. Consideration is to be given to friction factors, relaxation, stress corrosion cracking, bolt fatigue, brittle failure, and other factors or combinations that may be present. The allowable stresses to be employed in the design of foundation structure for steel gravity bases or piles are to be in accordance with 5/2.6.3. In the case that the design is based on the LRFD approach, the load and resistance factors to be employed in the design of structure-pile connections are to be in accordance with 5/2.6.4. The design of laterally loaded piles is in accordance with 5/4.5.

3.4.2 Structural Response to Hydrostatic Loads

Analyses of the structural stability are to be performed to demonstrate the ability of structural parts to withstand hydrostatic collapse at the expected water depth. Hydrostatic collapse is to be checked in accordance with the Class guidance for performing buckling and ultimate strength assessment for offshore structures.

3.4.3 Deflections

The substructure deflections that may affect proper operations of the RNA, the design of piles, and other structures in way of the substructure are to be considered. Where appropriate, the associated geometric nonlinearity is to be considered.

3.4.4 Local Structures

Structures that do not directly contribute to the overall strength of the substructure, i.e., their loss or damage does not impair the structural integrity of the substructure, are local structures. Local structures including the interface with the power cable system (I-tube and J-tube support structure and foundation) are to be adequate for the nature and magnitude of applied loads. The criteria of 5/2.6.1 apply in the design of local structural components, except for those structural parts whose primary function is to absorb energy, in which case sufficient ductility is to be demonstrated.

4 Concrete Structures

4.1 General

The requirements of this chapter are to be applied to the substructure constructed of reinforced and prestressed concrete. The contents of this chapter are to be consulted regarding general design requirements.

4.1.1 Materials

Unless otherwise specified, the requirements of this chapter are intended for the substructure constructed of materials manufactured and having properties as specified in 2/1. Use of materials having properties differing from those specified in 2/1 will be specially considered by Class. Specifications for alternative materials, details of the proposed methods of manufacture and, where available, evidence of satisfactory previous performance, are to be submitted for review. For structural lightweight concrete, the reference is made to ACI 213R, and lightweight aggregates are to conform to the requirements of ASTM C330.

4.1.2 Durability

Materials, concrete mix proportions, construction procedures and quality control are to be chosen to produce satisfactory durability for structures located in a marine environment. Issues to be specifically addressed include chemical deterioration of concrete, corrosion of the reinforcement and hardware, abrasion of concrete, freeze–thaw durability, and fire hazards as they pertain to the zones of exposure defined in 5/2.1.6. Test mixes are to be prepared and tested early in the design phase to verify that proper values of strength, creep, alkali resistance, etc. can be achieved.

4.1.3 Access for Inspection

The components of the structure are to be designed to enable their inspection during construction and, to the extent practicable, periodic survey after installation.

4.2 Design Method

The criteria specified in this chapter for concrete structures are based on the Load and Resistance Factor Design (LRFD) approach. Where the Survival Load Cases in accordance with 4/3.3 are assessed, all partial factors are to be taken as 1.0 and structural members can reach their ultimate capacities or beyond provided the global strength of the substructure is not compromised.

4.2.1 Load Magnitude

The magnitude of a design load for a given type of loading k is obtained by multiplying the load, F_k, by the appropriate load factor, c_k (i.e., design load $= c_k F_k$).

4.2.2 Design Strength

In the analysis of sections, the design strength of a given material is obtained by multiplying the material strength, f_k, by the appropriate strength reduction factor, φ (i.e., design strength $= \varphi f_k$). The material strength, f_k, for concrete is the specified compression strength of concrete ($f_c\prime$) after 28 days and for steel is the minimum specified yield strength (f_y). Refer to also 5/3.4.2.

4.2.3 Design Reference

Design considerations for concrete substructure not directly addressed in this book are to follow the requirements of the ACI 318 and ACI 357, or equivalent.

4.3 Design Requirements

4.3.1 General

The strength of the concrete substructure of an offshore wind turbine is to be such that adequate safety exists against failure of the structure or its components. Among the modes of failure to be considered are the following:

(1) Loss of overall equilibrium.
(2) Failure of critical section.
(3) Instability resulting from large deformation.
(4) Excessive plastic or creep deformation.

The serviceability of the substructure is to be assessed. The following items are to be considered in relation to their potential influences on the serviceability of the structure:

(1) Cracking and spalling.
(2) Deformation.
(3) Corrosion of reinforcement or deterioration of concrete.
(4) Vibration.
(5) Leakage.

4.3.2 Load Combinations

The load combination is to be in accordance with 5/2.1.1 in general. The load categories referred in this chapter (i.e., dead loads, live loads, deformation loads, and environmental loads) are defined in 4/2. The Design Load Cases (DLCs) in Tables 1 and 3 of chap. 4 and 4, if applicable, are to be assessed as a minimum requirement of design load conditions. DLCs for ice conditions are to be considered in accordance with 4/3 for the installation site where ice is expected to occur. For the design of the concrete substructure of an offshore wind turbine, the partial safety factors for loads are to follow 5/2.6.4. While the critical design loadings are to be identified from the load combinations given above, the other simultaneously occurring load combinations during construction, transport and installation phases are to be considered if they can cause critical load effects.

4.3.3 Strength Reduction Factors

The strength of a member or a cross-section is to be calculated in accordance with the provisions of 5/3.4 and it is to be multiplied by the following strength reduction factor, φ.

(1) In the case of bending without axial tension, $\varphi = 0.90$.
(2) In the case of axial compression or axial compression combined with bending:

- For reinforced members with spiral reinforcement, $\varphi = 0.70$.
- For other reinforced members (excluding slabs and shells), $\varphi = 0.65$.

The values given in the above for two types of members may be increased linearly to 0.9 as φP_u decreases from $0.1 f_{c'} A_g$ or φP_u, whichever is smaller, to zero, where:

$f_{c'}$ Specified compression strength of concrete.
A_g Gross area of section.

P_uAxial design load in compression member.
P_bAxial load capacity assuming simultaneous occurrence of the ultimate strain of concrete and yielding of tension steel.

- For slabs and shells, $\varphi = 0.70$

In the case of shear and torsion, $\varphi = 0.75$.
In the case of bearing on concrete, $\varphi = 0.65$, except for post-tensioning anchorage bearing. For bearing on concrete in post-tension anchorage, $\varphi = 0.85$.

Alternatively, the expected strength of concrete members can be determined by using idealised stress–strain curves and material factors (c_M) given in ACI 357R. The material factors applied to the stress–strain curves limit the maximum stress to achieve the desired reliability like using the strength reduction factors given above. The strength reduction factors (φ) and the material factors (c_M) are not to be used simultaneously.

4.3.4 Fatigue

The fatigue strength of the concrete substructure of an offshore wind turbine is considered satisfactory if under the unfactored fatigue loads (i.e., $c_k = 1$) the following conditions are satisfied. The fatigue analysis is based on the stress in a critical section. The stress range is to be derived from all fatigue load cycles with a 1-year return probability level.

(1) The stress range in reinforcing or prestressing steel does not exceed 138 MPa (20 ksi), or where reinforcement is bent, welded, or spliced, 69 MPa (10 ksi).
(2) There is no membrane tensile stress in concrete and not more than 1.4 MPa (200 psi) flexural tensile stress in concrete.
(3) The stress range in compression in concrete does not exceed $0.5 f_c\prime$ where $f_c\prime$ is the specified compressive strength of concrete.
(4) Where maximum shear exceeds the allowable shear of the concrete alone, and where the cyclic range is more than half the maximum allowable shear in the concrete alone, all shear is taken by reinforcement. In determining the allowable shear of the concrete alone, the influence of permanent compressive stress may be considered.
(5) In situations where fatigue stress ranges allow greater latitude than those under the serviceability requirements given in Table 3, the latter condition is to assume precedence.
(6) Bond stress does not exceed 50% of that permitted for static loads. If lap splices of reinforcement or pretensioning anchorage development are subject to cyclic tensile stresses greater than 50% of the allowable static stress, the lap length or prestressing development length is to be increased by 50%.

Where the above nominal values are exceeded, an in-depth fatigue analysis is to be performed. In such an analysis, the possible reduction of material strength is

Table 3 Allowable tensile stresses for prestress and reinforcing steel to control cracking

Stage	Loading	Allowable stress, MPa (ksi)	
		Reinforcing Steel, f_s	Prestressing tendons, D_{ps}
Construction: where cracking during construction would be detrimental to the completed structure	All loads on the structure during construction	160 (23.0)	130 (18.5)
Construction: where cracking during construction is not detrimental to the completed structure	All loads on the structure during construction	210 (30.0) or $0.6 f_y$, whichever is less	130 (18.5)
Transportation and installation	All loads on the structure during transportation and installation	160 (23.0)	130 (18.5)
At offshore site	Dead and live loads plus design environmental loads	$0.8 f_y$	

to be considered based on appropriate data (S–N curves) corresponding to the 95th percentile of specimen survival. In this regard, consideration is to be given not only to the effects of fatigue induced by normal stresses, but also to fatigue effects due to shear and bond stresses under unfactored load combinations. Particular attention is to be given to submerged areas subjected to the low-cycle, high-stress components of the loading history. In prestressed members containing unbonded reinforcement, special attention is to be given to the possibility of fatigue in the anchorages or couplers that may be subjected to corrosive action. Where a fatigue analysis is performed for the concrete substructure, the safety factors for fatigue life [i.e., fatigue design factors (FDFs)] are to be in accordance with Table 2, except that the calculated fatigue life is to be at least twice the design life of the Bottom-founded Offshore Wind Turbine. To estimate the cumulative fatigue damage under variable amplitude stresses, a recognised cumulative rule is to be used. Miner's rule is an acceptable method for the cumulative fatigue damage analysis.

4.3.5 Serviceability Requirements

Serviceability

The serviceability of the concrete substructure is to be checked using stress–strain diagrams, as depicted in Figs. 1 and 2. The strength reduction factor, φ, and partial safety factors for loads, ck, are to be taken as 1.0. The unfactored ($c_k = 1.0$) load

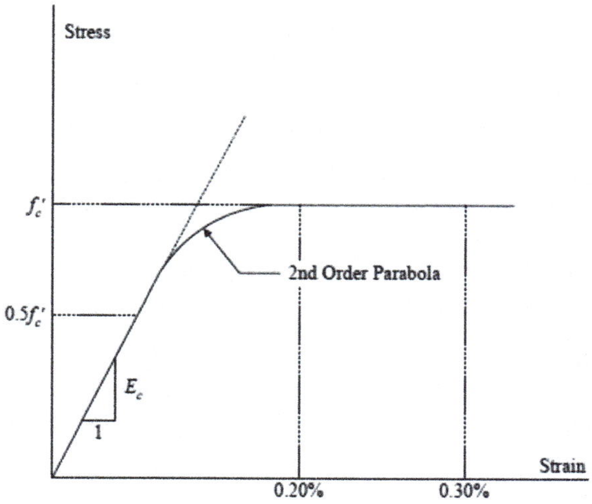

Fig. 1 Idealised stress–strain relation for concrete in uniaxial compression

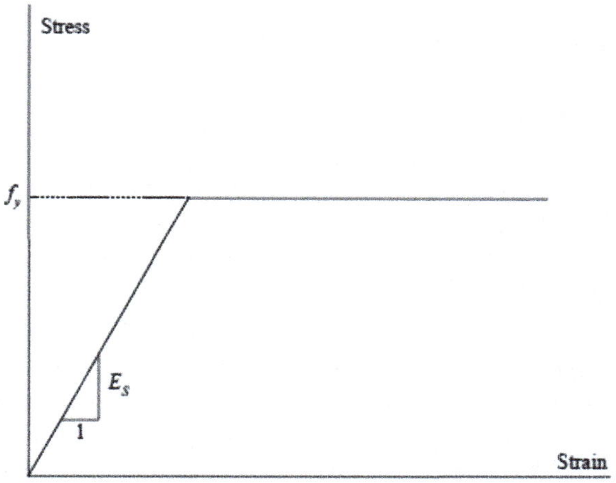

Fig. 2 Idealised stress–strain relation for non-prestressed steel in uniaxial tension

combination of most unfavourable dead loads, deformation loads and live loads as well as the design environmental loads is to be applied. Using this method, the reinforcing stresses are to be limited in compliance with Table 3. Additionally for hollow structural cross-sections, the maximum permissible membrane strain across the walls is not to cause cracking under any combination of unfactored loads. For structures prestressed in one direction only, tensile stresses in reinforcement transverse to the prestressing steel are to be limited so that the strains at the plane of the prestressing steel do not exceed $\frac{D_{ps}}{E_S}$, where D_{ps} is as defined in Table 3 and E_S is the modulus of elasticity of reinforcement (refer to 5/3.4.2). Alternative criteria such as those which directly limit crack width may also be considered.

Liquid-Containing Structures

The following criteria are to be satisfied for liquid-containing structures to verify adequate design against leakage.

(1) The reinforcing steel stresses are to be in accordance with 5/3.6.1.
(2) The compression zone is to extend over 25% of the wall thickness or 205 mm (8 in), whichever is less.
(3) There is to be no membrane tensile stress unless other construction arrangements are made, such as the use of special barriers to prevent leakage.

f_y yield stress of the reinforcing steel.
f_s allowable stress in the reinforcing steel.
D_{ps} *increase in tensile stress in prestressing steel with reference to the stress at zero strain in the concrete.*

4.4 Analysis and Design

4.4.1 General

The analysis of a concrete structure may be performed under the assumption of linearly elastic materials and linearly elastic structural behaviour in accordance with the requirements of ACI 318 and the additional requirements of this chapter. The material properties to be used in analysis are to conform to 5/3.4.2. However, the inelastic behaviour of concrete based on the true variation of the modulus of elasticity with stress and the geometric nonlinearities, including the effects of initial deviation of the structure from the design geometry, are to be considered whenever their effects reduce the strength of the substructure. The beneficial effects of the concrete's nonlinear behaviour may be accounted for in the analysis and design of the structure to resist dynamic loadings. Where required, the dynamic behaviour of a concrete structure may be investigated using a linear structural model, but soil-structural impedances are to be suitably considered. The analysis of the structure under earthquake conditions may be performed under the assumption of elastoplastic behaviour due to yielding, provided the requirements of 5/3.5.2 are satisfied.

5 Material Properties for Structural Analysis

5.1 Specified Compressive Strength

The specified compressive strength of concrete, $f_c{\prime}$, is to be based on 28-day tests performed in accordance with specifications ASTM C172, ASTM C31 and ASTM C39.

5.1.1 Early Loadings

For structures that are subjected to loadings before the end of the 28-day hardening period of concrete, the compressive strength of concrete is to be taken at the actual age of concrete at the time of loading.

5.1.2 Early Strength -Concrete

For early-strength concrete, the age for the tests for $f_c{\prime}$ may be determined based on the cement manufacturer's certificate.

5.1.3 Modulus of Elasticity—Concrete

For the purposes of structural analyses and deflection checks, the modulus of elasticity, E_c, of normal weight concrete may be assumed as equal to $4733(f_c{\prime})^{0.5}$ MPa ($57(f_c{\prime})^{0.5}$ ksi) or determined from stress–strain curves developed by tests (refer to Fig. 1). When the latter method is used, the modulus of elasticity is to be determined using the secant modulus for the stress equal to $0.50 f_c{\prime}$.

5.1.4 Uniaxial Compression—Concrete

In lieu of tests, the stress–strain relation shown in Fig. 1 may be used for uniaxial compression of concrete.

5.1.5 Poisson Ratio

The Poisson ratio of concrete may be taken equal to 0.20.

5.1.6 Modulus of Elasticity—Reinforcement

The modulus of elasticity, E_s of non-prestressed steel reinforcement is to be taken as 200 × 103 MPa (29 × 103 ksi). The modulus of elasticity of prestressing tendons is to be determined by tests.

5.1.7 Uniaxial Tension—Reinforcement

The stress–strain relation of non-prestressed steel reinforcement in uniaxial tension is to be assumed as shown in Fig. 2. The stress–strain relation of prestressing tendons is to be determined by tests or taken from the manufacturer's certificate.

5.1.8 Yield Strength—Reinforcement

If the specified yield strength, f_y, of non-prestressed reinforcement exceeds 420 MPa (60 ksi), the value of f_y used in the analysis is to be taken as the stress corresponding to a strain of 0.35%.

5.1.9 Analysis of Plates, Shells, and Folded Plates

In all analyses of shell structures, the theory employed in analysis is not to be based solely on membrane or direct stress approaches. The buckling strength of plate and shell structures is to be checked by an analysis that considers the geometrical imperfections of the structure, the inelastic behaviour of concrete and the creep deformations of concrete under sustained loading. Special attention is to be given to structures subjected to external pressure and the possibility of their collapse (implosion) by failure of concrete in compression.

5.1.10 Deflection Analysis

Immediate deflections may be determined by the methods of linear structural analysis. For the purposes of deflection analysis, the member stiffness is to be computed using the material properties specified in the design and is to consider the effect of cracks in tension zones of concrete. The effect of creep strain in concrete is to be considered in the computations of deflections under sustained loadings.

5.1.11 Analysis and Design for Shear and Torsion

The analysis and design of members subjected to shear or torsion or to combined shear and torsion is to adhere with the applicable requirements of ACI 318 or their equivalent.

5.1.12 Analysis and Design for Bending and Axial Loads

5.2 Assumed Conditions

The analysis and design of members subjected to bending and axial loads are to be based on the following assumptions:

(1) The strains in steel and concrete are proportional to the distance from the neutral axis.
(2) Tensile strength of the concrete is to be neglected, except in prestressed concrete members under unfactored loads, where the requirements in 5/3.3.5 apply.
(3) The stress in steel is to be taken as equal to *ES* (refer to 5/3.4.2) times the steel strain, but not larger than f_y (refer to refer to 5/3.4.2).
(4) The stresses in the compression zone of concrete are to be assumed to vary with strain according to the curve given in Fig. 1 or any other conservative rule. Rectangular distribution of compressive stresses in concrete specified by ACI 318 may be used.
(5) The maximum strain in concrete at the ultimate state is not to be larger than 0.30%.

5.2.1 Failure

The members in bending are to be designed in such a way that any section yielding of steel occurs prior to compressive failure of concrete.

5.2.2 Seismic Analysis

Dynamic Analysis

For the substructure to be located at sites known to be seismically active (refer to 5/3.5.3), dynamic analyses are to be performed to determine the response of the structure to design earthquake loading. The substructure is to be designed to withstand this loading without damage. In addition, a ductility check is also to be performed to verify that the structure has sufficient ductility to resist deflections more severe than those resulting from the design earthquake loading without the collapse of the substructure or any major structural component. Reference is made to 3/7.5 and 4/

5.2 for the Strength Level and Ductile Level earthquakes as well as the earthquake loads.

Design Conditions

The dynamic analysis for earthquake loadings is to be performed with consideration of:

(1) The interaction of all loads bearing or load carrying components of the structure.
(2) The compliance of the soil and the dynamic soil-structure interaction.
(3) The dynamic effects of the ambient and contained fluids.

Method of Analysis

The dynamic analysis for earthquake loadings may be performed by any recognised method, such as direct integration of the equations of motion or the response spectra method.

Ductility Check

In the ductility check, ground motions (e.g., spectral ordinates) at least twice those used for the design earthquake are to be assumed. If the ductility check is performed with the assumption of elastoplastic behaviour of the structure, the selected method of analysis is to be capable of considering the non-linearity of the structural model. The possibility of dynamic instability (dynamic buckling) of individual members and of the whole structure is to be considered.

5.2.3 Seismic Design

Compressive Strain

The compressive strain in concrete at critical sections (including plastic hinge locations) is to be limited to 0.30%, except when greater strain may be accommodated by confining steel.

Flexural Bending or Load Reversals

For structural members or sections subject to flexural bending or to load reversals, where the percentage of tensile reinforcement exceeds 70% of the reinforcement at which yield stress in the steel is reached simultaneously with compression failure in the concrete, special confining reinforcement (e.g., T-headed bars) and/or compressive reinforcement are to be provided to prevent brittle failure in the compressive zone of concrete.

Web Reinforcement

Web reinforcement (stirrups) of flexural members is to be designed for shear forces which develop at full plastic bending capacity of end sections. In addition:

(1) The diameter of rods used as stirrups is not to be less than 10 mm (#3 bar).
(2) Only closed stirrups (stirrup ties) are to be used. T-headed bars or other mechanically headed bars may be used if their effectiveness has been verified.
(3) The spacing of stirrups is not to exceed the lesser of $d/2$ or 16 bar diameters of compressive reinforcement, where d is the distance from the extreme compression fibre to the centroid of tensile reinforcement. Tails of stirrups are to be anchored within a confined zone (i.e., turned inward).

Splices

No splice is allowed within a distance, d, defined in 5/3.5.3, from a plastic hinge. Lap splices are to be designed in accordance with ACI 318. Mechanical and welded splices are permitted. Mechanical splices are to comply with ACI 349.

5.3 *Design Details*

5.3.1 Concrete Cover

General

The following minimum concrete cover for reinforcing bars is required:

(1) Atmospheric zone not subject to salt spray: 50 mm (2 in).
(2) Splash and atmospheric zones subjected to salt spray and exposed to soil: 65 mm (2.5 in).
(3) Submerged zone: 50 mm (2 in).
(4) Areas not exposed to weather or soil: 40 mm (1.5 in).
(5) Cover of stirrups may be 13 mm (0.5 in) less than covers listed above.

Tendons and Ducts

The concrete cover of prestressing tendons and post-tensioning ducts is to be increased by 25 mm (1 in) above the values listed in 5/3.6.6.

Sections Less Than 500 mm (20 in) Thick

In sections less than 500 mm (20 in) thick, the concrete cover of reinforcing bars and stirrups may be reduced below the values listed in 5/3.6.6. However, the cover is not to be less than the following:

(1) 1.5 times the nominal aggregate size.
(2) 1.5 times the maximum diameter of reinforcement, or 19 mm (0.75 in).
(3) Tendons and post-tensioning duct covers are to have 12.5 mm (0.5 in) added to the above.

5.3.2 Minimum Reinforcement

The minimum requirements of ACI 318 are to be satisfied. In addition, for loadings during all phases of construction, transportation, and operation (including design environmental loading) where tensile stresses occur on a face of the structure, the following minimum reinforcement is to be provided:

$$A_s = \left(\frac{f_t}{f_y}\right) b d_e$$

where:

A_s total cross-section area of reinforcement.
f_t mean tensile strength of concrete.
f_y yield stress of the reinforcing steel.
b width of structural element.
d_e effective tension zone, to be taken as $1.5c + 10d_b$ = cover of reinforcement.
d_b diameter of reinforcement bar.

d_e is to be at least 0.2 times the depth of the section, but not greater than $0.5(h-x)$, where x is the depth of the compression zone prior to cracking and h is the section thickness. At intersections between structural elements, where transfer of shear forces is essential to the integrity of the structure, adequate transverse reinforcement is to be provided.

5.3.3 Reinforcement Details

Lapped joints and mechanical splices are to be avoided in structural members subject to significant fatigue loading. Where lapped splices are used in members subject to fatigue, the development length of reinforcing bars is to be twice that required by ACI 318, and lapped bars are to be tied with tie wire. Where mechanical splices are used in members subject to fatigue, the coupled assembly of reinforcing bars and the mechanical coupler are to demonstrate adequate fatigue resistance by test.

Where lapped bars are expected to be subject to tension during operation, through-slab confinement reinforcement is to be considered at the splices. Where longitudinal bars are subject to tension during operation, special consideration is to be given to number of reinforcements with splices at a single location. Reinforcing steel is to comply with the chemical composition specifications of ACI 359 if welded splices are used. For anchorage of shear and main reinforcement, mechanically-headed bars (T-headed bars) may be used if their effectiveness has been verified by static and dynamic testing. Shear reinforcement is to be full length without splices. Entire close-up stirrups are to be anchored by hooks or bends of at least 90 degrees followed by a straight leg length of a minimum 12 bar diameters.

5.3.4 Post Tensioning Ducts

Ducting for post-tensioning ducts may be rigid steel or plastic (polyethylene or polystyrene). Steel tubing is to have a minimum wall thickness of 1 mm. Plastic tubing is to have a minimum wall thickness of 2 mm. Ducts may also be semi-rigid steel, spirally wrapped, of minimum thickness of 0.75 mm, and is to be grout-tight. All splices in steel tubes and semi-rigid duct are to be sleeved and the joints sealed with heat shrink tape. Joints in plastic duct are to be fused or sleeved and sealed. The inside diameter of ducts is to be at least 6 mm (0.25 in) larger than the diameter of the post-tensioning tendon to facilitate grout injection. Flexible ducts are to be used only in special areas where the rigid or semi-rigid duct is impracticable, such as at sharp bends. A mandrel is to be inserted into the ducts to prevent them from dislocating during concreting.

5.3.5 Post-tensioning Anchorages and Couplers

Anchorages for unbonded tendons and couplers are to develop the specified ultimate capacity of the tendons without exceeding the anticipated set. Anchorages for bonded tendons are to develop at least 90% of the specified ultimate capacity of the tendons, when tested in an unbonded condition without exceeding anticipated set. However, 100% of the specified ultimate capacity of the tendons is to be developed after the tendons are bonded in the member. Anchorage and end fittings are to be permanently protected against corrosion. Post-tensioning anchorages are to preferably be recessed in a pocket which is then filled with concrete. The fill is to be mechanically tied to the structure by reinforcements as well as bonded by epoxy or polymer. Anchor fittings for unbonded tendons are to be capable of transferring to the concrete a load equal to the capacity of the tendon under both static and cyclic loading conditions.

5.3.6 Embedded Metals in Concrete

Consideration is to be given to prevention of corrosion of exposed faces of steel embedment. These embedments are to be separated from the reinforcing steel. Effects of dimensional changes due to factors such as prestressing and temperature changes which may result in fractures near embedments may require provisions to prevent deformation.

5.4 Construction and Detailing

5.4.1 General

Construction methods and workmanship are to follow accepted practices as described in ACI 301, ACI 318, ACI 357 or other relevant standards. Additional requirements relevant to concrete substructure of an offshore wind turbine are included below.

5.4.2 Mixing, Placing, and Curing of Concrete

Mixing

Mixing of concrete is to conform to the requirements of ACI 318 and ASTM C94.

Cold Weather

In cold weather, concreting in air temperatures below 2°C (35°F) is to be carried out only if special precautions are taken to protect the fresh concrete from damage by frost. The temperature of the concrete at the time of placing is to be at least 4°C (40°F) and the concrete is to be maintained at this or a higher temperature until it has reached a strength of at least 5 MPa (700 psi). Protection and insulation are to be provided to the concrete where necessary. The aggregates and water used in the mix are to be free from snow, ice, and frost. The temperature of the fresh concrete may be raised by heating the mixing water or the aggregates or both. Cement is never to be heated, nor is it to be allowed to encounter water at a temperature greater than 60°C (140°F).

Hot Weather

During hot weather, proper attention is to be given to ingredients, production methods, handling, placing, protection and curing to prevent excessive concrete temperatures or water evaporation which will impair the required strength or serviceability of the member or structure. The temperature of concrete as placed is not to exceed 30°C (90°F) and the maximum temperature due to heat of hydration is not to exceed 65°C (145°F).

Curing

Special attention is to be paid to the curing of concrete to verify maximum durability and to minimise cracking. Concrete is to be cured with fresh water, whenever possible, to keep the concrete surface wet during hardening. Care is to be taken to avoid the rapid lowering of concrete temperatures (thermal shock) caused by applying cold water to hot concrete surfaces.

Sea Water

Sea water is not to be used for curing reinforced or prestressed concrete, although, if demanded by the construction programme, "young" concrete may be submerged in sea water provided it has gained sufficient strength to withstand physical damage. When there is doubt about the ability to keep concrete surfaces permanently wet for the whole curing period, a heavy-duty membrane curing compound is to be used.

Temperature Rise

The rise of temperature in the concrete, caused by the heat of hydration of the cement, is to be controlled to prevent steep temperature stress gradients which could cause cracking of the concrete. Since the heat of hydration may cause significant expansion, members must be free to contract, so as not to induce excessive cracking. In general, when sections thicker than 610 mm (2 ft) are concreted, the temperature gradients between internal concrete and external ambient conditions are to be kept below 20°C (68°F).

Joints

Construction joints are to be made and located in such a way as not to impair the strength and crack resistance of the structure. Where a joint is to be made, the surface of the concrete is to be thoroughly cleaned, and all laitance and standing water removed. Vertical joints are to be thoroughly wetted and coated with neat cement grout or equivalent enriched cement paste or epoxy coating immediately before placing of new concrete.

Watertight Joints

Whenever watertight construction joints are required, in addition to the above provisions, the heavy aggregate of the existing concrete is to be exposed, and an epoxide-resin bonding compound is to be sprayed on just before concreting. In this case, the neat cement grout can be omitted.

5.4.3 Reinforcement

The reinforcement is to be free from loose rust, grease, oil, deposits of salt or any other material likely to affect the durability or bond of the reinforcement. The specified cover to the reinforcement is to be maintained accurately. Special care is to be taken to correctly position and rigidly hold the reinforcement to prevent displacement during concreting.

5.4.4 Prestressing Tendons, Ducts, and Grouting

General

Further guidance on prestressing steels, sheathing, grouts, and procedures to be used when storing, making up, positioning, tensioning and grouting tendons can be found in the relevant paragraphs of ACI 318, Prestressed Concrete Institute (PCI) publications, *Federation Internationale de la Precontrainte* (FIP) Recommended Practices, and the specialist literature.

Cleanliness

All steel for prestressing tendons is to be clean and free from grease, insoluble oil, deposits of salt or any other material likely to affect the durability or bond of the tendons.

Storage

During storage, prestressing tendons are to be kept clear of the ground and protected from weather, moisture from the ground, sea spray and mist. No welding, flame cutting, or similar operations are to be carried out on or adjacent to prestressing tendons under any circumstances where the temperature of the tendons could be raised, or weld splash could reach them.

Protective Coatings

Where protective wrappings or coatings are used on prestressing tendons, they are to be chemically neutral so as not to produce chemical or electrochemical corrosive attack on the tendons.

Entry of Water

All ducts are to be watertight, and all splices carefully taped to prevent the ingress of water, grout, or concrete. During construction, the ends of ducts are to be capped and sealed to prevent the ingress of sea water. Ducts may be protected from excessive rust using chemically neutral protective agents such as vapor phase inhibitor powder.

Grouting

Where ducts are to be grouted, all oil or similar material used for internal protection of the sheathing is to be removed before grouting. However, water-soluble oil used internally in the ducts or on the tendons may be left on, to be removed by the initial portion of the grout.

Air Vents

Air vents are to be provided at all crests in the duct profile. Threaded grout entries, which permit the use of a screwed connector from the grout pump, may be used with advantage.

Procedures

For long vertical tendons, the grout mixes, admixtures, and grouting procedures are to be checked to verify that no water is trapped at the upper end of the tendon due to excessive bleeding or other causes. Suitable admixtures known to have no injurious effects on the metal or concrete may be used for grouting to increase workability and to reduce bleeding and shrinkage. The temperature of members is to be maintained above 10°C (50°F) for at least 48 h after grouting. General guidance on grouting can be found in the specialist literature. Holes left by unused ducts or by climbing rods of slipforms are to be grouted in the same manner as described above.

6 Foundations

6.1 General

This chapter addresses site investigations, design considerations for the supporting soil, and the influence of the soil and foundation on the Bottom-founded Offshore Wind Turbine. The degree of design conservatism is to reflect prior experience under similar conditions, the manner and extent of data collection, the scatter of design data, and the consequences of failure. For cases where the limits of applicability of any method of calculation employed are not well defined, or where the soil characteristics vary significantly, more than one method of calculation or a parametric study of the sensitivity of the relevant design data is to be used. If the design involves novel foundations or unique soil conditions, model testing may be required to verify design assumptions and methodologies.

6.2 Documentation

The design data and documentation to be submitted for Class review are summarised in 1/5. As a minimum, the following documents are to be submitted, as applicable.

(1) Site Investigation reports including geological survey, geophysical/geohazard survey and geotechnical investigation and reports.
(2) A site-specific seismic hazard report if the reference site is within a seismic zone.
(3) As applicable, the results of studies to assess the following effects are also to be submitted. In these studies, the structure is to be considered present:
    - Scouring potential of the sea floor.
    - Hydraulic instability and the occurrence of sand waves.
    - Instability of slopes in the area where the Support Structure is to be installed.
    - Liquefaction and other soil instabilities.
    - Geo-hazards such as faults, gas seeps etc.
    - For Arctic areas, degradation of subsea permafrost layers.
    - Soils conditions in the vicinity of footprints left by temporarily situated service units or other service units.
    - Effects of volcanic sands, organic matter, carbonate soil, calcareous sands and other substances which degrade the strength of the soil foundation.
(4) Design reports covering the following:
    - Holding capacity of the foundation.
    - Foundation structure design.
    - Fatigue analysis if applicable.
    - Installation analyses.

- Corrosion protection.

(5) Design drawings and specifications for material, welding, and fabrication.
(6) Foundation installation procedure.

6.3 Site Investigation

6.3.1 General

The site investigation programme is to consist of the following three phases.

(1) Sea Floor Survey to obtain relevant geophysical data.
(2) Geological Survey to obtain data of a regional nature concerning the site.
(3) Subsurface Investigation and Testing to obtain the necessary geotechnical data.

The results of these investigations are to be the bases for the additional site related studies which are listed in (3) above. A complete site investigation programme is to be accomplished. However, use of the complete or partial results of a previously completed site investigation as the design basis for another similarly designed and adjacent structure is permitted when the adequacy of the previous site's investigation for the new location is satisfactorily demonstrated. When deciding the area to be investigated, due allowance is to be given to the accuracy of positioning devices used on the vessel employed in the site investigation such that the data obtained are pertinent to the actual location of the Bottom-founded Offshore Wind Turbine.

6.3.2 Sea Floor Survey

Geophysical data for the conditions existing at and near the surface of the sea floor are to be obtained. The following information is to be obtained where applicable to the planned Bottom-founded Offshore Wind Turbine.

(1) Soundings or contours of the seabed.
(2) Position of bottom shapes which might affect scour.
(3) The presence of boulders, obstructions, and small craters.
(4) Gas seeps.
(5) Shallow faults.
(6) Slump blocks.
(7) Ice scour of sea floor sediments.
(8) Subsea permafrost or ice bonded soils.

6.3.3 Geological Survey

Data of the regional geological characteristics which can affect the design and siting of the Bottom-founded Offshore Wind Turbine are to be considered in planning the subsurface investigation, and they are also to be used to verify that the findings of the subsurface investigation are consistent with known geological conditions. Where necessary, an assessment of the seismic activity at the site is to be made. Emphasis is to be placed on the identification of fault zones, the extent and geometry of faulting and attenuation effects due to conditions in the region of the site.

6.3.4 Subsurface Investigation

The subsurface investigation is to obtain reliable geotechnical data concerning the stratigraphy and properties of the soil. These data are to be used to assess whether the desired level of structural safety and performance can be obtained and to assess the feasibility of the proposed method of installation. Consistent with the stated objective, the subsurface investigation programme is to consist of an adequate number of in-situ tests, borings, probing and geophysical investigation to examine all important soil and rock strata at each foundation location. The extent of the investigation is to be discussed and agreed upon with Class prior to the beginning of the subsurface investigation and may be adjusted based on the results of the Sea Floor Survey and Geological survey. Soil data are to be taken in the vicinity of the foundation locations, including seabed samples for evaluation of scour potential. An interpretation of such data is to be submitted by a recognised geotechnical consultant. To establish the soil characteristics at the foundation locations, borings or probing are to be taken at all foundation locations to a suitable depth of at least the anticipated depth of any pile penetrations plus a consideration for the soil variability as described below. If probing is used, they are to be calibrated with the data obtained from conventional boring in the vicinity of the probing. As an alternative, geophysical surveys may be carried out and correlated with at least two borings or probing in the vicinity to establish the soil profile at each foundation location. The integration of geophysical, geotechnical, and geological findings is to be carried out by recognised consultants. If the foundation location is at a significant distance from the boring/probing location, additional number of verification borings/probing may be required to validate the extrapolated data. To account for soil variability, a lower bound and an upper bound soil profile are to be established for use in pile holding capacity calculations and installation analyses, respectively.

For pile-supported structures, the minimum depth of a bore hole, for either individual or clustered piles, is to be the design penetration plus a zone of influence. The zone of influence is to be at least 15.2 m (50 ft) or 1.5 times the diameter of the cluster, whichever is greater, unless it can be demonstrated by analytical methods that a lesser depth is justified. Additional bore holes of lesser depth are required if discontinuities in the soil are likely to exist within the area of the structure. For a gravity-type foundation, the required depth of at least one boring is to be at least equal to the larger horizontal dimension or three times the smaller dimension of the base, whichever is greater. In-situ tests are to be carried out, where possible, to a depth that will include the anticipated shearing failure zone. A continuous profile is to be obtained during recovery of the boring samples. The recovery of the materials to a depth of 12 m (40 ft) below the mudline is to be as complete as possible. Thereafter, samples at significant changes in strata are to be obtained, at approximately 3 m (10 ft) intervals to 61 m (200 ft) and approximately 8 m (25 ft) intervals below 61 m (200 ft). The existence of carbonate soils is to be determined in frontier areas or areas known to contain carbonate materials during the subsurface investigation. Additional field and laboratory testing may be required when a soil profile shows the contained carbonate materials are more than 15% of the soil fraction.

6.3.5 Soil Testing Programme

The testing programme is to reveal the necessary engineering properties of the soil including strength, classification, and deformation properties of the soil. Testing is to be performed in accordance with recognised standards. At least one undrained strength test (vane, drop cone, unconfined compression, etc.) on selected recovered cohesive samples is to be performed in the field. Where practicable, a standard penetration test or equivalent on each significant sand stratum is to be performed, recovering samples where possible. Field samples for laboratory work are to be retained and carefully packaged to minimise changes in moisture content and disturbance. Samples from the field are to be sent to a recognised laboratory for further testing. They are to be accurately labelled, and the results of visual inspection recorded. The testing in the laboratory is to include at least the following.

(1) Perform unconfined compression tests on clay strata where needed to supplement field data.
(2) Determine water content and Atterberg limits on selected cohesive samples.
(3) Determine density of selected samples.
(4) As necessary, develop appropriate constitutive parameters or stress–strain relationships from either unconfined compression tests, unconsolidated undrained triaxial compression tests, or consolidated undrained triaxial compression tests.
(5) Perform grain size sieve analysis, complete with percentage passing 200 sieves, on each significant sand or silt stratum.
(6) Other advanced testing to determine soil parameters necessary for detailed geotechnical analyses.

For pile-supported structures, consideration is also to be given to the need for additional tests to adequately describe the dynamic, creep and set-up characteristics of the soil as well as the static and cyclic lateral properties of soil-pile system. For gravity structures, laboratory tests are also to include, where necessary, the following.

(1) Shear strength tests with pore pressure measurements. The shear strength parameters and porewater pressures are to be measured for the relevant stress conditions.
(2) Cyclic loading tests with deformation and pore pressure measurements to determine the soil behaviour during alternating stress.
(3) Permeability and consolidation tests performed as required.

6.4 Foundation Design Requirements

6.4.1 General

The loadings used in the design of foundations are to include those defined in 5/2.5 and those experienced during installation. Foundation displacements are to be evaluated to the extent necessary to conclude that they are within the limits that do not impair the intended function and safety of the Bottom-founded Offshore Wind Turbine. The soil and the offshore wind turbine are to be considered as an interactive system. The results of analyses, as required in subsequent paragraphs, are to be evaluated from this point of view. In addition to this paragraph, specific design requirements are described in 5/4.5 for pile foundations and 5/4.6 for gravity structures. Other types of foundations are subject to special considerations by Class. Where the Survival Load Cases in accordance with 4/2.5 are assessed, the ultimate strength analysis method provided in API RP 2SIM is to be applied for the foundation design, where applicable. For the pile foundation, if a global collapse mechanism in the foundation controls the ultimate strength of the Bottom-founded Offshore Wind Turbine, then a geotechnical engineer familiar with assessment should be involved in the assessment. Refer to API RP 2SIM for guidance.

6.4.2 Cyclic Loading Effects

The influence of cyclic loading on soil properties is to be considered and reduction of soil strength is to be investigated and employed in the design of the foundation. In particular, the effects of wind and wave induced forces on the soil properties under the following conditions are to be considered.

(1) Design storm during the initial consolidation phase.
(2) Short-term effects of the design storm.
(3) Long-term cumulative effects of several storms, including the design storm, and turbine operations

Reduced soil strength characteristics resulting from these conditions are to be employed in the design. In seismically active zones, similar deteriorating effects due to repeated loadings are to be considered. Other possible cyclic load effects, such as changes in load–deflection characteristics, liquefaction potential and slope stability are also to be considered when they are expected to affect the design.

6.4.3 Scour

Where scour is expected to occur, either effective protection is to be furnished soon after the installation of pile or gravity base, or the depth and lateral extent of scouring, as evaluated in the site investigation programme, are to be accounted for in the design of the foundation.

6.4.4 Deflections and Rotations

Tolerable limits of deflections and rotations are to be established based on the type of the substructure and the effects of those movements on the turbine operations. Maximum allowable values of pile or gravity base movements, as limited by the structural considerations, overall stability of the Support Structure and serviceability limit of the wind turbine, are to be considered in the design of the foundation. Pile installation tolerances are also to be considered in the design.

6.4.5 Soil Strength

The ultimate strength or stability of soil is to be determined using test results which are compatible with the analysis method selected. In a total stress approach, the total shear strength of the soil obtained from simple tests is used. A total stress approach ignores changes in the soil's pore water pressure under varying loads and the drainage conditions at the site. When an effective stress approach is used, effective soil strength parameters and pore water pressures are to be determined from tests which predict in-situ total stresses and pore pressures.

6.4.6 Dynamic and Impact Considerations

For dynamic and impact loading conditions, a realistic and compatible treatment is to be given to the interactive effects between the soil and structure. Dynamic analyses of the Support Structure of an offshore wind turbine may be accomplished by lumped parameter, foundation impedance functions, or by continuum approaches including the use of finite element methods. Such models are to include consideration of the internal and radiational damping provided by the soil and the effects of soil layering. Studies of the dynamic response of the structure are to include, where applicable, consideration of the nonlinear and inelastic characteristics of the soil, the possibilities of deteriorating strength and increased or decreased damping due to cyclic soil loading, and the added mass of soil subject to acceleration. Where applicable, the influence of nearby structures is to be included in the analysis.

6.4.7 Loading Conditions

Those loading conditions which produce the most unfavourable effects on the foundation during and after installation are to be considered. Post installation loadings to be checked are to include at least those relating to the design load conditions, as specified in 4/3, combined with dead loads, live loads and deformation loads appropriate to the function and operations of an offshore wind turbine. For areas with potential seismic activity, the foundation is to be designed for sufficient strength to sustain seismic loads. As appropriate, effects of liquefaction shear failure of soft mud and submarine slides are to be considered.

6.4.8 Loads and Soil Conditions Due to Temporarily Situated Structures

Changes in soil conditions due to temporarily situated structures, such as self-elevating wind turbine installation units placed near the Support Structure, are to be assessed. These changes and their influence on the structure are to be incorporated in the foundation design such that the intended function and safety of the Bottom-founded Offshore Wind Turbine are not impaired.

6.5 Pile Foundations

6.5.1 General

The effects of axial, bending, and lateral loads are to be accounted for in the design of pile foundations. The design of a pile is to reflect the interactive behaviour between the soil and the pile and between the pile and other components of the Support Structure. Piles are to be designed to withstand in-place and installation loads. Foundation displacements are to be evaluated to the extent necessary to conclude that they are within limits which do not impair the intended function and safety of the Bottom-founded Offshore Wind Turbine. The following is to be considered in the evaluations of the pile holding capacity.

(1) Cyclic, creep and soil set-up effects on the soil strength.
(2) Punch through failure in layered soils.
(3) Installation tolerances and soil disturbance due to protuberances as applicable.
(4) Potential of soil liquefaction and its impact if the reference site is within a seismic zone.
(5) Pile group effect if applicable.
(6) Bottom instability such as slope stability, hydraulic stability (see page, piping), scour potential, etc.

Where applicable, the effects of close spacing on the load and deflection characteristics of pile groups are to be determined. The allowable load for a group, both axial and lateral, is not to exceed the sum of the apparent individual pile allowable loads reduced by a suitable factor. The required load conditions and strength criteria of steel structural members are to be in accordance with 5/2.3. Depending on water depth, anticipated loads and the pile configuration, fatigue analysis may be required. The adequacy of the corrosion protection of the piles is to be evaluated. A corrosion protection system may be designed in accordance with the recognised industry standards, such as those published by NACE. The design life of the corrosion protection

Table 4 Factor of safety for axially loaded piles

loading condition	Factor of safety
Normal	2.0
Abnormal	1.5

system is to be equal or greater than the design life of the Bottom-founded Offshore Wind Turbine. Methods of pile installation are to be consistent with the type of soil at the site, and with the installation equipment available. Pile installation is to be carried out and supervised by qualified and experienced personnel and is to be witnessed by the Class Surveyor.

6.5.2 Slender Piles

Axial Loads

The axial capacity of slender piles in compression consists of the skin friction, Q_f, developed along the length of the pile, and the end bearing, Q_p at the tip of the pile. The axial capacity of a pile subjected to tension is to be equal to or less than the skin friction alone. Predictions of the various parameters needed to evaluate Q_f and Q_p are to be accomplished using a recognised analytical method, such as that recommended in the API RP 2A, or other methods shown to be more appropriate to the conditions at the site. When required, the acceptability of any method used to predict the components of pile resistance is to be demonstrated by showing satisfactory performance of the method under conditions like those existing at the actual site. The results of dynamic pile driving analysis alone are not to be used to predict the axial load capacity of a pile. When the ultimate capacities of a pile are evaluated using the API method cited above, the allowable axial pile bearing, and pullout loads are to be determined by dividing the ultimate capacities by the factor of safety specified in Table 4. For the Design Earthquake, the factor of safety is to be specially considered.

Lateral Loads

In the evaluation of the pile's behaviour under lateral loadings, the combined-load–deflection characteristics of the soil and pile, and of the pile and structure are to be considered. The representation of pile's lateral deflection when the pile is subjected to lateral loads is to adequately reflect the deterioration of the lateral bearing capacity when the soil is subjected to cyclic loading. The description of the lateral load versus deflection characteristics ($p - y$ curves) for the various soil strata is to be based on constitutive data obtained from suitable soil tests. Reference may be made to the API RP 2A for a procedure to evaluate the load–deflection characteristics of laterally loaded piles. However, the use of alternative methods is permitted when they are more appropriate for conditions at the site. Where applicable, the rapidly deteriorating cyclic bearing capacity of stiff clays, especially those exhibiting the presence of a secondary structure, is to be considered. The strength criteria for steel structural members of the pile subjected to lateral loads are to be in accordance with 5/2.6.1.

6.5.3 Large Diameter Piles

The design method in 5/4.5.2 slender piles may not be applicable to large diameter piles. Due to the geometry of the pile, the failure modes of the soil may be different than those applicable for slender piles as described in 5–4/4.5.2. A suitable analytical method, such as the finite element analyses, may be employed to evaluate the holding capacity of large diameter piles and the adequacy of the piles to withstand in-place loads. Soil properties necessary for these analyses are to be established in consultation with the geotechnical consultant. Where the limit equilibrium method is used, different failure modes, which are affected by various factors such as the pile load and load direction, are to be considered to establish the holding capacity of a pile. Model tests, centrifuge tests, or full-scale pile testing may be carried out to provide justification of the design method.

6.5.4 Pile Installation Analysis

Pile installation analyses including lifting, transportation and penetration analyses are to be carried out according to the relevant requirements in API RP 2A. For driven piles, the pile stick-up and driveability analysis are to be performed to evaluate the adequacy of the pile to withstand loads during driving conditions in accordance with API RP 2A. Under the stick-up condition, the pile's adequacy to withstand the self-weight, hammer weight and environmental loads acting on the pile section above the mudline, if applicable, is to be verified. Pile driveability study is to use acceptable methods and appropriate analysis parameters for operating efficiency, hammer data and accessories, etc. Driving stresses, number of blows per unit depth and total blow count required for driving are to be established using appropriate methods. During driving, the sum of the stresses in the pile due to the impact of the hammer (the dynamic stresses) and the stresses due to axial load and bending (static stresses) is not to exceed the minimum yield stress of the steel. Further, the dynamic stresses are not to exceed 80 to 90 percent of the minimum yield stress of the steel. If piles are installed by suction, the suction pressure required to reach the target penetration is to provide a minimum factor of 1.5 against soil heaving inside the pile. The pile capacity to resist hydrostatic collapse during installation under the differential pressure is to be evaluated using the procedures given in the Class guidance for performing buckling and ultimate strength assessment for offshore structures.

6.6 *Gravity Structures*

6.6.1 General

For foundation systems consisting of gravity structures, the stability of the foundation regarding the bearing and sliding failure modes is to be investigated using the soil shear strengths determined in accordance with 5/4.3.5 and 5/4.4.2. The effects of adjacent structures and the variation of soil properties in the horizontal direction are to be considered where relevant. Where levelling of the site is not carried out, the predicted tilt of the overall foundation is to be based on the average bottom slope of the sea floor and the tolerance of the elevation measuring device used in the site investigation programme. Differential settlement is also to be calculated, and the tilting of the structure caused by this settlement is to be combined with the predicted structural tilt. Any increased loading effects caused by the tilting of the structure are to be considered in the foundation stability requirements described in 5/5.2. When an under-pressure or over-pressure is experienced by the sea floor under the structure, provision is to be made to prevent piping which could impair the integrity of the foundation. The influence of hydraulic and slope instability, if any, is to be determined. Initial consolidation and secondary settlements, as well as permanent horizontal displacements, are to be calculated.

6.6.2 Stability

The bearing capacity and lateral resistance are to be calculated under the most unfavourable combination of loads. Possible long-term redistribution of bearing pressures under the base slab are to be considered to verify that the maximum edge pressures are used in the design of the perimeter of the base. The lateral resistance of the foundation is to be investigated with respect to various potential shearing planes. Special consideration is to be given to any layers of soft soil. Calculations for overturning moment and vertical forces induced by the passage of a wave are to include the vertical pressure distribution across the top of the foundation and along the sea floor. The capacity of the foundation to resist a deep-seated bearing failure is to be analysed. In lieu of a more rigorous analysis, the capacity of the foundation to resist a deep-seated bearing failure can be calculated by standard bearing capacity formulas applicable to eccentrically loaded shallow foundations, provided that.

(1) Uniform soil conditions are present or conservatively chosen soil properties are used to approximate a non-uniform soil condition.
(2) A trapezoidal distribution of soil pressure is a reasonable expectation. Alternatively, slip-surface methods, covering a range of kinematically possible deep rupture surfaces can be employed in the bearing capacity calculations.

The maximum allowable shear strength of the soil is to be determined by dividing the ultimate shear strength of the soil by a factor of safety. When the ultimate soil

Table 5 Factor of safety for allowable shear strength of soil

Loading condition	Factor of safety
Normal	2.0
Abnormal	1.5

strength is determined by an effective stress method, the safety factor is to be applied to both the cohesive and frictional terms. If a total stress method is used, the safety factor is to be applied to the undrained shear strength. The minimum safety factors to be applied when using a standard bearing capacity formulation and various trial sliding failure planes are specified in Table 5. The safety factors to be applied when considering the earthquake conditions will be specially considered.

Where present, additional effects of penetrating walls or skirts which transfer vertical and lateral loads to the soil are to be investigated for their contribution to the bearing capacity and lateral resistance.

6.6.3 Soil Reaction on the Base

For conditions during and after installation, the reaction of the soil against all structural members seated on or penetrating the sea floor is to be accounted for in the design of these members. The distribution of soil reactions is to be based on the results obtained according to 5/4.3.4. Calculations of soil reactions are to account for any deviation from a plane surface, the load–deflection characteristics of the soil, and the geometry of the base of the structure. Where applicable, effects of local soil stiffening, nonhomogeneous soil properties, as well as the presence of boulders and other obstructions, are to be accounted for. During installation, consideration is to be given to the possibility of local contact pressures due to irregular contact between the base and the sea floor; these pressures are additive to the hydrostatic pressure. An analysis of the penetration resistance of structural elements projecting into the sea floor below the foundation structure is to be performed. The design of the ballasting system is to reflect uncertainties associated with achieving the required penetration of the structure. Since the achievement of the required penetration of the platform and its skirts is of critical importance, the highest expected values of soil strength are to be used in the calculation of penetration.

6.7 Marine Operations

6.7.1 General

In this book, marine operations include the following activities as appropriate to the planned Bottom-founded Offshore Wind Turbine.

(1) Load-out and transport.

(2) Pile installation.
(3) Final field erection and levelling.
(4) Installation of the Rotor-Nacelle Assembly (RNA).

The unfavourable effects that may be induced in the Support Structure during marine operations are to be accounted for. This chapter is intended to address the requirements on considering the influence that marine operations may have on the safety and integrity of the Bottom-founded Offshore Wind Turbine.

6.7.2 Documentation

The extent of documentation and analysis of marine operations is to be adequate for the size and type of the Support Structure involved, the operation being considered, the extent of experience with similar operations, and the severity of the expected environmental conditions. A report on the marine operations planned to install the Support Structure is to be developed and submitted for use in association with the review of the analyses required in 5/5. The purpose of this report is to demonstrate that the strength and integrity of the Support Structure are not reduced or otherwise jeopardised by the marine operations. This report is to contain the following information.

(1) Description of the marine operations to be performed and the procedures to be employed.
(2) Pre-installation verification procedures for the sea-bed condition and contingency procedures.
(3) For operations which do not govern the design of the Support Structure, a description of the engineering logic, experience, or preliminary calculations supporting this conclusion.
(4) For operations which govern design of the offshore wind turbine Support Structure, the assumptions, calculations, and results of the analyses required in 5/5.

7 Analyses

7.1 Loads

Analyses are to be performed to determine the type and magnitude of the loads and load combinations to which the Support Structure is exposed during the performance of marine operations. The design conditions (DLC 8.1 in Table 1 of Chap. 4) for marine operations are to be defined as appropriate by the Fabricator and/or Owner. Particular attention is to be given to inertial, impact, and local loads that are likely to occur during marine operations. In addition, DLC 8.2 and DLC 8.3 in Table 1 of

Chap. 4 are also to be considered in accordance with paragraph 7.4.9 of IEC 61,400-3-1 (2019). Where significant fatigue damage occurs during marine operations, it is to be included in calculating the total fatigue life.

7.2 Stability

Analyses are to be performed to verify that the Support Structure, or its means of support where such exist, has sufficient hydrostatic stability and reserve buoyancy to allow for successful execution of all phases of marine operations. For the large or unusual Support Structure, an experimental determination of the centre of gravity of the structure and its means of support, where such exist, is to be performed.

Chapter 6
Equipment and Systems for Bottom-Founded Offshore Wind Turbines

1 General

The requirements for the onboard equipment and systems for the Bottom-founded Offshore Wind Turbine are given in this chapter. In addition, equipment and systems used for Self-Elevating Units are to comply with the MOU Rules, where applicable.

2 Fire Fighting Systems and Equipment

Firefighting systems and equipment are to be designed with consideration of the size, type, and intended service of the Bottom-founded Offshore Wind Turbine. In all cases, the selection of the fire extinguishing medium is to be based on the fire hazard for the space protected. The fire extinguishers are to be visible and readily accessible. Attention is drawn to the relevant requirements of the Coastal State or other governmental authorities having jurisdiction over the Bottom-founded Offshore Wind Turbine to be installed.

3 Safety Outfitting

Safety outfitting and measures are to be designed with suitable consideration of the nature of unmanned installations. It is suggested that the Coastal State or other governmental authorities having jurisdiction over the Bottom-founded Offshore Wind Turbine are to be installed be contacted to obtain applicable requirements.

F. Aderouane, *Design, Construction, and Global Performance Analysis of Offshore Wind Turbines*, Green Energy and Technology, https://doi.org/10.1007/978-3-031-86010-2_6

4 Identification Marks

A unique name or number is to be assigned to each Bottom-founded Offshore Wind Turbine, which is to conform to requirements of the Coastal State or other governmental authorities having jurisdiction over the Bottom-founded Offshore Wind Turbine. This name or number is to be permanently displayed on the structure and will be entered in the Class Record.

Chapter 7
Surveys for Bottom-Founded Offshore Wind Turbines

1 General

This chapter pertains to surveys during the construction and installation of the Bottom-founded Offshore Wind Turbine. The requirements of 7/1 are to apply to all Bottom-founded Offshore Wind Turbines covered by this book regardless of type. Additional specific requirements are contained in 7/2.2 for steel substructures and in 7/2.5 for concrete substructures. The phases of construction covered by this chapter include material manufacture, fabrication, installation and final field erection of the foundation, substructure, tower, and RNA. For the RNA and tower, the survey scope includes the following two situations:

- The optional RNA notation is requested by the Owner. The RNA and tower including its end connections are type approved by Class and are within the scope of classification and survey.
- The optional RNA notation is not requested by the Owner. However, the tower and its connection to the RNA and/or to the substructure are not included in the wind turbine type certificate. In this scenario, the tower and its connection are within the scope of classification and survey. The RNA with a wind turbine type certificate is not within the scope of classification.

2 Quality Control Programme

A quality control programme compatible with the type and size of the planned Bottom-founded Offshore Wind Turbine is to be developed and submitted to Class for review. Class will review, approve and, as necessary, request modification of this programme. The Fabricator is to work with the attending Surveyor to establish the required hold points on the quality control programme to form the basis for all future surveys at the fabrication yard. As a minimum, the items enumerated in the various

F. Aderouane, *Design, Construction, and Global Performance Analysis of Offshore Wind Turbines*, Green Energy and Technology, https://doi.org/10.1007/978-3-031-86010-2_7

applicable paragraphs below are to be covered by the quality control programme. Surveyors will be assigned to monitor the fabrication of items within the scope of classification, and to verify that competent personnel are carrying out all tests and inspections specified in the quality control programme. It is to be noted that the monitoring provided by Class is a supplement to and not a replacement for inspections to be carried out by the Fabricator or Owner.

2.1 Access and Notification

During construction, the Surveyor is to have access to structures at all reasonable times. The attending Surveyor is to be notified as to when and where parts of the structure may be examined. If, at any visit, the Surveyor finds occasion to recommend repairs or further inspection, notice is to be made to the Fabricator or its representatives.

2.1.1 Identification of Materials

The Fabricator is to maintain a system of material traceability to the satisfaction of the attending Surveyor, for all Special and Primary Application Structures. Data as to place of origin and results of relevant material tests for structural materials are to be retained and made readily available during all stages of construction (refer to 7/2.2.12 and 7/2.3.8). Such data are to be available to the Surveyor upon request.

2.2 Steel Structures

2.2.1 Quality Control Programme

The quality control programme (refer to 7/2.2.1) for the construction of the steel Support Structure is to include the following items, as appropriate.

(1) Material quality and traceability.
(2) Steel Forming.
(3) Welder qualification and records.
(4) Welding procedure specifications and qualifications.
(5) Weld inspection.
(6) Tolerances alignments and compartment testing.
(7) Corrosion control systems.
(8) Tightness and hydrostatic testing procedures.
(9) Non-destructive testing.
(10) Installation of the Support Structure.

The items which are to be considered for each of the topics mentioned above are indicated in 7/2.2.

2.2.2 Material Quality and Traceability

The properties of the material are to be in accordance with 2/1 of this book. Manufacturer's certificates are to be supplied with the material. Verification of the material's quality is to be done by the Surveyor at the plant of manufacture in accordance with the prescribed Class Rules for materials and welding. Alternatively, material manufactured to recognised standards may be accepted in lieu of the above steel requirements provided the substitution of such materials is approved by Class. Materials used are to be in accordance with those specified in the approved design. All materials required for classification purposes are to be tested in the presence of the Class Surveyor. The Fabricator is to maintain a material traceability system for all the Primary and Special Application Structures.

2.2.3 Steel Forming

When forming changes base plate properties beyond acceptable limits, appropriate heat treatments are to be carried out to re-establish required properties. Unless approved otherwise, the acceptable limits of the re-established properties are to meet the minimums specified for the original material before forming. As required, Class will survey formed members for their compliance with the forming dimensional tolerances required by the design.

2.2.4 Welder Qualification and Records

Welders who will work on the Bottom-founded Offshore Wind Turbine are to be qualified in accordance with the welder qualification tests specified in a recognised code to the satisfaction of the attending Surveyor. Certificates of qualification are to be prepared to record evidence of the qualification of each welder qualified by an approved standard/code, and such certificates are to be made available to the Surveyor. If welders have been previously tested in accordance with the requirements of a recognised code and provided that the period of effectiveness of the previous testing has not lapsed, these previous welder qualification tests may be accepted.

2.2.5 Welding Procedure Specifications and Qualifications

Welding procedures are to be approved in accordance with the Class Rules for materials and welding. Welding procedures conforming to the provisions of a recognised code may, at the Surveyor's discretion, be accepted. A written description of all procedures previously qualified may be employed in the structure's construction provided it is included in the quality control programme and made available to the Surveyor. When it is necessary to qualify a welding procedure, this is to be accomplished by employing the methods specified in the recognised code, and in the presence of the Surveyor.

2.2.6 Weld Inspection

As part of the overall quality control programme, a detailed plan for the inspection and testing of welds is to be prepared. This plan is to include the applicable provisions of this chapter.

2.2.7 Tolerances and Alignments

The overall structural tolerances, forming tolerances, and local alignment tolerances are to be commensurate with those considered in developing the structural design. Inspections are to be carried out to verify that the dimensional tolerance criteria are being met. Particular attention is to be paid to the out-of-roundness of members for which buckling is an anticipated mode of failure. Structural alignment and fit-up prior to welding are to be monitored to verify consistent production of quality welds.

2.2.8 Corrosion Control Systems

The details of any corrosion control systems employed for the structure are to be submitted for review. Installation and testing of the corrosion control systems are to be carried out to the satisfaction of the attending Surveyor in accordance with the approved plans.

2.2.9 Tightness and Hydrostatic Testing Procedures

Compartments, which are designed to be permanently watertight or to be maintained watertight during installation, are to be tested by a procedure approved by the attending Surveyor. The testing is also to be witnessed by the attending Surveyor.

2.2.10 Non-Destructive Testing

A system of non-destructive testing is to be included in the fabrication specification of the structures. The minimum extent of non-destructive testing is to be in accordance with this book or recognised design Code. All non-destructive testing records are to be reviewed and approved by the attending Surveyor. Additional non-destructive testing may be requested by the attending Surveyor if the quality of fabrication is not in accordance with industry standards.

2.2.11 Testing and Trials of a Self-Elevating Unit to Be Used as the Substructure

A Self-Elevating Mobile Offshore Unit intended to be converted to the site dependent substructures of an offshore wind turbine is to be subject to the applicable surveys during construction and installation specified in this book in addition to the testing and trials required by the Class Rules for design, construction, and classification of mobile offshore units.

2.2.12 Records

A data book of the records of construction activities is to be developed and maintained to compile a record as complete as practicable. The pertinent records are to be adequately prepared and indexed to assure their usefulness, and they are to be stored so that they may be easily recovered. For a steel structure, the construction record is to include, as applicable, the following:

(1) Material traceability records including mill certificates.
(2) Welding procedure specification and qualification records.
(3) Shop welding practices.
(4) Welding inspection records.
(5) Construction specifications.
(6) Structural dimension check records.
(7) Non-destructive testing records.
(8) Records of completion of items identified in the quality control programme.
(9) Towing and pile driving records.
(10) Position and orientation records.
(11) Levelling and elevation records, etc.

The compilation of these records is a condition of classification. After fabrication and installation, these records are to be retained by the Owner or Fabricator for future references. The minimum time for record retention is not to be less than the greatest of the following: the warranty period, the time specified in construction agreements, or the time required by statute or governmental regulations.

2.3 Concrete Structures

2.3.1 Quality Control Programme

The quality control programme (refer to 7/2.2.1) for a concrete structure is to cover the following items, as appropriate.

(1) Inspections prior to concreting.
(2) Inspection of batching, mixing, and placing concrete.
(3) Inspections of form removal and concrete curing.
(4) Inspection of prestressing and grouting.
(5) Inspection of joints.
(6) Inspection of finished concrete.
(7) Installation of the Support Structure.
(8) Tightness and hydrostatic testing as applicable (refer to 7/2.2.9).

The items which are to be considered for each of the topics mentioned above, except for (8), are indicated in 7/2.3.2 through 7/2.3.7.

2.3.2 Inspections Prior to Concreting

Prior to their use in construction, the manufacturers of cement, reinforcing rods, prestressing tendons and appliances are to provide documentation of the pertinent physical properties. These data are to be made available to the attending Surveyor for verification of conformity with the properties specified in the approved design. As applicable, at the construction site, the Surveyor is to be satisfied that proper consideration is being given to the support of the structure during construction, the storage of cement and prestressing tendons in weathertight areas, the storage of admixtures and epoxies to manufacturer's specifications, and the storage of aggregates to limit segregation, contamination by deleterious substances and moisture variations within the stockpile. Forms and shores supporting the forms are to be inspected to verify that they are adequate in number and type, and that they are in accordance with the approved plans. The dimensions and alignment of the forms are to be verified by the attending Surveyor. The measurements are to be within the allowable finished dimensional tolerances specified in the approved design. Reinforcing steel, prestressing tendons, post-tensioning ducts, anchorages and any included steel are to be checked, as appropriate to the planned structure, for size, bending, spacing, location, firmness of installation, surface condition, vent locations, proper duct coupling, and duct capping.

2.3.3 Inspection of Batching, Mixing, and Placing Concrete

The production and placing of the concrete are to employ procedures which will provide a well-mixed and well-compacted concrete. Such procedures are also to limit segregation, loss of material, contamination, and premature initial set during all operations. Mix components of each batch of concrete are to be measured by a method specified in the quality control programme. The designer is to specify the allowable variation of mix component proportions, and the Fabricator is to record the actual proportions of each batch. Testing during the production of concrete is to be carried out following the procedures specified in the quality control programme. As a minimum, the following concrete qualities are to be measured by the Fabricator:

(1) Consistency.
(2) Air content.
(3) Density or Specific Gravity.
(4) Strength.

Field testing of aggregate gradation, cleanliness, moisture content, and unit weight is to be performed by the Fabricator following standards and schedules specified in the quality control programme. The frequency of testing is to be determined with consideration of the uniformity of the supply source, volume of concreting, and variations of atmospheric conditions. Mix water is to be tested for purity following the methods and schedules specified in the quality control programme.

2.3.4 Inspections of Form Removal and Concrete Curing

The structure is to have sufficient strength to bear its own weight, construction loads and the anticipated environmental loads without undue deformations before forms and form supports are removed. The schedule of form removal is to be specified in the quality control programme, giving due account to the loads and the anticipated strength. Curing procedures for use on the structure are to be specified in the quality control programme. When conditions at the construction site cause a deviation from these procedures, justification for these deviations is to be fully documented and included in the construction records. Where the construction procedures require the submergence of recently placed concrete, special methods for protecting the concrete from the effects of salt water are to be specified in the quality control programme. Concrete is not to be submerged until 28 days after placing (refer to also 5/3.7.2).

2.3.5 Inspection of Prestressing and Grouting

A schedule indicating the sequence and anticipated elongation and stress accompanying the tensioning of tendons is to be prepared. Any failures to achieve proper tensioning are to be immediately reported to the designer to obtain guidance on needed remedial actions. Pre- or post-tensioning loads are to be determined by measuring both tendon elongation and tendon stress. These measurements are to be compared. In the case that the variation of measurements exceeds the specified amount, the cause of the variation is to be determined, and any necessary corrective actions are to be accomplished. The grout mix is to conform to that specified in the design. The Fabricator is to keep records of the mix proportions and ambient conditions during grout mixing. Tests for grout viscosity, expansion and bleeding, compressive strength, and setting time are to be made by the Fabricator using methods and schedules specified in the quality control programme. Employed procedures are to verify that ducts are filled. Anchorages are to be inspected to verify that they are located and sized as specified in the design. Anchorages are also to be inspected to verify that they will be provided with adequate cover to mitigate the effects of corrosion.

2.3.6 Inspection of Joints

Where required, leak testing of construction joints is to be carried out using procedures specified in the quality control programme. When deciding which joints are to be inspected, consideration is to be given to the hydrostatic head on the subject joint during normal operation, the consequence of a leak at the subject joint, and the ease of repair once the Bottom-founded Offshore Wind Turbine is in service.

2.3.7 Inspection of Finished Concrete

The surface of the hardened concrete is to be comprehensively inspected for cracks, honeycombing, popouts, spalling and other surface imperfections. When such defects are found, their extent is to be reported to the Surveyor and to the designer for guidance on any necessary repairs. The structure is to be examined using a calibrated rebound hammer or a similar non-destructive testing device. Where the results of surface inspection, cylinder strength tests or non-destructive testing do not meet the design criteria, the designer is to be consulted regarding remedial actions to be taken. The completed sections of the structure are to be checked for compliance to specified design tolerances for thickness, alignment, etc., and to the extent practicable, the location of reinforcing and prestressing steel and post-tensioning ducts. Variations from the tolerance limits are to be reported to the designer for evaluation and guidance on any necessary remedial actions.

2.3.8 Records

Refer to 7/2.2.12 for information on compiling construction records. For the concrete Support Structure of an offshore wind turbine, the construction records are to include, as applicable, the following:

(1) All material certificates and test reports.
(2) Tensioning and grouting records.
(3) Concrete records including weight.
(4) Moisture content and mix proportions.

(5) A list of test methods and results.
(6) Ambient conditions during the pours.
(7) Calibration data for test equipment.
(8) Towing records.
(9) Data on initial structural settlements.
(10) Inspector's logs.

These records are to be retained by the Owner.

2.4 *Installation and Commissioning Surveys*

2.4.1 Installation of Steel Substructure and Foundation

Installation procedures are to be submitted to Class for review and approval as described in 5/4.7. The Surveyor is to witness the following activities, as applicable to the planned structure, to ascertain whether they have been accomplished in a manner conforming to the approved procedures:

(1) Installation of the Substructure and Foundation.
(2) Piling and grouting.
(3) Welding and non-destructive testing.
(4) Final field erection and levelling.
(5) Pre-tensioning.

Significant deviations from approved plans and procedures or any incidents such as excessive tilting of the foundation or abnormal vibrations during pile driving may require resubmission of supporting documentation to provide an assessment of the significance of deviation and any necessary remedial actions to be taken. To verify that overstressing of the structure during transportation has not occurred, Class is to be provided access to towing records to ascertain if conditions during the towing operations exceeded those employed in the design. Results are to be submitted to demonstrate compliance with the reviewed design analysis.

2.4.2 Installation of Concrete Substructure and Foundation

Installation procedures are to be submitted to Class for review and approval as described in 5/4.7. The Surveyor is to witness the following operations, as applicable to the planned structure, to verify that they have been accomplished in a manner conforming to plans or drawings covering these operations:

(1) Installation.
(2) Final field erection and levelling.
(3) Pre-tensioning.

Significant deviations from approved plans and procedures may require resubmission of supporting documentation to provide an assessment of the significance of the deviation and the remedial actions to be taken. To verify that overstressing of the structure during transportation has not occurred, Class is to have access to towing records to ascertain if conditions during the towing operations exceeded those employed in the design. Results are to be submitted to demonstrate compliance with the reviewed design analysis.

2.4.3 Installation of Rotor-Nacelle Assemblies (RNAs) and Tower

Where the RNA and tower are not within the scope of classification, the valid type certificate of the RNA and tower is to be examined by the attending Surveyors. The installation of the RNA and tower and the hook-up of the power cable system is not to damage the interface structure to the Substructure and the Foundation. Where the RNA and tower are within the scope of classification, the Surveyor is to witness the installation of at least one (1) RNA and one (1) tower per each wind turbine type. Where there are more than fifty (50) RNAs and towers of the same type, at least one (1) RNA and one (1) tower installation per every fifty (50) wind turbines of the same type are to be witnessed by the Surveyor. The selection of the RNA and tower installations to be witnessed is to reflect having Surveys at the start and end of RNA and tower installation periods and the rate of installation within an installation period. The attending Surveyor is to verify that the RNA and tower to be installed comply with the relevant design documents for the Bottom-founded Offshore Wind Turbine. Deviations from approved design documents and plans or any incidents such as damage or overstress to the Bottom-founded Offshore Wind Turbine during the installation may require resubmission of supporting documentation to provide an assessment of the significance of deviation and any necessary remedial actions to be taken.

2.4.4 Commissioning Surveys

The commissioning date will be the date on which the Surveyor issues the Interim Classification Certificate for the Bottom-founded Offshore Wind Turbine. Commissioning of all Class Rules/guideline required systems are to be verified by the attending Class Surveyor. The commissioning is to be in accordance with the approved step-by-step commissioning procedures. The Surveyor is to be permitted access to critical/hold points to verify that the procedures are satisfactorily accomplished. The Surveyor is to observe the Bottom-founded Offshore Wind Turbine operating under various capacities and conditions. Where the RNAs are within the scope of classification, the Surveyor is to witness the commissioning of at least one (1) RNA per each wind turbine type. Where there are more than fifty (50) RNAs of the same type, the commissioning of at least one (1) additional RNA per every fifty (50) turbines of the same type is to be witnessed by the Surveyor. The selection of

RNA commissioning to be witnessed is to reflect having Surveys at the start and end of RNA commissioning periods. Approved turbine RNA operations including emergency procedures are to be verified to the extent deemed necessary by the attending Surveyor. Overall performances of the turbine RNA are to be verified for compliance with the design parameters used in the design of the Bottom-founded Offshore Wind Turbine. Records of all these performances are to be maintained and made available to Class.

2.4.5 Personnel Safety

Verification of personnel safety features which include checks of operational readiness of all lifesaving, fire detection and firefighting equipment, emergency shutdown systems, unobstructed escape routes and establishment of communication procedures are to be taken during commissioning and are required to be verified by the attending Surveyor.

3 Surveys After Construction

3.1 Condition for Surveys After Construction

3.1.1 Damage, Failure, and Repair

Examination and Repair

Damage, failure, deterioration, or repair to a classed Bottom-founded Offshore Wind Turbine, which affects or may affect classification, is to be submitted by the Owner or their representatives for examination by the Surveyor at the first opportunity. All repairs found necessary by the Surveyor are to be carried out to the Surveyor's satisfaction.

Repairs on Site

Where repairs to the structure, which affect or may affect classification, are intended to be carried out at site, complete repair procedure including the extent of proposed repair and the need for Surveyor's attendance on site is to be submitted to and agreed upon by the Surveyor in advance. The above is not intended for routine maintenance.

Representation

Nothing contained in this chapter or in a regulation of any government or other administration, or the issuance of any report or certificate pursuant to this chapter or such a rule or regulation, is to be deemed to enlarge upon the representations expressed in the Class Rules relating to the conditions for the classification of offshore units and structures.

3.1.2 Notification and Availability for Survey

The Surveyor is to be provided access to a classed Bottom-founded Offshore Wind Turbine at all reasonable times. The Owners or their representatives are to notify the Surveyor on all occasions when the Bottom-founded Offshore Wind Turbine can be examined on site. The Surveyor is to undertake all surveys on the classed Bottom-founded Offshore Wind Turbine upon request, with adequate notification, from the Owners or their representatives and is to report thereon to Class. Should the Surveyor find occasion during any survey to recommend repairs or further examination, notification is to be given immediately to the Owners or their representatives in order that appropriate actions may be taken. The Surveyor is to avail himself of every convenient opportunity for performing periodical surveys in conjunction with surveys of damages and repairs to avoid duplication of work. Also refer to the Class Rules relating to the conditions for the classification of offshore units and structures.

3.1.3 In-Service Inspection Plan

The surveys are to be carried out in accordance with the reviewed In-Service Inspection Plan to confirm that the Bottom-founded Offshore Wind Turbine remains in compliance with the applicable classification requirements and other relevant standards. The In-Service Inspection Plan as required by 1/5.3 is to cover all surveys for the design life of the Bottom-founded Offshore Wind Turbine. As an alternative, a Risk Based Inspection (RBI) plan approved in accordance with 7/3.10 of this book may be considered so long as it includes all components of the regular inspection programme. The RBI plan is to be submitted and agreed by Class.

3.1.4 Annual Surveys

Annual Surveys of the Bottom-founded Offshore Wind Turbine are to be made within three months either way of each annual anniversary date of crediting of the previous Special Periodical Survey or original construction date. Where the Surveyor is engaged in the survey of a grouping of structures of similar design and location, and where requested by the Owner, special consideration will be given to the timing of Annual Surveys and Special Periodical Surveys such that all periodical survey due dates can be harmonised. For the Annual Survey of a wind farm with a group of wind turbines having a similar design, the group of wind turbines can be considered as a set and examined in accordance with the approved In-Service Inspection Plan.

3.1.5 Special Periodical Surveys

Special Periodical Surveys of the Bottom-founded Offshore Wind Turbine are to be carried out at least once every five years. If a Special Periodical Survey is not completed at one time, it will be credited as of the completion date of the survey provided the due date of the Special Periodical Survey is not overdue by more than six months. Where the Surveyor is engaged in the survey of a grouping of structures of similar design and location, and where requested by the Owner, special consideration will be given to the timing of Special Periodical Surveys so that the due dates for all periodical surveys can be harmonised. For Special Periodical Surveys of an offshore wind farm with a group of wind turbines having a similar design, the group of wind turbines can be surveyed as a set, in accordance with the In-Service Inspection Plan. Continuous Surveys can be applied (refer to 7/3.4).

3.1.6 Continuous Surveys

At the request of the Owner, and upon approval of the proposed arrangement, a system of Continuous Surveys may be undertaken whereby all the Special Periodical Survey requirements are carried out in regular rotation and completed within the normal Special Periodical Survey interval. For Continuous Surveys, a suitable notation will be entered in the Class Record and the date of completion of the cycle published. If any defects are found during these surveys, they are to be examined and dealt with to the satisfaction of the Surveyor.

3.1.7 Reactivation Surveys

When the Bottom-founded Offshore Wind Turbine has been out of service for an extended period, the requirements for surveys on reactivation are to be specially considered in each case. Due regard is to be given to the status of surveys at the time of the commencement of the deactivation period, the length of the period, and conditions under which the Bottom-founded Offshore Wind Turbine had been maintained during that period.

3.1.8 Incomplete Surveys

When a survey is not completed, the Surveyor is to report immediately upon the work done in order that the Owner and Class may be advised of the remaining parts to be surveyed.

3.1.9 Alterations

No major alterations, which affect classification of the installation, are to be made to a classed Bottom-founded Offshore Wind Turbine unless plans of the proposed alterations are submitted and approved by Class before the alterations are undertaken. Such alterations are to be carried out to the satisfaction of the Surveyor. Nothing contained in this chapter or in a rule or regulation of any government or other administration, or the issuance of any report or certificate pursuant to this chapter or such a rule or regulation, is to be deemed to enlarge upon the representations expressed in the Class Rules relating to the conditions for the classification of offshore units and structures and the issuance and use of any such reports or certificates are to be, in all respects, governed by the Class Rules relating to the conditions for the classification of offshore units and structures.

3.1.10 Survey of a Self-Elevating Unit to Be Used as the Substructure

A Self-Elevating Mobile Offshore Unit that has been converted to the site-dependent substructure of an offshore wind turbine is to be subject to the surveys after construction required in this book in addition to the applicable surveys of structures required by the Class Rules relating to the conditions for the classification of offshore units and structures.

3.1.11 Survey for Extension of Use

Existing installations to be used at the same location for an extended period beyond their original design life are subject to additional surveys, as deemed necessary, to determine the actual condition of the Bottom-founded Offshore Wind Turbine. The extent of the survey will depend on the completeness of the existing survey documents. Class will review and verify maintenance manual, logs, and records. Any alterations, repairs, or installation of equipment since installation are to be included in the records. Those survey requirements in 7/3.1.5 for the Special Periodical Survey must be included in the survey for extension of use. The surveys cover examination of splash zone, inspection of above water and underwater structural members and welds for damages and deteriorations, examination and measurements of corrosion protection systems and marine growth, sea floor condition survey, examination of secondary structural attachments such as J-Tube, service decks, etc. Special attention is to be given to the following critical areas:

(1) Areas of high stress.
(2) Areas of low fatigue life.
(3) Damage incurred during installation or while in service.
(4) Repairs or modifications made while in service.
(5) Abnormalities found during previous surveys.

An inspection report of the findings is to be submitted to Class for review and evaluation of the condition of the Bottom-founded Offshore Wind Turbine. The need for more frequent future periodical surveys is to be determined based on the calculated remaining fatigue life described in the Class Rules for the design, construction, and classification of offshore installations, with the exception that the load, strength, and fatigue calculations are to follow Chaps. 4 and 5 of this book as well as past inspection results.

3.2 Annual Surveys

Each Annual Survey is to include a thorough visual examination of above water structure. Special attention will be given to the splash zone for damage or deterioration from corrosion. Additionally, where significant deterioration or damage has occurred to an installation since the last survey, a general examination, by diver, underwater camera, submersible, or other suitable means, of the underwater structure, the scour protection, the sea floor, and the corrosion control system is to be carried out. Underwater examinations are to be contracted by the Owner and monitored by the Surveyor. Any novel features incorporated in the design are to be given special attention according to procedures agreed to during review of the design. Particular attention is to be given to significant modifications or repairs made because of findings at the previous survey. The Annual Survey is also to include verification that the approved design life has not been exceeded. The Surveyor is to confirm the design life limits. The Class technical office is to be consulted for verification. If the end of the design life has been reached, the provisions of the Class Rules for the design, construction, and classification of offshore installations, with the exception that the load, strength, and fatigue calculations are to follow Chaps. 4 and 5 of this book, are to be applied and specific requirements for maintaining the class of the Bottom-founded Offshore Wind Turbine are to be obtained from Class.

3.3 Special Periodical Surveys

The requirements of the Annual Survey are to be met during the Special Periodical Survey. Additionally, underwater inspection of selected areas is to be carried out.

In addition, non-destructive testing is to be carried out on representative joints of the structures and if found necessary, structural supports of J-Tubes. The extent and methods to be employed in such testing, cleaning, and inspection of the structure are to be in accordance with an approved inspection plan. The inspection plan, which is to be submitted for approval, is to cover all Special Periodical Surveys for the design life of the Bottom-founded Offshore Wind Turbine. It is to enumerate in detail the items to be surveyed, the testing and inspection procedures to be employed, and where necessary, cleaning, and non-destructive testing procedures. The plan is to include sufficiently detailed drawings which can be used by the Surveyor to reference and locate the items to be surveyed. The testing, cleaning, and inspection services are to be provided by the Owner and monitored by the Surveyor. Divers carrying out structural inspections and non-destructive testing on the structures are to be suitably qualified. In addition, the Special Periodical Survey is to include monitoring of the effectiveness of the corrosion protection system. The effectiveness of the corrosion protection system is to be monitored by taking measurements of the potential voltages generated by such systems. Scour in way of foundation legs and tilt of the Bottom-founded Offshore Wind Turbine are also to be checked and witnessed by the attending Surveyor. The Special Periodical Survey is also to include verification that the approved design life has not been exceeded. The Surveyor is to confirm the design life limits and the Class technical office is to be consulted for verification. If the end of the design life has been reached, the provisions of the Class Rules for the design, construction, and classification of offshore installations, with the exception that the load, strength, and fatigue calculations are to follow Chaps. 4 and 5 of this book, are to be applied and specific requirements for maintaining the class of Bottom-founded Offshore Wind Turbine are to be obtained from Class.

3.4 Continuous Surveys

As an alternative arrangement for the Special Periodical Surveys, the Continuous Surveys are to be applied annually to a minimum of 20% of the Bottom-founded Offshore Wind Turbines selected from the offshore wind farm. All the Special Periodical Survey requirements as specified in 7/3.1.5 are to be fulfilled during the Continuous Surveys. Each Bottom-founded Offshore Wind Turbine in the offshore wind farm is to be subject to at least once Special Periodical Survey within any five consecutive years. Where significant deterioration or damage has occurred to the installations or specific problems are identified as an indication of a serial defect, the Continuous Surveys are to be carried out to the satisfaction of the Surveyor for an increased number of the Bottom-founded Offshore Wind Turbines.

3.5 Gauging

Thickness gauging is required to be taken at each Special Periodical Survey. Suspect areas including structures in way of the splash zone are to be tested for thickness and results submitted to the attending Surveyor for review.

3.6 Structural Deterioration

Where thickness measurement and visual examination show evidence of significant structural deterioration, the structural integrity of the structure for continuous use is to be justified by engineering analyses. Deteriorated structural members are to be modelled such that they will add hydrodynamic, or wind loads to the Bottom-founded Offshore Wind Turbine but will not contribute to the strength of the Bottom-founded Offshore Wind Turbine. Results of these analyses are to be submitted to Class for review and approval. If the results show that structural integrity is inadequate, the deteriorated structural components are to be suitably reinforced or replaced with new materials having required dimensions and properties in accordance with approved procedure.

3.7 Marine Growth

During any Annual or Special Periodical Survey, assessment of the degree of marine growth is to be carried out. Marine growth is to be removed where it is found to be thicker than the original approved design. If the Owner decides to leave the marine growth greater than what is allowed in the approved design, the Owner is to justify that the higher hydrodynamic loading due to the additional marine growth will not affect the structural integrity of the structure. The Owner is to submit as a minimum, an in-place analysis to justify that the installation is capable of withstanding environmental wave loads resulting from the maximum marine growth that the Owner is prepared to maintain.

3.8 Maintenance of Rotor-Nacelle Assembly

The Owner is to submit an annual report for review by the attending Class Surveyor attesting to the following:

(1) Maintenance has been carried out by authorised and qualified personnel in accordance with the maintenance manual.

(2) The control settings have been checked regarding conformance with the limiting values specified in design documentation of the classed Bottom-founded Offshore Wind Turbine.
(3) All repair, modification and replacement have been carried out in accordance with the manufacturers' recommendations and have not altered values of RNA parameters specified in design documentation of the classed Bottom-founded Offshore Wind Turbine.

3.9 Statutory Certification

When Class is authorised to perform surveys on behalf of a Coastal State or a governmental authority, and when requested by the Owner, items as specified by the Coastal State, the governmental authority or Owner will be surveyed. Reports indicating the results of such surveys will be issued accordingly. Where the periodicity and types of surveys on behalf of a Coastal State or a governmental authority differ from those required by the applicable portions of this chapter, Coastal State or governmental authority's requirements take precedence.

3.9.1 Risk-Based Surveys for Maintenance of Class

A properly developed risk-based inspection plan or reliability centred maintenance plan may be credited as satisfying requirements of surveys for maintenance of Class for the Bottom-founded Offshore Wind Turbine. The plan is to be developed in accordance with the Class guidance for performing risk-based inspection for floating offshore installations and/or the Class guidance for performing surveys based on machinery reliability and maintenance techniques. The application of the books referenced above in this paragraph does not cover any statutory survey requirements that may apply to the Bottom-founded Offshore Wind Turbine being considered. Only when the governmental authorities accept and authorise Class for certification based on risk-based inspection techniques, Class will carry out such surveys, as authorised. The Owner is to ensure that in developing the inspection plan, consideration is given to applicable requirements external to Class. If the governmental authorities do not accept a risk-based approach, surveys will be carried out in a conventional, prescriptive manner in compliance with 7/3.1 through 7/3.10.

Appendix 1: Wind Spectra and Coherence Functions

Kaimal Wind Spectrum and Exponential Coherence Model

A modified version of the Kaimal wind spectrum is provided in IEC 61,400–1 (2019). The power spectral densities for the wind fluctuations in three dimensions are given as follows:

$$\frac{f \cdot S_k(f)}{\sigma_k^2} = \frac{\frac{4fL_k}{V_{ub}}}{(1 + 6fL_k/V_{hub})^{\frac{5}{3}}}$$

where:

$S_k(f)$ spectral energy density at frequency f, in m^{2}s^{-2}/Hz (ft^{2}s^{-2}/Hz).
f frequency, in Hz.
k index referring to the direction of wind speed component (i.e., 1 = longitudinal, 2 = lateral, and 3 = upward, as depicted in Fig. 1 of Chap. 3).
σ_k standard deviation of turbulent wind speed component (refer to Table A1.1)

$$\left[\int_0^{\infty} S_k(f)df\right]^{0.5}$$

L_k integral parameter of turbulent wind speed component (refer to Table A1.1).

Along with the Kaimal spectrum, an exponential coherence model is provided in IEC 61,400–1 (2019) to account for the spatial correlation of the longitudinal wind speed:

$$Coh(r,f) = exp\left[\sqrt{\left(f \cdot \frac{r}{V_{hub}}\right)^2 + \left(0.12 \cdot \frac{r}{L_c}\right)^2}\right]$$

where:

$Coh(r,f)$coherence function at frequency f.

f frequency, in Hz.
r magnitude of the projection of the separation vector between the two points on to a plane normal to the average wind direction, in m (ft).
V_{hub} 10-min mean wind speed at hub height, in m/s (ft/s).
L_c coherence scale parameter, in m (ft).

$8.1\Lambda_1$, where $8.1\Lambda_1$, is specified in 3/3.6.2

NPD (Frøya) Wind Spectrum and Coherence Model

For the extratropical storm wind, the NPD (Frøya) wind spectrum in conjunction with the two-point coherence function recommended by API RP 2MET (2019) may be applied.

Wind Spectrum

The following wind spectrum is defined for the energy density of the longitudinal wind speed fluctuations:

$$S(f) = \frac{320\phi^2\left(\frac{U_0}{10\phi}\right)^2\left(\frac{z}{10\phi}\right)^{0.45}}{\left(1+\tilde{f}^n\right)^{\left(\frac{5}{3n}\right)}}$$

$$\tilde{f} = 172f\left(\frac{z}{10\phi}\right)^{\frac{2}{3}}\left(\frac{U_0}{10\phi}\right)^{-0.75}$$

where:

$S(f)$ Spectral energy density at frequency f, in m^2s^{-2}/Hz (ft^2s^{-2}/Hz).
f Frequency, in Hz.
U_0 1-h mean wind speed at 10 m (32.8 ft) above the SWL, in m/s (ft/s).
n 0.468.
z Height above the SWL, in m (ft).
ϕ Unit conversion factor.

1 when using SI units (m, m/s)
3.28 when using US Customary units (ft, ft/s)

Spatial Coherence

The squared correlation between the spectral energy densities $S(f)$ of the longitudinal wind speed fluctuations between two points $(xj, yj, zj), j = 1, 2$ in space is described by the two-point coherence function as follows:

$$Coh(f) = exp\left\{-\frac{1}{U_0/\phi}\cdot\left[\sum_{i=1}^{3}A_i^2\right]^{\frac{1}{2}}\right\}$$

Table A1.1 Spectral parameters for the kaimal model

	Wind speed direction		
	k = 1 (longitudinal)	k = 2 (lateral)	k = 3 (upward)
Standard deviation (σ_k)	σ_1	$0.8\sigma_1$	$0.5\sigma_k$
Integral scale (L_k)	$8.1\Lambda_1$	$2.7\Lambda_1$	$0.66\Lambda_1$

Note σ_1 standard deviation of longitudinal turbulent wind speed. Λ_1 scale parameter of the turbulence as specified in 3/3.6.3

Table A1.2 Coefficients and distances for the three-dimensional coherence function

i	Δ_i	q_i	p_i	r_i	α_i
1	$\|x_2 - x_1\|$	1.00	0.4	0.92	2.9
2	$\|y_2 - y_1\|$	1.00	0.4	0.92	45.0
3	$\|z_2 - z_1\|$	1.25	0.5	0.85	13.0

$$A_i = \alpha_i f^{ri} \left(\frac{\Delta_i}{\phi} \right)^{qi} \left(\frac{\sqrt{z_1 z_2}}{10\phi} \right)^{-pi}$$

where the coefficients α_i, p_i, q_i, r_i and the distances Δ_i are specified in Table A1.2, and

$Coh(f)$ coherence function at frequency f.
f frequency, in Hz.
U_0 1-h mean wind speed at 10 m (32.8ft) above the SWL, in m/s (ft/s).
(x_j, y_j, z_j) spatial coordinates of two points ($j = 1, 2$) where x_j is in the longitudinal (along wind) direction, y_j is in the lateral (across wind) direction and z_j is the height above the SWL in the upward direction, in m (ft).
ϕ unit conversion factor.

1 when using SI units (m, m/s)
3.28 when using US Customary units (ft, ft/s)

Tropical Cyclone Wind Spectrum

For tropical storm wind, the following wind speed spectrum for the energy density of the longitudinal wind speed fluctuations is recommended by API RP 2MET (2019):

$$S(f, z) = \frac{4I_u^2 U_{w,1h}(z) L_{u,x}(z)}{\left[1 + 70.8\left(\frac{fL_{u,x}(z)}{U_{w,1h}(z)}\right)^2\right]^{\frac{5}{6}}}$$

where:

$S(f, z)$ Spectral energy density at frequency, f, and height z, in m^2s^{-2}/Hz (ft^2s^{-2}/Hz).
f Frequency, in Hz.
z Height above the SWL, in m (ft).
$U_{w,1h}(z)$ 1-h average wind speed at a height z above the MSL, in m/s (ft/s).
$I_{u,x}(z)$ Turbulence intensity at the height z above the MSL (refer to A2/3), (dimensionless).
$L_{u,x}(z)$ Integral length scale at the height z above the MSL, in m (ft)

$$L_{u,x}(z) = \frac{50\phi\left(\frac{z}{\phi}\right)^{0.35}}{\left(\frac{z_0}{\phi}\right)^{0.063}}$$

Z_0 surface roughness length for tropical cyclone winds (refer to A2/3), in m (ft).
ϕ unit conversion factor, (dimensionless).

1 when using SI units (m, m/s)
3.28 when using US Customary units (ft, ft/s)

Appendix 2: Tropical Cyclone Wind Speed Profile, Standard Deviation, Turbulence Intensity and Gust Factor

General

Extensive dropsonde measurements of major tropical cyclones in the Gulf of Mexico indicate that the tropical cyclone wind speed profile is on average steeper than the exponential wind speed profile specified in IEC 61,400–1 for Extreme Wind Speed Model (EWM). The measured wind profiles are also steeper than those NPD (Frøya) wind model originally developed based on the North Sea data. Tropical cyclone wind data collected on the platforms in the Gulf of Mexico indicate that the turbulence intensity is insensitive to the change of mean wind speed. The average observed value of turbulence intensity is slightly smaller than 0.11 as defined in IEC 61,400–1 for the Extreme Wind Speed Model (EWM) but could differ noticeably from the value calculated with the NPD (Frøya) wind model. In addition, the measured gust factors of tropical cyclone winds are found to be less sensitive to the change of mean wind speed in comparison to the calculated gust factors in the NPD (Frøya) wind model. The formulation of the wind speed profile, wind speed standard deviation, turbulence intensity and gust factors presented in this appendix is found to provide good representations of the measurements of tropical cyclone wind in the Gulf of Mexico. In absence of site data, the formulation in this appendix may be used to model tropical cyclone (also known as hurricane or typhoon) wind conditions over the open ocean.

Tropical Cyclone Wind Speed Profile

Tropical cyclone wind boundary layers within approximately 200 m above the sea surface of the open ocean can be represented by an equilibrium form of the logarithmic boundary layer profile under neutral stability conditions, i.e.:

$$V_{1hr}(z) = \frac{u*}{k}\ln\left(\frac{z}{z_0}\right)$$

where:

$V_{1hr}(z)$ One hour mean longitudinal wind speed at the height z above the sea surface, in m/s (ft/s).
$u*$ Friction velocity, in m/s (ft/s).
z Height above the sea surface, in m (ft).
z_0 Surface roughness length, in m (ft).
k Von Karman constant, (dimensionless).

0.4

For mean wind speed $V_{1hr}(z_{ref})$ lower than approximately 27 m/s (52.5 knots), the surface roughness length z_0 over the open ocean may be calculated using the Charnock equation:

$$z_0 = \frac{\alpha u^2*}{g} = \frac{\alpha}{g}\left[\frac{kV_{1hr}(z)}{\left(ln\frac{z}{z_0}\right)}\right]^2$$

where:

α 0.011, as suggested in IEC 61,400–3, for the open ocean.
z_{ref} 10 m (33 ft) reference height above the sea surface, in m (ft).
$V_{1hr}(z_{ref})$ 1-h mean wind speed at the reference height z_{ref} above the sea surface, in m/s (ft/s).
g Acceleration of gravity, in m/s^2 (ft/s^2).

Alternatively, the surface roughness length z_0 over the open ocean can be determined from the boundary layer wind profile and expressed in terms of the surface drag coefficient $C_d(z_{ref})$ at the reference height $z_{ref} = 10$ m (33 ft) through the surface shear stress $\tau_0 = pu_*^2 = \rho C_d(z_{ref})V_{1hr}^2(z_{ref})$, where ρ is the air density and $C_d(z_{ref})$ is the surface drag coefficient at the reference height z_{ref} above the sea surface, such that:

$$u* = \sqrt{C_d(z_{ref})V_{1hr}(z_{ref})}$$

$$z_0 = 10 \times exp\left(-\frac{k}{\sqrt{C_d(z_{ref})}}\right)$$

where the surface drag coefficient $C_d(z_{ref})$ may be calculated by:

$$C_d(z_{ref}) = [0.49 + 0.065V_{1hr}(z_{ref})] \times 10^{-3}$$

The values of $C_d(z_{ref})$ and z_0 determined by the equations given above will increase along with the mean wind speed. However, it is found that the $C_d(z_{ref})$ levels off for the wind speed $V_{1hr}(z_{ref})$ exceeding approximately 22 m/s ~ 30 m/s (43.8 knots ~ 58.3 knots) and may even decrease slightly at very high wind speeds, probably due to wave breaking and increasingly frothy sea surface. The maximum $C_d(z_{ref})$ over the open ocean is estimated in the range of 0.0019 ~ 0.0025 depending on the size of the tropical cyclone and the maximum wind speed. As a results, the surface roughness length z_0 over the open ocean is in the range of 0.001 m ~ 0.0034 m (0.0033 ft ~ 0.011 ft). For typical tropical cyclones over the open ocean, the following limiting value should be applied:

$$C_d(z_{ref}) \leq 0.0023$$

Standard Deviation of Tropical Cyclone Wind Speed and Turbulence Intensity

The standard deviation of instantaneous tropical cyclone wind speed within one hour time duration is determined using the formulation suggested in ESDU 83,045 Strong Winds in the Atmospheric Boundary Layer, Part 2: Discrete Gust Speeds (2002), in conjunction with the friction velocity $u*$ and the surface roughness length z_0 determined in accordance with this appendix (refer to A2/2):

$$\sigma_u(z) = \frac{7.5\eta\left[0.538 + 0.09ln\left(\frac{z}{z_0}\right)\right]^{\eta 16}}{1 + 0.156ln\left(\frac{u*}{fz_0}\right)}u*$$

where:

$\sigma_u(z)$ standard deviation of the instantaneous longitudinal wind speed at the height z above the sea surface, in m/s (ft/s).

η scaling parameter, (dimensionless)

$$1 - \frac{6fz}{u*}$$

f Coriolis parameter, in rad/s

$$1.458 \times 10^{-8} sin\phi$$

ϕ local angle of latitude, in degrees.

The turbulence intensity, defined as the standard deviation of the wind speed divided by the mean wind speed at the same height, is given by:

$$I_u(z) = \frac{\sigma_u(z)}{V_{1hr}(z)} = \frac{\sigma_u(z)}{u*}\frac{u*}{V_{1hr}(z)}$$

$$= \frac{7.5\eta\left[0.538 + 0.09ln\left(\frac{z}{z_0}\right)\right]^{\eta 16}}{1 + 0.156ln\left(\frac{u*}{fz_0}\right)}\frac{k}{ln\left(\frac{z}{z_0}\right)}$$

where the frication velocity $u*$ can be determined either by the surface wind profile, or the relation derived from surface shear stress. The turbulence intensity $I_u(z)$ is weakly dependent on the mean wind speed and, for strong winds, can be considered independent of the mean wind speed.

Tropical Cyclone Wind Gust Factor

The longitudinal gust wind speed, $V_\tau(z)$, which represents the maximum value of longitudinal wind speed occurring in any given observation period averaged over a shorter period, is related to the mean longitudinal wind speed in the same observation period through the gust factor as defined below:

$$V_\tau(z) = G(z, \tau, T_0)V_{\tau o}(z)$$

$$= \left[1 + g(z, \tau, T_0)I_u(z)\right]V_{To}(z)$$

where:

$V_\tau(z)$ Longitudinal gust wind speed at the height z above the sea surface with averaging time duration of τ,, in m/s (ft/s).

$V_{\tau o}(z)$ Mean longitudinal wind speed at the height z above the sea surface in any given reference observation period T_o,, in m/s (ft/s).
$G(z, \tau, T_0)$ Gust factor, (dimensionless).
$g(z, \tau, T_0)$ Peak factor, (dimensionless).
z Height above the sea surface, in m (ft).
τ Gust averaging time duration, in seconds.
T_o Reference observation time duration ($T_o > \tau$), in seconds.
$I_u(z)$ Turbulence intensity at the height z above the sea surface, (dimensionless).

For the one-hour observation period, the peak factor can be calculated by:

$$g(z, \tau, T_0 = 3600s) = \frac{\sigma_u(z, \tau, T_o = 3600s)}{\sigma_u(z)}\left[\sqrt{2\ln(3600v)} + \frac{0.577}{\sqrt{2\ln(3600v)}}\right]$$

$$\frac{\sigma_{u(z,\tau,T_0=3600s)}}{\sigma_u(z)} = 1 - 0.193\left(\frac{T_u}{\tau} + 0.1\right)^{-0.68}$$

where:

$\sigma_{u(z,\tau,T_0)}$ standard deviation of the wind speed at the height z above the sea surface with averaging time duration of τ observed during a period of T_o,, in m/s (ft/s).
$\sigma_u(z)$ standard deviation of the instantaneous wind speed at the height z above the sea surface, in m/s (ft/s).
υ zero up-crossing frequency of wind speed, in Hz

$$\frac{\left[0.007 + 0.213\left(\frac{T_u}{\tau}\right)^{0.654}\right]}{T_u}$$

T_u Longitudinal integral length time scale, in seconds

$$3.12z^{0.2}$$

The gust factor associated with one hour reference period is relate to the peak factor through:

$$G(z, \tau, T_0 = 3600s) = 1 - g(z, \tau, T_0 = 3600s)I_u(z)$$

As an example, the maximum 10-min mean wind speed observed in a one-hour time duration can be obtained by:

$$V_{10min}(z) = G(z, \tau = 600s, T_0 = 3600s)V_{1hr}(z)$$

For $z = z_{ref} = 10$ m (33 ft) above the sea surface,

$$V_{10min}(z_{ref}) = 1.03V_{1hr}(z_{ref})$$

For a reference period $T_o(> \tau)$ smaller than $3600s$, the gust factor can be calculated by:

$$G(z, \tau, T_0) = [0.2193ln(log_{10}T_0) - 0.7242] \times G(z.\tau, T_0 = 3600s)$$

Appendix 3: Fatigue Analysis for Bottom-Founded Offshore Wind Turbine Support Structures

General

Spectral fatigue analysis method is widely used for designing fixed offshore platforms in the offshore oil and gas industry. A basic assumption of the spectral fatigue analysis method is that the offshore structure acts linearly in response to wave loading to the structural responses are proportional to the wave height through the transfer functions. For bottom-founded offshore wind turbines, however, such assumption may no longer be valid when aerodynamic loads become the primary source of fatigue damage in the turbine Support Structure. Such aerodynamic loads are mostly the result of nonlinear responses of the turbine RNA and are determined by nonlinear aeroelastic responses of the blades and actions of the turbine control and protection systems. Time-domain fatigue analysis method is therefore considered as a more rational approach and should in general be employed in the design of the turbine support structures. Figure A3.1 depicts a typical flowchart for the time-domain fatigue analysis for bottom-founded offshore wind turbine support structures.

Scatter Diagram Lumping

One of the main difficulties of using the time-domain fatigue analysis method in the practical design arises from the time-consuming effort of numerical simulations in the time domain. This could be a costly process for those support structures having complex geometries and connections, such as jacket-type support structures. The number of simulations is proportional to the number of combinations of wind, sea state and direction as defined in the scatter diagram. For a given direction, a typical scatter diagram for the fatigue analysis of a turbine support structure includes three

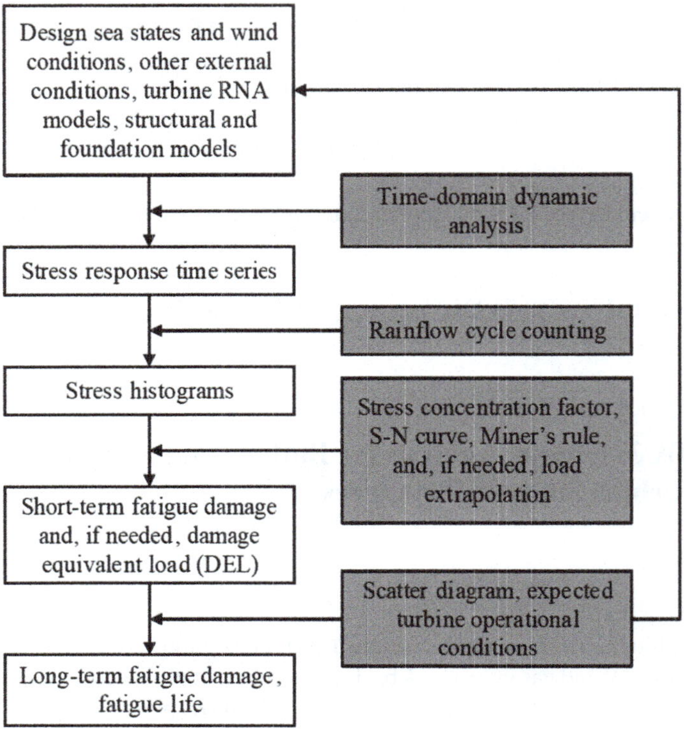

Fig. A3.1 Flowchart of time domain fatigue analysis method

variables [i.e., the significant wave height (H_s), the wave peak period or zero up-crossing period (T_p or T_z), and the hub-height wind speed (V_{hub})]. To reduce the computational effort, the scatter diagram lumping method may be used to reduce the total number of combinations of sea state and wind speed. The hub height wind speed bin and directionality should be verified in accordance with the requirement of 4/3.2. In general, the hub height wind speed bin should not be lumped. With the assumption that the scatter diagram is expressed in the form of a series of T_p vs. H_s tables for a given hub height wind speed at a given direction, the main steps of the suggested scatter diagram lumping method are:

(1) Divide a scatter diagram into up to five lumping partitions depending on the range of structural natural period: lumping partition A of the scatter diagram with associated wave period near the first structural nature period; lumping partition B and C of the scatter diagram with associated wave periods adjacent to structural natural period within the range of plus and minus 2 s, respectively; and lumping partition D and E of scatter diagram with associated wave periods outside the range of the structural natural period plus and minus 2 s, respectively. For a typical bottom-founded support structure with the first natural period

around 3 s and a scatter diagram with a lower bound wave period of 3 s, there would be only three partitions (i.e., A, B and D).

(2) Calculate the equivalent significant wave height and the equivalent wave period for each partition using the following equations. As a result, a full $T_p - H_s$ scatter diagram can be condensed to no more than 5 equivalent sea states.

$$(H_s)_{eq} = \left[\frac{\sum_{i=1}^{N} P_i (H_s)_{i^m}}{\sum_{i=1}^{N} P_i} \right]^{\frac{1}{m}}$$

$$\left(T_p\right)_{eq} = \left[\frac{\sum_{i=1}^{N} P_i}{\sum_{i=1}^{N} \frac{P_i}{\left(T_p\right)_i}} \right]$$

where:

$(H_s)_{eq}$	Equivalent significant wave height, in m (ft).
H_{si}	Significant wave height of sea state i,, in m (ft).
$\left(T_p\right)_{eq}$	Equivalent wave peak period, in seconds.
$\left(T_p\right)_i$	Wave peak period of sea state i, in seconds.
p_i	Occurrence number or probability of sea state i.
m	Slope of a selected S–N curve (the smaller slope in the case of a two-segment S–N curve).
N	Number of non-zero entries of p_i in any given partition of the scatter diagram.

Part II
Floating Offshore Wind Turbines

Chapter 8
General

1 General

This Part 2 provides criteria for the design, construction, installation, and survey of permanently sited Floating Offshore Wind Turbines. It addresses three principal areas: the floating substructure, the station keeping system, and onboard machinery, equipment and systems including applicable marine systems and associated equipment and machinery, safety systems and associated equipment, and lifesaving appliances and machinery. The criteria applicable to bottom-founded offshore wind turbine substructures and foundations are provided in Part 1. Requirements for ancillary offshore wind farm installations such as meteorological measuring towers, accommodation units, and transformer platforms are not addressed in this book. For the requirements for bottom-founded ancillary offshore wind farm installations, refer to the Class Rules for the design, construction, and classification of offshore installations. The requirements applicable to floating ancillary offshore wind farm installations are provided in the Class Rules for the design, construction, and classification of mobile offshore units and the Class Rules for the design, construction, and classification of floating production installations.

2 Classification

The general requirements for conditions of classification are contained in the Class Rules relating to the conditions for the classification of offshore units and structures. Additional requirements specific to the Floating Offshore Wind Turbine are contained in the following chapters.

F. Aderouane, *Design, Construction, and Global Performance Analysis of Offshore Wind Turbines*, Green Energy and Technology, https://doi.org/10.1007/978-3-031-86010-2_8

2.1 *Classification Boundaries*

The classification of the Floating Offshore Wind Turbine addresses three principal areas which are subject to the requirements of this book:

(1) The floating substructure, which carries the wind turbine Rotor-Nacelle Assembly (RNA) and tower.
(2) The station keeping system.
(3) Applicable marine systems and associated equipment and machinery, safety systems and associated equipment, and lifesaving appliances and machinery.

Additionally, at the request of the Owner, the RNA and the tower may be included in the classification. Optional notations for the RNA and the tower will be added to Class Record subject to the satisfactory compliance with the requirements given in 8/2.3. Where the RNA and tower are not included in the classification, the RNA and tower installed on the Floating Offshore Wind Turbine are required to have a type certificate in accordance with IECRE OD-501 or other recognised standards. The type certificate will be reviewed by Class solely to confirm it is current and verify that the information of the installed RNA and tower are consistent with the design information, criteria and limitations considered in the classification of the Floating Offshore Wind Turbine. Class will not review or be responsible for the accuracy of the type certificate. Where the tower and its connection to the RNA are not covered by the wind turbine type certificate and the optional class notation is not requested, the tower and its connection to the RNA are to be included in the scope of classification. Where interface structures of the tower connecting to the substructure are not covered by the wind turbine type certificate and the optional class notation is not requested, these interface structures are to be included in the scope of classification. RNAs lacking a type certification or the wind turbines possessing a non-horizontal-axis configuration are subject to special consideration by Class. Design documents are to be submitted for review. Scope of the design review is to be defined on a case-by-case basis. In this book, the Floating Offshore Wind Turbine is considered as an unmanned structure.

2.2 *Offshore Wind Turbines Built Under Class Survey*

The following class notations apply to the Floating Offshore Wind Turbine, as defined in 8/2.1:

2.2.1 Offshore Wind Turbines Built Under Class Survey

Offshore wind turbines built and constructed to the satisfaction of the Class Surveyors and to the requirements of this book or to their equivalent, where approved by Class, may be classed, and distinguished in the Class Record by an appropriate notation, such as:

✠ A1 Offshore Wind Turbine (Floating)

2.2.2 Offshore Wind Turbines not Built Under Class Survey

Offshore wind turbines not built under Class survey but submitted for classification are subject to special classification survey. Where found satisfactory and thereafter approved by Class, such offshore wind turbines may be classed with the notation:

✠ A1 Offshore Wind Turbine (Floating)

Minus the Class identifier (in this example, the Maltese Cross ✠).

2.3 Additional Class Notation

2.3.1 Notation for Rotor-Nacelle Assembly (RNA) and Tower

The RNA and tower may be considered for the classification at the Owner's request. A Floating Offshore Wind Turbine may be classed and distinguished in the Record by the optional class notation RNA if:

(1) The RNA and tower comply with the Class Type Approval requirements in the Class Rules relating to the conditions for classification of offshore units and structures.
(2) The site-specific assessment demonstrates, as a minimum, which loads and deflections calculated for the Floating Offshore Wind Turbine under the site-specific conditions do not exceed those calculated for the RNA and tower approved by the Class Type Approval.

2.3.2 Strength Criteria for Site-Specific Conditions (S)

For the Floating Offshore Wind Turbine designed and built to the requirements for pre-service conditions and strength criteria for in-service site-specific conditions in accordance with this book and maintained in accordance with the applicable Class requirements, the site-specific environmental data and electric network condition data for in-service conditions will be indicated by the (S) qualifier following the basic notation of 8/2.2 and 8/2.3 as described in the following paragraphs. The (S) qualifier followed by the definition of the site is mandatory to be assigned and published in the Class Record.

New Construction

This qualifier (S) will be followed by the definition of the site. For example:

✠ A1 Offshore Wind Turbine (Floating) (S) Gulf of Maine
A1 Offshore Wind Turbine (Floating) (S) Gulf of Maine

Life Extension of Floating Offshore Wind Turbines at the Same Site

For life extension of a Floating Offshore Wind Turbine remaining at the same site, the intended period of operation is extended. The strength of the Floating Offshore Wind Turbine is to be reassessed to demonstrate that it remains in compliance with applicable requirements. If the environmental criteria for the site or operating area have been revised since the original approval due to new environmental data or changing environmental conditions, the Coastal State may require the use of new environmental criteria for the life extension. For the life extension of a Floating Offshore Wind Turbine remaining at the same site, the following actions are to be taken:

- Structural strength analysis and fatigue life re-evaluation reports, as required in 8/2.3.3.
- Design review and surveys related to any major modifications affecting class items.
- Special survey, including Underwater Inspection, to determine the structural condition of the Floating Offshore Wind Turbine at the time of the life extension. Survey requirements for extension of use are given in 17/2.12. After the life extension is completed in accordance with this book, the qualifier (S) notation can be retained for the site.

Relocation of Floating Offshore Wind Turbines

A Floating Offshore Wind Turbine is designed and classed for a specific site location where it will be operated for an intended period. When the Floating Offshore Wind Turbine is relocated to a new site, either within the same offshore wind farm or in a different operating area, the strength of the Floating Offshore Wind Turbine is to be reassessed to confirm that it remains in compliance with applicable requirements.

Relocation Within the Same Offshore Wind Farm

If the environmental conditions for the offshore wind farm have been revised since the original approval due to new environmental data or changing environmental conditions, the Coastal State may require the use of new environmental conditions for the relocation, in which case the same requirements as those required for relocation to a different operating area will apply. It is expected that relocation within the same offshore wind farm will require at least a new station keeping system for the new site. When the original station keeping system or part of the components are to be re-used, the condition of the station keeping system, or the components is to be assessed. The design assessment is to be performed. For the relocation within the same offshore wind farm, without exceeding the original design life of the Floating Offshore Wind Turbine, the following actions are to be taken:

- Design review and surveys related to the new station keeping system.
- Design review and surveys related to any other modifications affecting class items.
- Survey reports, including gauging to bring the Floating Offshore Wind Turbine to a satisfactory condition to complete the remaining design life at the specific site.

The expected operating life in the new location may be within the original design life or otherwise it may extend beyond the original design life. In the latter case, in addition to the requirements for relocation, the requirements for life extension will apply. The design and assessment reports as required in 8/2.3.2 are to be submitted for review. Survey requirements for relocation are given in 17/2.13. After the relocation is completed in accordance with this book, the qualifier (S) notation can be updated for the new site.

Relocation to a Different Operating Area

For a Floating Offshore Wind Turbine to be relocated to an operating area where the site conditions are different than those at the original site, the structural strength and fatigue life of the Floating Offshore Wind Turbine are to be reassessed for the new conditions. However, if the new location has milder environmental conditions than the current site and comparable electrical network conditions, the reassessment may not be necessary, provided that the Floating Offshore Wind Turbine maintains the same structural condition as at the original site and the design fatigue life of the Floating Offshore Wind Turbine is not extended. Relocation to a different operating area will require a new station keeping system for the new site and the following actions:

- Structural strength analysis and fatigue life re-evaluation reports, as required in 8/2.3.3.
- Design review and surveys related to the new station keeping system.
- Design review and surveys related to any other modifications affecting class items.
- Survey reports, including gauging to bring the Floating Offshore Wind Turbine to a satisfactory condition to complete the remaining design life at the specific site.

The expected operating life in the new location may be within the original design life or otherwise it may extend beyond the original design life. In the latter case, in addition to the requirements for relocation, the requirements for life extension will apply. The design and assessment reports as listed in 8/2.3.2 are to be submitted for review. Survey requirements for relocation are given in 17/2.13. After the relocation is completed in accordance with this book, the qualifier (S) notation can be updated for the new site.

2.3.3 Design Fatigue Life

New Construction

The design fatigue life value notation is to be assigned and published in the Record. The "design fatigue life" refers to the target value set by the owner or designer, not the value calculated in the analysis. The design fatigue life is to be verified to comply with the fatigue criteria in this book. The class notation FL (number of years), Year will be assigned to identify the design fatigue life in years and the year of maturation of fatigue life in the defined site. For example, FL (25), 2045 for a Floating Offshore Wind Turbine built in 2020 if the minimum design fatigue life specified is 25 years. Where the design fatigue life for the floating substructure differs from that of the station keeping system, the notation FL(number of years), Year will be followed by the notation FLM (number of years), Year, where (number of years) refers to the design fatigue life for the station keeping system and Year refers to the year of maturation associated with the station keeping system. The class notation FL (number of years), Year and/or FLM (number of years), Year is mandatory to be assigned and published in the Class Record.

Life Extension of Floating Offshore Wind Turbines at the Same Site

When the operating life of a Floating Offshore Wind Turbine exceeds the design fatigue life specified in either FL (number of years), Year, or in FLM (number of years), Year notation for which it was classed, an evaluation is to be made and appropriate actions are to be taken to extend the use of the Floating Offshore Wind Turbine up to the new operating life under the site-specific conditions. For the life extension of the Floating Offshore Wind Turbine remaining in the same location, the following actions are to be taken:

- Verification from the original fatigue analysis that the actual fatigue values of all the structural elements of the Floating Offshore Wind Turbine are still higher than the proposed extended fatigue life.
- New fatigue analysis covering all the structural elements in accordance with FL or FLM requirements, as applicable.
- Identification of structural elements or details with a fatigue life below the new intended.
- design fatigue life of the Floating Offshore Wind Turbine and proposed actions to increase the fatigue life of those elements or details.
- Design review and surveys of structural modifications proposed because of the fatigue analysis.
- Enhanced survey programme to monitor those structural elements or details with lower fatigue life which cannot be modified or renewed on site.
- Special survey, including Underwater Inspection, to determine the structural condition of the Floating Offshore Wind Turbine at the time of the life extension.

Survey requirements for extension of use are given in 7/2.12. After the life extension is completed in accordance with this book, the existing FL or FLM notation with total number of fatigue life and year of maturation is to be updated accordingly.

Relocation of Floating Offshore Wind Turbines

When a Floating Offshore Wind Turbine is relocated to a new site, either within the same offshore wind farm or in a different operating area, the fatigue life of the Floating Offshore Wind Turbine is to be reassessed to demonstrate that the remaining fatigue life for the new operating conditions is within the design fatigue life of the Floating Offshore Wind Turbine. Survey requirements for relocation are given in 17/2.13. After the relocation is completed in accordance with this book, the existing FL or FLM notation can be retained for the new site.

2.3.4 Other Class Notations

When requested by the Owner, applicable notations listed in the MOU Rules and FPI Rules may be assigned upon agreement from Class.

3 Rules for Classification

3.1 Application

The requirements of this book are applicable to Floating Offshore Wind Turbines. This book is applicable to those features that are permanent in nature and can be verified by plan review, calculation, physical survey, or other appropriate means. Any statement in this document regarding other features is to be considered as guidance to the designer, Fabricator, Owner, et al.

3.2 Reference

References are made in this book to Class Rules and other criteria issued by Class and other organisations. Unless otherwise noted, the applicable edition of a reference is the one officially issued and available on the date the Agreement for Classification is accepted by Class. Where a particular edition or date associated with a reference is given, it means that edition is relevant to the topic being presented in this book. Class may consider at its discretion, upon the request of the Owner, the application of other appropriate alternative methods and recognised codes of practice.

3.3 *Alternatives*

Any departure from the requirements of this book may be considered by Class based on suitable engineering analyses or risk assessments. In the case of such departures, classification is subject to Class's approval upon a demonstration of acceptable levels of safety in line with the principles of this book (refer to 9/2) and recognised and accepted current offshore wind industry practice. Risk acceptance criteria are subject to approval by Class. Refer to the Class guidance on risk assessment applications for the marine and offshore industries for an overview of risk assessment techniques and additional information. Using a risk assessment approach to justify alternatives may be applicable either to the Floating Offshore Wind Turbine as a whole or to individual systems, subsystems, or components. As appropriate, remote hazards outside of the bounds of the system under consideration are to be considered. Such account must include incidents relating to remote hazards directly affecting or being influenced by the system under consideration. Class will consider the application of risk-based techniques in the design of the Floating Offshore Wind Turbine as well as surveys during construction and surveys for maintenance of class. Portions of the Floating Offshore Wind Turbine not included in the risk assessment are to comply with the applicable parts of the Class Rules. The following are the responsibility of the Owner:

(1) Risk acceptance criteria.
(2) Hazard identification.
(3) Risk assessment.
(4) Risk management.
(5) Compliance with the applicable requirements of the Coastal State or other governmental authorities.

4 Design Documentation to Be Submitted

The design documentation to be submitted is to describe the data, tools, procedures and methodologies of design and analysis which are employed to establish the design of the Floating Offshore Wind Turbine. The intended design life is also to be stated. It is recommended that a list of drawings which are planned for Class approval is to be developed and submitted for review prior to detailed plans are submitted. Where model testing is used as a basis for a design, the applicability of test results depends on the demonstration of the adequacy of the methods employed, including enumeration of sources of error, limits of applicability, and methods of extrapolation to full scale. Preferably, procedures are to be reviewed and agreed upon before model testing is done. As required in the subsequent paragraphs of this chapter, calculations are to be submitted to demonstrate the sufficiency of a proposed design. Such calculations are to be presented in a logical and well-referenced fashion employing a consistent system of units. Where the calculations are in the form of computer analysis, the submission is to provide input and output data with computer generated plots for the analysis model. A programme description (not code listings), user manuals, and the results of programme verification sample problems may be required to be submitted. The design documentation to be submitted is to include the reports, calculations, plans, specifications, and other documentation where applicable. The extensiveness of the submitted documentation is to reflect:

(1) The uniqueness of a specific design of the Floating Offshore Wind Turbine within an offshore wind farm.
(2) The level of experience with conditions in the area where the Floating Offshore Wind Turbine is to be located.
(3) Plans or requests related to statutory conventions or exemptions requested by the Flag State or Coastal State, where applicable. Owner is to develop a statutory Class deliverable matrix for review and concurrence in projects early stages.

Design documentation should be submitted electronically to Class. However, hard copies will also be accepted.

4.1 Reports

Reports by consultants and other specialists used as a basis for design are to be submitted for review. The contents of reports on offshore wind farm conditions, environmental considerations, foundation data, and materials are, in general, to comply with the recommended list of items given below.

4.2 Offshore Wind Farm Conditions

A report on offshore wind farm conditions is to present the configuration of an offshore wind farm and the exact locations of all individual floating offshore wind turbines, subsea cables, transformer platform, service and accommodation units and any other supporting structures and facilities in the offshore wind farm where applicable. The report is also to contain information on wind turbines, particularly those properties that are used as the input or as the basis of the input for the design of the Floating Offshore Wind Turbine.

4.3 Environmental Considerations

Reports on environmental considerations, as described in 4/1, are to describe all environmental phenomena appropriate to the areas for the pre-service (load-out, transportation, installation, and commissioning) and in-service (operation, maintenance, and repair) phases. The types of environmental phenomena to be accounted for, as appropriate to the type and location of the Floating Offshore Wind Turbine, include wind, waves, currents, temperature, tide, marine growth, chemical components and density of air and water, snow and ice, earthquake, and other pertinent phenomena. The establishment of environmental conditions is to be based on appropriate original data or, when permitted, data from analogous areas. Demonstrably valid statistical models are to be employed to perform the extrapolation to long-term values. Any calculations required to establish the pertinent environmental conditions are to be submitted for review. The report on environmental considerations is also to contain the calculations which quantify the effects or loadings on the Floating Offshore Wind Turbine where these are not provided in other documentation.

4.4 Soil Data

Reports on soil data are to present the findings of investigations or, where applicable, data from analogous areas on geophysical, geological, and geotechnical considerations existing at and near the installation site of anchoring structures of the station keeping system. As appropriate to the planned anchoring structure, the way such data is established and the specific items to be assessed are in general to comply with 5/4 and 5/4.3 of Part 1. The report is to contain a listing of references to cover the investigation, sampling, testing, and interpretive techniques employed during and after the site investigation. Where applicable, the report is to include a description of the predicted interaction between the soil and the intended anchoring structure of the station keeping system. As appropriate to the planned anchoring structure, the items which may need to be covered in the reports are axial and lateral pile capacities and response characteristics, the effects of cyclic loading on soil strength, scour, settlements and lateral displacements, dynamic interaction between soil and structure, the capacity of pile groups, slope stability, bearing and lateral stability, soil reactions on the structure, and penetration resistance. Recommendations relative to any special anticipated problem regarding installation are to be included in the report on soil data. Items such as the following are to be included, as appropriate: hammer sizes, soil erosion during installation, bottom preparation, and procedures to be followed in the case that pile anchor installation procedures significantly deviate from those anticipated.

4.5 Materials and Welding

Reports on structural materials and welding may be required for structures or welding procedures where materials or welding procedures do not conform to those provided in Chap. 9 of this book. For metallic structures intending to employ new alloys not defined by a recognised specification, reports are to be submitted indicating the adequacy of the material's metallurgical properties, fracture toughness, yield and tensile strengths, and corrosion resistance, with respect to their intended application and service temperatures. For a concrete structure, where it is not intended to test or define material properties in accordance with Chap. 9, a report is to be provided indicating the standards to be employed and their relative adequacy with respect to the corresponding standards referenced in Chap. 9.

4.6 Model Test

If model testing is performed (for example as per 12/5 for the assessment of global performance, 11/4 for the determination of environmental loads, or 15/2 for establishing the dynamic-response-based intact stability criteria), a model test report containing the information required in the relevant chapters are to be submitted.

5 Design Data and Calculations

Design and analysis calculations are to be submitted for items relating to loadings and responses for operations during the pre-service and in-service phases. Calculations are in general to comply with the items listed below. Calculations to be provided in association with environmental considerations and soil data are described in 8/4.

5.1 Loadings

Calculations for loadings are to be submitted in accordance with 9/5 and Chap. 11.

5.2 Stability

Stability analyses are to be performed, and the results are to be submitted in accordance with 15/2.

5.3 Dynamic Properties

Calculations of natural periods of the Floating Offshore Wind Turbine are to be submitted for review. A resonance diagram (Campbell diagram) depicting the relationship between the rotor speeds and the natural periods of turbine components, and the Floating Offshore Wind Turbine is to be submitted. As applicable, the calculation of vibration amplitudes, velocities and accelerations of the Floating Support Structure may also be required.

5.4 Global Performance

Global performance analyses are to be carried out in accordance with 9/6 and Chap. 12. Descriptions of analysis methods and calculation results of the parameters listed in 12/1 are to be submitted for review. The design documentation for the mooring lines and anchors (or tendons and foundation) and other station keeping system components is to be submitted in accordance with Chap. 14.

5.5 Structural Responses

Calculations to verify structural adequacy of the floating substructure and the tower (if classed), are to be submitted for review. The appropriate extent and types of analyses and the sophistication of such analyses are dependent on one or a combination of the following factors:

(1) The design basis of the structure relative to the conditions to be encountered at the site.
(2) The relative lack of experience with the structure's arrangement, local details, loading patterns, and failure mode sensitivities.
(3) Potential deleterious interactions with other subsystems of the Floating Offshore Wind Turbine.

The required structural analyses are to employ the loads associated with the design load conditions determined in accordance with Chap. 11. More specific information on required structural analyses is given in Chap. 13 for the Floating Support Structure.

5.6 Other Calculations

For the Floating Offshore Wind Turbine carrying the optional RNA notation, a site-specific assessment is to be performed to calculate the loads and deflections of the type-approved RNA and tower under site-specific conditions. Calculations are to include those performed in the design of the corrosion protection system. Additional calculations which demonstrate the adequacy of an overall design may also be required.

6 Design Plans of Floating Support Structure

Plans showing the scantlings, arrangements, and details of the principal parts of the floating substructure to be built under survey are to be submitted and approved before the work of construction is commenced. These plans are to clearly indicate the scantlings, joint details and welding, or other methods of connection. In general, plans to be submitted are to include the following, where applicable:

- General arrangement.
- Body plan, lines, offsets, curves of form, inboard and outboard profile.
- Layout plans indicating the locations, dimension, and weights of turbine components (e.g., blade, hub, nacelle, etc.) and the components (e.g., electrical, mechanical and control systems, etc.) in nacelle housing.
- Layout plans of secondary structures, fenders, ladders, access platform, boat landing, power cable support, etc.
- Wind heeling moment curves or equivalent data for the Floating Support Structure carrying the RNA
- Thrust curve of the turbine rotor.
- Arrangement plan of watertight compartmentation.
- Diagrams showing the extent to which the watertight and weathertight integrity are intended to be maintained, as well as the location, type, and disposition of watertight and weathertight closures
- Capacity plan and tank sounding tables.
- Summary of distributions of weights (fixed, variable, ballast, etc.) for various conditions.
- Estimations for additional accumulated masses due to sea sediment resting on structures prone to hold such masses i.e. water entrapment plates.
- Type, location and quantities of permanent ballast, if any.
- Loadings for all decks.
- Tower scantlings and tower-RNA connection details, where applicable.
- Tower-hull connection details.
- Tower pedestal and foundation (hull structure supporting the tower) details.
- Hull transverse section showing scantlings.
- Hull longitudinal sections showing scantlings.
- Decks, including helicopter deck if applicable.
- Framing, shell plating, watertight bulkheads and flats, structural bulkheads and flats, tank bulkheads and flats with location of overflows and air pipes.
- Pillars, girders, diagonals, and struts.
- Stability columns, intermediate columns, hulls, pontoons, superstructure, and deck houses.
- Arrangement and details of watertight doors and hatches.
- Foundations for anchoring equipment, industrial equipment, etc., where attached to hull structure, superstructures, or deckhouses.
- Mooring turrets and yoke arms, including mechanical details, if applicable.
- Corrosion control arrangements.

- Specification of floating substructure internal and external coatings/antifouling.
- Structure strength analysis and corrosion consideration due to additional sea sediment mass trapped on structures prone to hold such masses i.e. water entrapment plates.
- Welding details and procedures.
- Methods and locations for non-destructive testing.
- Standard details of hull fabrication.
- Information in support of novel features utilised in the design, where applicable.
- Documentation to facilitate the survey of concrete quality, as applicable to the Quality Control Programme (QCP) (refer to 17/1.3 for details), is to be submitted for approval.
- For concrete structures, plans indicating general notes about materials and workmanship, arrangements and details of reinforcement, typical details of concrete cover, the location and detail of construction joints, water stops, etc.

7 Design Documentation for Station Keeping Systems

The design documentation for the station keeping system is to include the following, where applicable:

- Mooring arrangement or pattern.
- Details of winching equipment.
- Details of anchoring system.
- Details of mooring line or tendon segments.
- Connections at anchors and between mooring line segments.
- Details of in-line (spring) buoys.
- Details of buoy of catenary anchor leg mooring (CALM) system.
- Details of single anchor leg mooring (SALM) structures.
- Details of turret system to show turret structure, swivel, and turntable.
- Details of yoke (hard or soft) connecting the floating substructure and the CALM/SALM structure.
- Reports on wind farm conditions, environmental considerations, and soil data, as required in 8/4.
- Global performance analysis report, as required in 8/5.
- Model test report where the design loads are based on model tests in a wave basin (refer to 8/4.1).

8 Design Plans of Machinery and Systems

Design plans of the onboard machinery and systems addressed in Chap. 16 are to be submitted for review and approval by Class.

9 Design Documentation of Electrical Installations

The design documentation for the electrical installation is to include the following, where applicable:

- Electrical one-line diagrams.
- Short-circuit current calculations.
- Coordination study.
- Specifications and data sheets for generators and motors.
- Specifications and data sheets for distribution transformers.
- Details of storage batteries.
- Details of emergency power source.
- Standard details of wiring cable and conduit installation practices.
- Switchboards and distribution panel.
- Panel board.

10 Design Documentation of Instrumentation and Control Systems

The design documentation for the instrumentation and control system is to include the following, where applicable:

- General arrangements.
- Data sheet.
- Schematic drawings—electrical systems.
- Schematic drawings—hydraulic and pneumatic systems.
- Programmable electronic systems.

11 Fire Protection and Personnel Safety Design Plans

The applicability of the following requirements to the submission of design documentation may vary, depending upon the nature of a specific design of the Floating Offshore Wind Turbine:

- Portable or semi-portable extinguishers.
- Fire detection and alarm systems.
- Fire cause and effect chart.
- Heating, ventilation, and air conditioning (HVAC) plan [including air handling unit (AHU)], location, duct layout, duct construction and bulkhead and deck penetration details
- Guard rails.
- Escape routes (may be included on the fire control plan or separate plan).
- Lifesaving appliances and equipment plan (escape routes must be indicated).

12 Design Plans for Other Machinery and Systems

Submission of design plans for other machinery and systems which are described in Chap. 16 but not specified in 8/8 through 1–1/11.9.3 of this book is to follow the applicable provisions of the Class Rules for the design, construction, and classification of offshore installations.

13 Additional Plans

Where additional class notations or certification under the other Rules, guides, or regulations, as described in Class Rules for the design, construction, and classification of offshore installations, are requested, submission of additional plans and calculations may be called for by Class.

14 Manuals and Procedures

14.1 Operating Manual

The Operating Manual pertaining to the safe operation of the Floating Offshore Wind Turbine is to be prepared and submitted for review. The Operating Manual is to be retained by the owner or operator and made readily available to the Class Surveyor and service personnel prior to conducting any maintenance or inspection. As far as classification is concerned, the Operating Manual is to include, as appropriate, the following information:

(1) A general description of the Floating Offshore Wind Turbine, including major dimensions, lightship characteristics.
(2) A general description and the maintenance record of the RNA and the tower.
(3) Summaries of approved modes of operation (refer to 11/3), including for each mode of operation:

- Design environmental conditions, including wave height and period, wind speed, current velocity, minimum air and sea temperatures, air gap, and water depth.
- Turbine RNA operating mode in conjunction with the condition of turbine control, protection and electrical systems and the status of electrical network connection.
- Design deck loadings, mooring loads, icing loads, variable load, cranes and, if applicable, types of helicopters for which the helideck is designed.

- Draft or draft range, disposition of movable equipment such as crane booms, etc.
- Maximum allowable KG versus draft curve or equivalent and associated limitations or assumptions upon which the allowable KG is based (refer to 15/2).
- Disposition (open or closed) of watertight and weathertight closures (refer to 15/6).
- Maximum allowable offset and heeling angle of the Floating Support Structure.

(4) Information showing:

- General arrangements.
- Type, location and quantities of permanent ballast.
- Allowable deck loadings.
- Information related to stability and watertight/weathertight integrity, as required in 15/4.2, 15/5, and 15/6.
- Capacity, centres of gravity and free surface correction for each tank.
- Capacity and centres of gravity of each void provided with sounding arrangements but not provided with means of draining.
- Location and means of draining voids.
- Hydrostatic curves or equivalent.
- Simplified electrical one-line diagrams of main power and emergency power systems.
- Schematic diagrams of the bilge, ballast, and ballast control system.

(5) Ballasting and deballasting procedure.

The Operating Manual is to be reviewed by Class to verify the operational procedures and conditions are consistent with the design information, criteria and limitations considered in the Floating Offshore Wind Turbine's classification. Class is not responsible for the operation of floating offshore wind turbines. The Operating Manual outlined in this paragraph does not need to be in addition to that required by the Coastal State or other governmental authorities. However, these administrations may require that additional information be included in the Operating Manual.

14.2 Procedures

Procedures are to be submitted for the following:

- Installation Procedures.
- Hook-Up Procedures for connecting the station keeping system to the floating substructure.
- Procedure for closure weathertightness integrity testing i.e. leak, chalk, or hose testing.

- Startup and Commissioning Procedures.
- In-Service Inspection Programme (ISIP), as required in 17/2.3.
- Underwater Inspection Procedure or, if applicable, Drydocking Survey Procedure.

14.3 Information Memorandum

An information memorandum is to be prepared and submitted to Class. Class will review the contents of the memorandum to establish consistency with other data submitted for the purpose of obtaining classification. Class will not review the contents of the memorandum for its accuracy, or the features described in the memorandum for their adequacy. An information memorandum is to contain, as appropriate to the Floating Offshore Wind Turbine, the following:

- Specifications of the RNA and tower suitable to be installed for the site.
- Valid wind turbine type certificate for the RNA and tower.
- Description of the wind turbine control system concept and the control system (structure of the control system, sequences of the start and stop procedures, behaviour of the turbine during normal operation and on detection of malfunctions, statement of trigger criteria).
- Description of turbine control system fault modes and frequency and duration of fault events.
- Description of the procedure for manual intervention after the activation of any wind turbine protection functions.
- Description of the wind turbine braking systems and their behaviour (structure of the braking systems, mode of operation, characteristic quantities, time constants).
- Functional description of the wind turbine locking devices.
- Description of the wind turbine control software used for load simulation. (interfaces to the load analysis such as programme modules for blade pitching as well as identification of critical load cases).
- Description of the process that confirms that the control system (including software) on the wind turbine has the same behaviour as modelled in the load simulation.
- Computer software interface of the turbine control system linked to the load analysis programme if an independent analysis is requested.
- Site plan indicating the general features at the site and the layout of the offshore wind farm.
- Environmental and soil design criteria, including the recurrence interval used to assess environmental phenomena.
- Plans showing the general arrangement of the Floating Offshore Wind Turbine.
- Description of the safety, protective, security and trespass avoidance systems provided.
- Description of the modes of operation.
- Description of the export or inter-array power cable termination at the floating substructure.

- Listing of governmental authorities having cognisance over the installation.
- Listing of any novel features.
- Brief description of any monitoring proposed for use on the installation.
- Description of pre-service load-out, transportation, installation, and commissioning procedures.
- Description of onsite maintenance and repair procedures for the Floating Offshore Wind Turbine.
- Description of disconnecting, transportation and reconnecting procedures for maintenance and repair operations, if applicable.
- A plan to control the procurement of equipment.
- An inventory of hazardous materials (IHM) plan.

Chapter 9
Materials and Welding Used in the Construction and Maintenance of Floating Offshore Wind Turbines

1 Metallic Materials

The properties of metallic materials using in the construction and maintenance of offshore floating wind turbines are to be in accordance with the applicable Class mandated MOU Rules. Guidance on determining the material application category of a hull structural component is provided in the FPI Rules.

2 Welding

Requirements for welding are to be in accordance with the Class Rules for materials and welding. The MOU Rules may be used to establish weld designs for the steel Floating Support Structure.

3 Concrete

Properties of the concrete are to comply with 2/1.3 which makes further reference to applicable standards of the American Society for Testing and Materials (ASTM) and the American Concrete Institute (ACI).

F. Aderouane, *Design, Construction, and Global Performance Analysis of Offshore Wind Turbines*, Green Energy and Technology, https://doi.org/10.1007/978-3-031-86010-2_9

4 Mooring Chains and Accessories

Materials and welding of offshore mooring chains and accessories for application in the station keeping system are to be in accordance with the Class guidance relating to the certification of offshore mooring chain.

5 Fibre Ropes

For materials of synthetic fibre ropes for application in the station keeping system, the requirements specified in the Class guidance relating to the application of fibre rope for offshore mooring are to be satisfied.

Chapter 10
General Design Requirements

1 General

The floating substructure, the station keeping system and onboard machinery, equipment and systems including applicable marine systems and associated equipment and machinery, safety systems and associated equipment, and lifesaving appliances and machinery are to be designed in compliance with the standards and requirements contained in this book with intention that the Floating Offshore Wind Turbine can fulfil its intended functions during its design life and meet specified minimum requirements for serviceability and operability. Compliance of the tower with the design requirements in this book is required where (refer to 8/2.1):

- The optional RNA notation is requested by the Owner.
- The optional RNA notation is not requested by the Owner, but the tower and its connection to the RNA and/or to the floating substructure are not included in the wind turbine type certificate.

2 Safety Level

The design criteria specified in this book are intended for the unmanned Floating Offshore Wind Turbine having a safety level equivalent to the medium (L2) exposure level as defined in ISO 19904-1 for unmanned floating offshore structures. The Floating Support Structure and the station keeping system can be categorised at various safety levels based on the consideration of life-safety and consequences of failure. Life-safety considers the most unfavourable combination of environmental event and turbine operating mode which would be expected to occur while personnel are on the Floating Offshore Wind Turbine. Consequences of failure deal with loss of life, environmental pollution, financial loss and potential negative impact to the industry. Since the Floating Offshore Wind Turbine is typically unmanned

F. Aderouane, *Design, Construction, and Global Performance Analysis of Offshore Wind Turbines*, Green Energy and Technology, https://doi.org/10.1007/978-3-031-86010-2_10

and considered as having moderate consequences of failure, the Floating Support Structure can in general be designed to have a safety level equivalent to the medium (L2) exposure level as defined in ISO 19904-1. A higher safety level equivalent to the high (L1) exposure level as defined in ISO 19904-1 may be warranted when:

(1) There is limited design and operation experience with floating offshore wind turbines.
(2) The Floating Support Structure is likely to have a low level of redundancy.
(3) The same or a similar design may be used in an entire offshore wind farm, which becomes vulnerable to "common-cause" failures.

For an offshore wind farm having multiple Floating Offshore Wind Turbines located in proximity, failure of an individual station keeping system could potentially cause extensive damage to many other floating offshore wind turbines and power cables as well as other offshore installations in the offshore wind farm. The risk analysis may be used to determine the intended safety level of the station keeping system. However, the safety level of the station keeping system is not to be lower than that of the Floating Support Structure.

3 Design Life

The design life of the Floating Offshore Wind Turbine is not to be less than 20 years. Upon the request by the Owner, a shorter design life may be accepted by Class for the Floating Offshore Wind Turbine for the purpose of demonstrating viability of a new design concept or conducting pilot operations. Continuance of classification beyond the design life is subject to the survey requirements specified in 17/2.12 and additional engineering analyses (refer to 8/2.3.2 and 8/2.3.3).

4 Environmental Conditions

The Floating Offshore Wind Turbine is to be designed to withstand specified operational and environmental conditions at the installation site while the RNA is in various operating conditions. The environmental conditions are defined in this chapter. The Floating Offshore Wind Turbine is also to be designed for all pre-service operations such as load-out, transportation, installation, and commissioning. Environmental conditions having a low probability of being exceeded where a mooring line or tendon of the station keeping system is broken or removed are also to be considered. Joint statistics may be used to determine a return period which, combined with the probability of occurrence of damage, produces a risk level consistent with the intended safety level. Additionally, environmental conditions are to be specified for verifying the survivability of the Floating Offshore Wind Turbine. Survival environmental conditions are those that produce responses having an exceptionally low probability of being exceeded during the design life, and that the Floating Offshore Wind Turbine can endure such responses without causing catastrophic consequences.

5 Design Load Conditions and Load Calculations

The Floating Support Structure's life cycle phases, including the pre-service (load-out, transportation, installation, and commissioning) and in-service (operation, maintenance, and repair) phases, are to be investigated using anticipated loads. Permanent and variable loads together with relevant environmental loads because of wind, waves, currents, water level variations and, where deemed necessary, the effects of earthquake, temperature, fouling, ice, and other potential impacts are to be considered. Combinations of these loads as well as the turbine operating conditions that produce the most unfavourable local and global effects on the Floating Support Structure and the station keeping system, as determined by the requirements of the pre-service and in-service phases, are to be applied. Dynamic analysis models are to be developed to predict design load effects for all relevant combinations of external site conditions and turbine operating conditions. A minimum set of such combinations, which is termed as Design Load Cases (DLCs), is defined in 11/3. In addition to the Design Load Cases, other load cases are also to be considered for the situations where the Floating Offshore Wind Turbine is in the survival condition. The adequacy of air gap and station keeping capacity, as well as structural integrity if considered relevant, are to be verified by the robustness check, in which the Floating Offshore Wind Turbine is subjected to survival environmental conditions that are more severe than extreme design environmental conditions. The robustness check provides a direct indicator of the survivability of a specific design. A minimum set of Survival Load Cases (SLCs) is defined in 11/3.3. When establishing design criteria for the Floating Offshore Wind Turbine, all applicable aspects contributing to the safety level need to be considered collectively. These include, but are not limited to.

(1) Prescribed safety factors.
(2) Return period of extreme design environmental conditions.
(3) Statistical variation of environmental conditions at different installation sites.
(4) Load models of environmental and operational loading specific to offshore wind turbines.
(5) Strength and fatigue capacity models of materials, structural members, joints, and foundation elements.
(6) Characteristics of the responses of the offshore wind turbine.

In contrast to bottom-founded offshore wind turbines, the Floating Offshore Wind Turbine having the typical hull types shows a different pattern of sensitivity to the return period of environmental conditions. The minimum return period for the extreme storm condition, as specified in 11/3.2, is 50 years.

6 Global Performance Analyses

Global performance analyses of the Floating Offshore Wind Turbine are used to determine the global effects of environmental loads and other loads on the Floating Offshore Wind Turbine. Global response analyses are to be performed for each of the critical design phases. Global performance analyses using various design load cases are required. It is recommended to include the turbine RNA and the tower, the floating substructure, the station keeping system, and, where relevant, the power cable system in an integrated (also known as "coupled") simulation model for global performance analyses. For those global loads and responses that are deemed as having weak coupling effects, global performance analyses may also be performed using a non-integrated model. Either frequency or time domain methods, or a combination of both, may be used in global performance analyses. However, for those cases that have transient or highly nonlinear effects, time-domain analyses are normally required. Methods and models employed in analyses are to account for the relevant nonlinear and dynamic coupling effects of the RNA, the tower, the floating substructure, the station keeping system, and, where relevant, the power cable system. Chapter 12 outlines specific requirements for global performance analyses of the Floating Offshore Wind Turbine.

7 Structural Designs

The Floating Support Structure is to be designed in accordance with the requirements specified in Chap. 13. In the case where the structure's configuration or loading is not specifically addressed in this book, other recognised design standards may be used. Where alternative standards are followed, it is to be demonstrated that the safety level specified in this book has been adequately satisfied.

8 Station Keeping System

The floating substructure is restrained by the station keeping system, which may be either a passive or an actively controlled system, or a combination of both. A passive station keeping system may be in the form of.

- Spread mooring (catenary, taut-line, or semi-taut-line).
- CALM consisting of a buoy and several catenary anchor legs.
- Turret mooring.
- SALM, such as an articulated leg.
- Tension leg system.

An active system could be a mooring system with the ability of changing mooring line tensions.

9 Stability and Watertight/Weathertight Integrity

Adequacy of stability of the Floating Offshore Wind Turbine is to be verified for all relevant pre-service and in-service phases. The assessment of stability is to include the consideration of both intact and damaged conditions. The design requirements of Chap. 15 are to be satisfied.

10 Structural Material Selection, Welding and Connection

Structural materials are to be selected with consideration of the requirements for performance, welding, and inspection. The materials used for the construction of the Floating Support Structure and the station keeping system are to be in accordance with the requirements of Chap. 9. The welding of steel for the Floating Support Structure is to follow the requirements of Chap. 9. Special attention is to be given to the weld details for fatigue sensitive areas, whenever relevant. For further information, refer to the Class Rules relating to fatigue assessment of offshore structures. For connections other than welded joints, such as clamps, connectors and bolts that are used to join diagonal braces to the column or the turbine tower to the support foundation, the strength and fatigue resistance are to be assessed by analytical methods or testing following established industry practices such as those specified in the AISC Steel Construction Manual. For concrete floating substructures, material and welding of steel reinforcement used for concrete construction are to comply with 2/1.5.8.

11 Corrosion Protection and Control

A corrosion protection and control system utilising sacrificial anodes and/or coating in accordance with the recognised industry standards such as those published by NACE are to be provided. The design life of the corrosion protection and control system is to be in general the design life of the Floating Offshore Wind Turbine unless a monitoring and repair plan is established. In the splash zone, corrosion allowance is to be added to the external shell plating. Reference may be made to the guidance on corrosion allowance in the relevant industry standards for floating offshore structures.

12 Considerations for Identical Floating Substructure Design

For the floating substructures of an identical design and supporting an identical RNA and tower combination, the station keeping system for these floating substructures could differ even in the same offshore wind farm. For design approval of these floating substructures, the extent of required design documentation submission and review may be adjusted by Class with consideration of:

- Differences in environmental conditions and electrical power network conditions.
- Difference in the load calculation for structural strength and fatigue assessment.
- Difference in the motion and acceleration calculation.
- Effect of modifications, if any, affecting class items.
- Effect of variations in construction.

13 Operating Manual

The Operating Manual, as required in 8/14.1, for the floating wind turbine is to be developed to specify the operating procedures and conditions that are consistent with the design information, criteria and limitations considered in the design of the Floating Offshore Wind Turbine.

Chapter 11
Environmental Conditions

1 General

Environmental conditions to which the Floating Offshore Wind Turbine is expected to be exposed during its design life are to be determined using adequate data for the areas in which the Floating Offshore Wind Turbine is to be transported and installed. Environmental phenomena that influence the pre-service and in-service phases of the Floating Offshore Wind Turbine are to be described in terms of relevant characteristic parameters. Statistical data and realistic statistical and mathematical models which describe the range of expected variations of environmental conditions are to be employed. All data used are to be fully documented with the sources and estimated reliability of data noted. Methods employed in developing available data into design criteria are to be described and submitted in accordance with 8/4.1. Probabilistic methods for short-term, long-term, and extreme-value prediction are to employ statistical distributions appropriate to environmental phenomena under consideration, as evidenced by relevant statistical tests, confidence limits and other measures of statistical significance. Directional data and angular separation for wind, waves and currents are to be established. Hindcasting methods and models are to be fully documented if they are used to derive the environmental data. Suitable data and analyses supplied by recognised consultants will be accepted as the basis for design. For installations in areas where published design standards and data exist, such standards and data may be cited in the design documentation.

2 Environmental Factors to Be Considered

In general, the design of the Floating Offshore Wind Turbine requires investigation of the following environmental factors.

F. Aderouane, *Design, Construction, and Global Performance Analysis of Offshore Wind Turbines*, Green Energy and Technology, https://doi.org/10.1007/978-3-031-86010-2_11

(1) Wind.
(2) Waves.
(3) Currents.
(4) Tides, storm surges, and water levels.
(5) Air and sea temperatures.
(6) Air density.
(7) Ice and snow accumulation.
(8) Marine growth.
(9) Seismicity.
(10) Sea ice or lake ice.

Other phenomena, such as tsunamis, submarine slides, seiche, abnormal composition of air and water, air humidity, salinity, ice drift, icebergs, ice scouring, etc., may require investigation depending upon specific conditions at an installation site and a specific design of the Floating Offshore Wind Turbine. The investigation of seabed and soil conditions is described in 11/7.7.

3 Wind

3.1 General

Statistical wind data are normally to include information on the frequency of occurrence, duration, and direction of various wind speeds at the location where offshore wind turbines are to be installed. If onsite measurements are taken, the duration of individual measurements and the height above sea-level of measuring devices are to be stated. A wind speed value is only meaningful when qualified by its elevation and time-averaging duration. In the absence of site data, published data and data from nearby land and sea stations may be used upon the agreement with Class.

3.2 Wind Properties

3.2.1 Wind Speed and Turbulence

A wind condition is typically represented by a mean wind speed and a standard deviation of wind speed. The turbulence intensity, which measures the variation of wind speed relative to the mean wind speed, is defined as the ratio of the wind speed standard deviation to the mean wind speed (i.e., coefficient of variance of wind speed). In this book, the mean wind speed, denoted as V_{hub}, at turbine hub height with 10-min averaging duration is employed to define the design load conditions in 12/3 and to calculate the rotor generated aerodynamic loads and thrust forces as described in 12/4.2. Mean wind speeds with 1-min or 1-h averaging time duration are used in

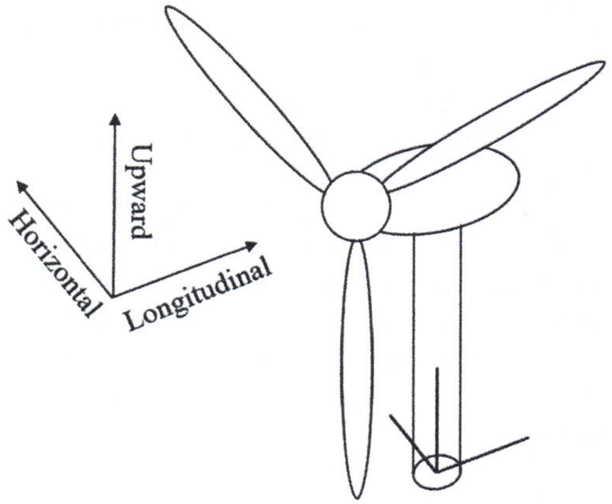

Fig. 1 Vector components of turbulent wind velocity

the calculation of steady and dynamic wind forces exerted on exposed components of the Floating Support Structure, as described in 12/4.2. Conversion of a mean wind speed from one averaging duration to another is to be based on site-specific wind conditions. Use of conversion factors in recognised industry standards appropriate to the installation site may be acceptable and is subject to approval by Class. For wind speeds given in terms of the "fastest mile of wind", V_f, the corresponding time-averaging period t in seconds is given by $t = \frac{3600}{V_f}$, where V_f is the fastest mile of wind at a reference height of 10 m (32.8 ft), in miles per hour. The turbulence of wind within 10 min is considered stationary and can be modelled by a power spectral density function and a coherence function. The turbulence model is to include the effects of varying wind speed, shears and directions and allow rotational sampling through varying shears. The three vector components of turbulent wind velocity, as depicted in Fig. 1 are defined as:

(1) Longitudinal—Along the direction of the mean wind speed.
(2) Lateral—Horizontal and normal to the longitudinal direction.
(3) Upward—Normal to both the longitudinal and lateral directions and pointing upward.

3.2.2 Wind Profile

The mean wind speed profile (vertical wind shear) is to be defined by the power law:

$$V(z) = V_{hub}(z/z_{hub})^{\alpha}$$

where

$V(z)$ = wind profile of the 10-min mean wind speed as a function of height, z, above the SWL, in m/s (ft/s)
V_{hub} = 10-min mean wind speed at hub height, in m/s (ft/s)

α = power law exponent, whose value is given in 11/3.5 and 11/3.5.3, (dimensionless)
z = height above the SWL, in m (ft)
z_{hub} = hub height above the SWL, in m (ft)

The power law wind profile can also be applied to the mean wind speeds with 1-min or 1-h averaging duration, except that the exponent is to be determined based on site-specific conditions or using published data, such as those in the FPI Rules, as appropriate to the installation site. For extratropical cyclone storm wind conditions, the mean wind speed profile may be represented by the following logarithmic wind shear law. It is expressed using the 1-h mean wind speed at 10 m (32.8 ft) above the SWL:

$$V(z,t) = V(z,t_0)\left[1 - 0.41 I_u(z)\ln\left(\frac{t}{t_0}\right)\right] \quad \text{for} \quad t < t_0$$

where

$V(z,t)$ = mean wind speed at height z and corresponding to an averaging time, t,, in m/s (ft/s)
z = height above the SWL, in m (ft)
t = averaging time shorter than $t_0 = 3600$ s, in seconds
$V(z,t_0)$ = 1-h mean wind speed at height z, in m/s (ft/s)
$= V_o\left[1 + C \ \ In\left(\frac{z}{10\phi}\right)\right].$
t_0 = reference averaging time (1 h), in seconds
= 3600 s
C $= 0.0573\sqrt{1 + \frac{0.15V_0}{\phi}}$ (dimensionless)
V_0 = 1-h mean wind speed at 10 m (32.8 ft) above the SWL, in m/s (ft/s)
$I_u(z)$ = turbulence intensity (i.e., the ratio of the wind speed standard deviation to the mean wind speed) at height z, (dimensionless)
$= 0.06\left[1 + \frac{0.043V_0}{\phi}\right]\left(\frac{z}{10\phi}\right)^{-0.22}.$
ϕ = unit conversion factor, (dimensionless).
= 1 when using SI units (m, m/s)
= 3.28 when using US Customary units (ft, ft/s)

For tropical cyclones (also termed as hurricanes or typhoons), the wind profile provided in Appendix A5/1 is to be applied. For squalls, the wind profile recommended in API RP 2MET may be applied. Other wind profile models may also be used if they can be justified by site-specific data.

3.2.3 Wind Spectrum and Spatial Coherence

Site-specific spectral density of wind speed and spatial coherence are preferable to be determined based on measured wind data. When the site assessment is not available, the Mann uniform shear turbulence model or the Kaimal spectrum in combination with the exponential coherence model, as recommended in Annex C of IEC 61400-1 (2019), may be applied. For the extratropical storm wind, the NPD (Frøya) wind spectrum in conjunction with the two-point coherence function provided in A4/1 may be applied. For the tropical cyclone storm wind, the wind spectrum provided in A4 may be applied.

3.3 Long-Term and Extreme-Value Predictions

Long-term and extreme-value predictions for sustained and gust winds are to be based on recognised techniques and clearly described in the design documentation. Preferably, the statistical data used for the long-term distributions of wind speed are to be based on the same averaging periods of wind speeds as are used for the determination of loads.

3.4 Wind Conditions

A wind condition for the design of the Floating Offshore Wind Turbine is represented by a steady mean flow and an associated turbulence, as well as a mean wind direction and a change of the wind direction. The design wind conditions are further categorised into:

(1) The normal wind conditions, which occur more frequently than once per year.
(2) The extreme wind conditions representing rare wind conditions with a given return period.
(3) The survival wind conditions having an exceptionally low probability of being exceeded during the design life of the Floating Offshore Wind Turbine.

The design wind conditions for the Floating Offshore Wind Turbine are specified in 11/3.5 through 11/3.6. The load case descriptions in 11/3 specify which wind condition is to be applied.

3.5 Normal Wind Conditions

3.5.1 Normal Wind Profile Model (NWP)

The normal mean wind speed profile (vertical wind shear) is to be defined by the power law specified in 10/3.3.2, where the power law exponent $\alpha = 0.14$.

3.5.2 Normal Turbulence Model (NTM)

The normal turbulence model (NTM) is to be applied together with the normal wind profile model (NWP) as defined in 11/3.5.1. The standard deviation of turbulence of the normal turbulence model, denoted as σ_{NTM},, is defined as the 90% quantile in the probability distribution of wind speed standard deviation conditioned upon a given 10-min mean wind speed at hub height (V_{hub}). The value of the turbulence standard deviation is to be determined using appropriate statistical techniques applied to measured and preferably de-trended data. Where the site assessment is not available, the recommended approach provided in paragraph 6.4.3 of IEC 61400-3-1 (2019) may be used to estimate the standard deviation.

3.5.3 Extreme Wind Conditions

The extreme wind conditions are represented by peak wind speeds due to storms, extreme operating gust and turbulence, extreme wind shear events, and rapid changes in wind speed and direction.

Extreme Wind Speed Model (EWM)

The EWM is defined as a turbulent wind model with a specified return period. Both the 1-year and 50-year return extreme wind conditions are considered in the design load conditions as specified in 18/1.1. When site data are not available, the wind profile of 10-min mean wind speeds for the EWM with return periods of 1 year and 50 years, respectively, is to be represented by the power law model as follows:

$$V(z) = V_{hub}(z/z_{hub})^{0.11}$$

where:

V_{hub}	= 10-min mean wind speed at hub height, in m/s (ft/s)
	= $V_{10\ \min,1-yr}$ for 1-year return extreme wind condition
	= $V_{10\ \min,50-yr}$ for 50-year return extreme wind condition
$V_{10\ \min,1-yr}$	= 10-min mean wind speed at hub height with a return period of 1 year, in m/s (ft/s)
$V_{10\ \min,50-yr}$	= 10-min mean wind speed at hub height with a return period of 50 years, in m/s (ft/s)
z	= height above the SWL, in m (ft)
z_{hub}	= hub height above the SWL, in m (ft)

The standard deviation of longitudinal turbulent wind speed of extreme wind condition, σ_1 at hub height is to be calculated as:

$$\sigma_1 = 0.11 \times V_{hub}$$

For the extratropical storm wind, the logarithmic wind shear law given in 11/3.3.2 may be used to calculate the wind profile and standard deviation of extreme wind conditions with a 50-year return period. For the tropical cyclone storm wind, the wind profile, the standard deviation, the turbulence intensity, and the gust factor provided in Appendix A5/1 are to be applied. Where applicable, extreme squall events with 1-year and 50-year return periods in time domain are to be considered as additional extreme wind conditions.

Extreme Operating Gust (EOG)

The EOG is represented by the hub height gust magnitude, V_{gust}, as defined in the following equation:

$$V_{gust} = \min\left\{1.35\left(V_{3s,1-\text{yr}} - V_{hub}\right); 3.3\left(\frac{\sigma_{NTM}}{1 + \frac{0.1D}{\Lambda_1}}\right)\right\}$$

where

$V_{3\,s,1-\text{yr}}$ = 3-s mean wind speed at hub height with a return period of 1 year, in m/s (ft/s)
V_{hub} = 10-min mean wind speed at hub height, as defined in Table 11.2, in m/s (ft/s)
σ_{NTM} = longitudinal turbulence standard deviation defined in 11/4.5.2, in m/s (ft/s)
Λ_1 = longitudinal turbulence length scale, in m (ft)
= 0.7 z when $z \leq 60$ m (196.8 ft)
= 42 m (137.8 ft) when $z \geq 60$ m (196.8 ft)
D = rotor diameter, in m (ft)

To account for resonance and interaction of gusts with the Floating Offshore Wind Turbine, gust events with longer durations are to be defined unless site-specific data shows that these events are unlikely to occur within the given return period. The time history of transient wind speed at height z is to be defined by:

$$V(z,t) = \begin{cases} V(z) - 0.37V_{gust}\sin\left(\frac{3\pi}{T}\right)\left[1 - \cos\left(\frac{2\pi}{T}\right)\right] & 0 \leq t \leq T \\ V(z) & \text{otherwise} \end{cases}$$

where

$V(t)$ = normal wind profile defined in 10/3.5.1, in m/s (ft/s)
z = height above the SWL, in m (ft)
T = 10.5 s
= 1.5 T_n.

$T_n =$ the natural period in seconds of the surge, sway, heave, roll, pitch, and yaw modes of the Floating Offshore Wind Turbine, in seconds. The natural period of the mode less than 7 s can be omitted

The amplitude for the EOG cannot be assumed to be constant with different time periods. When site data is not available, the hub-height gust magnitude, V_{gust}, may be replaced by the following:

$$V_{gust} = \min\left\{1.35\left(V_{3\,\mathrm{s},1-\mathrm{yr}} - V_{hub}\right); 0.9Ln(T+1.18)\left(\frac{\sigma_{NTM}}{1+\frac{0.1D}{\wedge_1}}\right)\right\}$$

3.5.4 Extreme Turbulence Model (ETM)

The ETM is to be represented by the normal wind profile (NWP) model specified in 11/3.5.1 and the turbulence whose standard deviation of longitudinal component is given by:

$$\sigma_1 = cI_{ref}\left[0.072\left(\frac{V_{ave}}{c}+3\right)\left(\frac{V_{hub}}{c}-4\right)+10\right]$$

where

c = 2 m/s (6.56 ft/s)

V_{hub} = 10-min mean wind speed at hub height, as defined in Table 11.1, in m/s (ft/s)

V_{ave} = site-specific annual mean wind speed at hub height, in m/s (ft/s)

I_{ref} = expected value of turbulence intensity at hub height when $V_{hub} = 15$ m/s (49.2 ft/s)

3.5.5 Extreme Direction Change (EDC)

The extreme direction change magnitude, θ_e, is to be calculated by:

$$\theta_e = \pm 4 arctan\left(\frac{\sigma_{NTM}}{V_{hub\left(1+\frac{0.1D}{\wedge_1}\right)}}\right) - 180^\circ \le \theta_e \le 180^\circ$$

where

σ_{NTM} = longitudinal turbulence standard deviation defined in 11/3.5.2, in m/s (ft/s)

V_{hub} = 10-min mean wind speed at hub height, as defined in Table 11.1, in m/s (ft/s)

$\wedge_1$ = longitudinal turbulence length scale, defined in 11/3.5.3.2, in m (ft)
D = rotor diameter, in m (ft)

The time history of transient extreme direction change, $\theta(t)$, is defined by:

$$\theta(t) = \begin{cases} 0^\circ & t < 0 \\ \pm 0.5\theta_e\left[1 - \cos\left(\frac{\pi t}{T}\right)\right] & 0 \le t \le T \\ \theta_e & t > Y \end{cases}$$

where $T = 6$ s is the duration of the extreme direction change. The sign in the equation is to be chosen such that the most unfavourable transient loading occurs. At the end of the time history of direction change, the direction is assumed to remain a constant value (θ_e). The wind speed is to follow the normal wind profile (NWP) model in 11/3.5.1. In addition, time periods coinciding with the floating substructure yaw natural period are to be considered.

3.5.6 Extreme Coherent Gust with Direction Change (ECD)

The extreme coherent gust with direction change is to have a magnitude of:

$$V_{cg} = 15\,\text{m/s or } 49.21\,\text{ft/s}$$

The time history of transient wind speed at height z, is defined by:

$$V(z,t) = \begin{cases} V(z) & t < 0 \\ (V(z) + 0.5V_{cg}\left[1 - \cos\left(\frac{\pi t}{T}\right)\right] & 0 \le t \le T \\ V(z) + V_{cg} & t > Y \end{cases}$$

where

$V(z)$ = normal wind profile defined in 11/3.5.1, in m/s (ft/s)
T = rise time of gust wind, in seconds
= 10 s
= height above the SWL, in m (ft)

The rise in wind speed is assumed to occur simultaneously with the time history of direction change:

$$\theta(t) = \begin{cases} 0^\circ & t < 0 \\ \pm 0.5\theta_{cg}(V(z) + 0.5V_{cg}\left[1 - \cos\left(\frac{\pi t}{T}\right)\right] & 0 \le t \le T \\ +\theta_{cg} & t > Y \end{cases}$$

where

θ_{cg} = magnitude of direction change, in degree

$$= \begin{cases} 180^\circ & V_{hub} \le 4\,\text{m/s (13.12 ft/s)} \\ 720^\circ (V_{hub}/\phi)^{-1} & V_{hub} \ge 4\,\text{m/s (13.12 ft/s)} \end{cases}$$

V_{hub} = 10-min mean wind speed at hub height, in m/s (ft/s).
T = rise time of gust wind, in seconds.
= 10 s
ϕ = unit conversion factor, (dimensionless).
= 1 when using SI units (m, m/s)
= 3.28 when using US Customary units (ft, ft/s)

In addition, time periods coinciding with the floating substructure yaw natural period are to be considered.

3.5.7 Extreme Wind Shear (EWS)

The extreme wind shear is to be applied in both vertical and horizontal directions. The two extreme wind shears are considered independent events and therefore they are not to be applied simultaneously. The time history of transient positive and negative vertical shear is given by:

$$V(z,t) = \begin{cases} V_{hub}\left(\frac{z}{z_{hub}}\right)^{\alpha} \pm \left(\frac{z-z_{hub}}{D}\right)\left(2.5\phi + 0.2\beta\sigma_{NTM}\left(\frac{D}{\Lambda_1}\right)^{\frac{1}{4}}\right)\left[1-\cos\left(\frac{2\pi t}{T}\right)\right] & 0 \le t \le T \\ V_{hub}\left(\frac{z}{z_{hub}}\right)^{\alpha} & \textit{otherwise} \end{cases}$$

The time history of transient horizontal shear is given by:

$$V(z,t) = \begin{cases} V_{hub}\left(\frac{z}{z_{hub}}\right)^{\alpha} \pm \left(\frac{y}{D}\right)\left(2.5\phi + 0.2\beta\sigma_{NTM}\left(\frac{D}{\Lambda_1}\right)^{\frac{1}{4}}\right)\left[1-\cos\left(\frac{2\pi t}{T}\right)\right] & 0 \le t \le T \\ V_{hub}\left(\frac{z}{z_{hub}}\right)^{\alpha} & \textit{otherwise} \end{cases}$$

where

V_{hub} = 10-min mean wind speed at hub height, as defined in Table 11.1, in m/s (ft/s)
σ_{NTM} = longitudinal turbulence standard deviation defined in 11/3.5.2, in m/s (ft/s)
Λ_1 = longitudinal turbulence length scale, defined in 11/3.5.3.2, in m (ft)
z = height above the SWL, in m (ft)
z_{hub} = hub height above the SWL, in m (ft)
y = horizontal distance from hub in the crosswind direction, in m (ft)
D = rotor diameter, in m (ft)
α = 0.14, (dimensionless)

β = 6.4, (dimensionless)
T = 12 s
ϕ = unit conversion factor, (dimensionless)
= 1 when using SI units (m, m/s)
= 3.28 when using US Customary units (ft, ft/s)

The sign for the transient wind shear is to be determined such that the most unfavourable transient loading occurs.

3.6 *Survival Wind Conditions*

The survival wind conditions are described by the Survival Wind Speed Model (SurWM).

3.6.1 Survival Wind Speed Model (SurWM)

The Survival Wind Speed Model (SurWM) is like the Extreme Wind Speed Model (EWM) defined in 11/3.5.3.1, but with a return period longer than 50 years, as specified in 12/3.3. When site data are not available, the wind profile and turbulence spectrum for the SurWM is to be defined according to 11/3.2.2 and 11/3.2.3. When the wind profile is assumed to follow the power law, the power law exponent $\alpha = 0.11$ and the standard deviation of longitudinal turbulent wind speed is to be taken as $\sigma_1 = 0.11 \times V_{hub}$.

4 Waves

4.1 *General*

The development of wave data is to reflect conditions at the installation site and the type of Floating Offshore Wind Turbine. Statistical wave data used to determine design parameters are normally to include the frequency of occurrence of various wave height groups, associated wave periods and directions. Published data and previously established design criteria for specific areas may be used where such exist. Hindcasting techniques that adequately account for shoaling and fetch limited effects on wave conditions at the site may be used to augment available data. Analytical wave spectra employed to augment available data are to reflect the shape and width of the data and to be appropriate to the general site conditions. As applicable, wave data are to be developed to determine the following:

(1) Dynamic responses of the Floating Support Structure and the station keeping system.
(2) Maximum responses of structures and station keeping system components.
(3) Fatigue.
(4) Air gap.
(5) Wave impact on the local structure.

All long-term and extreme value predictions employed for the determination of design wave conditions are to be fully described and based on recognised techniques. Because the wave-induced global responses may be increased due to the change of wave period and direction, consideration is to be given to waves of less than the maximum height but with a different period and/or direction. Waves that cause the most unfavourable effects on the overall structure may also differ from waves having the most severe effects on individual structural components. In addition to the most severe wave conditions, frequent waves of smaller heights are to be investigated to assess their effect on the fatigue and dynamic responses. Swells can be of importance in conjunction with low-frequency responses of the Floating Offshore Wind Turbine. The potential bi-directional wave loading for the Floating Offshore Wind Turbine is to be considered. The design wave conditions for the design of the Floating Offshore Wind Turbine are described below in 11/4.2 through 11/4.5. The design load conditions defined in 12/3 specify how these sea states are to be applied in combination with other design environmental conditions.

4.2 Normal Sea State (NSS)

The Normal Sea State (NSS) is represented by a significant wave height, a peak spectral period, and a wave direction. It is to be determined based on the site-specific long-term joint probability distribution of metocean parameters. The Normal Sea State (NSS) is used in Table 11.3 to define several Design Load Cases (DLCs) requiring either strength analysis or fatigue analysis. For strength analyses, the Normal Sea State (NSS) can be characterised by the expected value of significant wave height, $H_{S,NSS}$, conditioned upon a given value of V_{hub} (i.e., $H_{S,NSS} = E[H_S|V_{hub}]$). A range of peak period, T_p, associated with each significant wave height is to be determined for load calculations. The resultant highest loads are to be used in the design of the Floating Support Structure. For fatigue analyses, the number and resolution of sea states are to be determined in such a manner that the fatigue damage associated with the full long-term distribution of metocean parameters can be sufficiently accounted for.

4.3 Severe Sea State (SSS)

The Severe Sea State (SSS) condition is to be applied in combination with the normal wind condition, as specified in 11/3.5, and the turbine operating mode assumed in power production. The Severe Sea State (SSS) is represented by a significant wave height, $H_{S,SSS}$, a peak spectral period and a wave direction. It is to be determined by extrapolation of site-specific long term joint probability distribution of metocean parameters to the extent that the joint occurrence of $H_{S,SSS}$ and a given 10-min mean wind speed, V_{hub}, at hub height has a return period of 50 years. A series of V_{hub} is to be selected between the cut-in and cut-out wind speed for power production. As a conservative estimation, the 50-year return significant wave height independent of wind speed may be used to approximate $H_{S,SSS}$. In the case that maximum operating sea states for the turbine in power production are pre-defined in the Operating Manual, those sea states may be used to determine $H_{S,SSS}$ in lieu of the requirements specified above. A range of peak period associated with each significant wave height is to be determined for load calculations. The resultant most unfavourable responses are to be used in the design of the Floating Offshore Wind Turbine.

4.4 Extreme Sea State (ESS)

The Extreme Sea State (ESS) is to represent a 1-year return or 50-year return wave condition. The significant wave height of the ESS model is denoted either as $H_{s,1-\mathrm{yr}}$ or $H_{s,50-\mathrm{yr}}$ for the extreme significant wave height with a return period of 1 year or 50 years, respectively. The values of $H_{s,1-\mathrm{yr}}$ and $H_{s,50-\mathrm{yr}}$ are to be determined from onsite measurements, hindcast data, or both for a specific installation site. Ranges of peak spectral periods appropriate to site-specific $H_{s,1-\mathrm{yr}}$ and $H_{s,50-\mathrm{yr}}$ respectively are to be determined for load calculations. The resultant most unfavourable responses for 1-year return conditions and 50-year return conditions, respectively, are to be used in the design of the Floating Offshore Wind Turbine. The Extreme Sea State (ESS) is to be applied in combination with the Extreme Wind Model (EWM) defined in 11/3.5.3.1, the Extreme Current Model (ECM) defined in 11/5.3, and the Extreme Water Level Range Model (EWLR) defined 11/6.3, with consideration of their joint occurrence probabilities as required in 12/3.2.1(8).

4.5 Survival Sea State (SurSS)

The Survival Sea State (SurSS) condition is like the Extreme Sea State (ESS) defined in 11/4.4, but with a return period longer than 50 years, as specified in 12/3.3.

4.6 Breaking Waves

Where breaking waves are likely to occur at an installation site, the loads exerted by those breaking waves are to be assessed in the design. Breaking wave criteria are to be appropriate to the installation site and based on recognised methods. In shallow water, the empirical limit of wave height is approximately 0.78 times the local water depth. In deep water, the theoretical limit of wave steepness prior to breaking is 1/7. Further guidance on breaking wave hydrodynamics can be found in IEC 61400-3-1 (2019), Annex B.

5 Currents

5.1 Currents

Data for currents are to include information on current speed, direction, and variation with depth. The extent of information needed is to be commensurate with the expected severity of current conditions at the site in relation to other load causing phenomena, experience in adjacent or analogous areas, and the type of the Floating Support Structure and the station keeping system to be installed. Onsite data collection may be required for previously unstudied areas or areas expected to have unusual or severe conditions. Consideration is to be given to the following types of current, as appropriate to the installation site:

(1) Wind-generated current.
(2) Tide, density, circulation, and river-outflow generated sub-surface current.
(3) Near shore, breaking wave induced surface currents running parallel to the coast.

The direction of wind generated surface current velocity is assumed to be aligned with the wind direction. Current velocity profiles with depth are to be based on site-specific data or recognised empirical relationships. Unusual profiles due to bottom currents and stratified effects due to river out-flow currents are to be accounted for. For the design of offshore wind turbines in US offshore regions, the current profile is to be determined in accordance with Annex H, Annex I and Annex J of API RP 2MET. The current models for the design of the Floating Offshore Wind Turbine are described below in 11/5.2 through 11/5.4. The design load conditions defined in 12/3 specify how the current models are to be applied in combination with other design environmental conditions.

5.2 Normal Current Model (NCM)

The Normal Current Model (NCM) is to be determined based on the long-term joint probability distribution of metocean parameters at the installation site. For strength analyses, the Normal Current Model (NCM) is defined to represent the site-specific wind-generated current conditioned upon a given 10-min mean wind speed at hub height (i.e., V_{hub}). Tide and storm-generated sub-surface currents are not included. For fatigue analyses, the Normal Current Model (NCM) is to be determined in such a manner that the fatigue damage associated with the full long-term distribution of metocean parameters can be sufficiently accounted for.

5.3 Extreme Current Model (ECM)

The Extreme Current Model (ECM) is defined as the site-specific current with a return period of 1 year or 50 years. The Extreme Current Model (ECM) is to be applied in combination with Extreme Wind Model (EWM) defined in 11/3.5.3.1, the Extreme Sea State (ESS) defined in 10/4.4, and the Extreme Water Level Range Model (EWLR) defined in 11/6.3, with consideration of their joint occurrence probabilities as required in 12/3.2.

5.4 Survival Current Model (SurCM)

The Survival Current Model (SurCM) is like the Extreme Current Model (ECM) defined in 11/5.3, but with a return period longer than 50 years, as specified in 12/3.

6 Tides, Storm Surges, and Water Levels

6.1 General

Tides can be classified as lunar or astronomical tides, wind tides, and pressure differential tides. The combination of the latter two is commonly called the storm surge. The water depth at any location consists of the mean depth, defined as the vertical distance between the sea floor and an appropriate near-surface datum, and a fluctuating component due to astronomical tides and storm surges. Astronomical tide variations are bounded by the highest astronomical tide (HAT) and the lowest astronomical tide (LAT). Storm surge is to be estimated from available statistics or by mathematical storm surge modelling. The still water level (SWL) referred in the air gap criterion is to be taken as the highest still water level (HSWL), which is defined

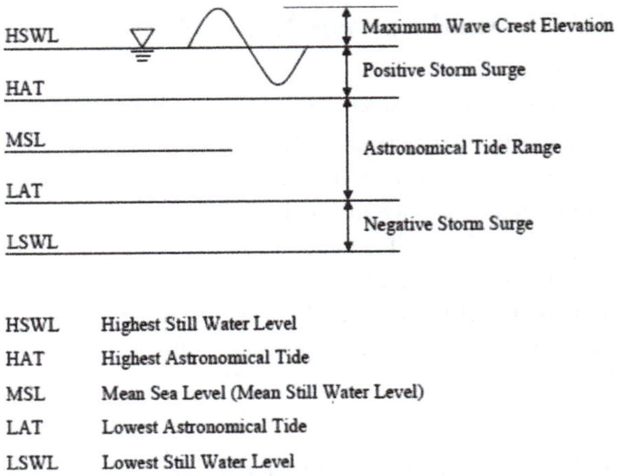

Fig. 2 Definition of water levels

as the sum of the highest astronomical level and the positive storm surge. Definitions of various water levels referred in this book are illustrated in Fig. 2. For the TLP-type floating substructure, variations in the elevation of the daily tide may be used in determining the elevations of boat landings, barge fenders and the top of the splash zone for corrosion protection of structure. Water depths assumed for various types of analysis are to be clearly stated. The water level models for the design of the Floating Offshore Wind Turbine are defined in 11/6.2 through 11/6.4. The design load conditions defined in 12/3 specify how the water level models are to be applied in combination with other design environmental conditions.

6.2 *Normal Water Level Range (NWLR)*

The Normal Water Level Range (NWLR) is defined as the variation in water level with a return period of one year. Load calculations for strength load cases are to be performed based on the water level within the Normal Water Level Range (NWLR) to determine the most unfavourable responses. The influence of water level variation on fatigue loads is also to be considered, if deemed necessary.

6.3 *Extreme Water Level Range (EWLR)*

The Extreme Water Level Range (EWLR) with a return period of 50 years is to be determined. It is to be applied in the Design Load Cases (DLCs, refer to 12/3), where the Extreme Wave Model (EWM) with a return period of 50 years is applied. Load calculations for strength load cases are to be performed based on the water level within the Extreme Water Level Range (EWLR) to determine the most unfavourable responses of the Floating Offshore Wind Turbine. In the absence of the long-term joint probability distribution of metocean parameters including water level, the following water levels are to be considered as a minimum:

- The Mean Sea Level (MSL).
- The highest still water level (HSWL), defined as a combination of highest astronomical tide (HAT) and positive storm surge, with a return period of 50 years.
- The lowest still water level (LSWL), defined as a combination of lowest astronomical tide (LAT) and negative storm surge, with a return period of 50 years.
- The water level associated with the highest breaking wave load, where applicable.

6.4 Survival Water Level Range (SurWLR)

The Survival Water Level Range (SurWLR) is defined in a manner like the Extreme Water Level Range (EWLR) in 11/6.3, but with a return period longer than 50 years, as specified in 12/3.

7 Other Conditions

7.1 Temperature

Extreme values of air, sea and seabed temperatures are to be expressed in terms of return periods and associated highest and lowest values. Wind speed data are typically presented with respect to a specific reference temperature. Temperature data is also to be used to evaluate the selection of air density, structural materials, ambient ranges and conditions for machinery and equipment design, and for determination of thermal stresses, as relevant to the Floating Offshore Wind Turbine.

7.2 Air Density

The air density is to be measured in conjunction with the wind conditions at the installation site. Where there are no site data for the air density, the value of air density is to be determined according to ISO 2533 and corrected as appropriate for annual average temperature at the installation site.

7.3 Ice and Snow Accumulation

For an installation site where ice and snow may accumulate, estimates are to be made of the extent to which ice and snow may accumulate. Data are to be derived from actual field measurements, laboratory data or data from analogous areas.

7.4 Marine Growth

Marine growth is to be considered in the design of the Floating Offshore Wind Turbine. Estimates of the rate and extent of marine growth may be based on experience and available field data. Particular attention is to be paid to increases in hydrodynamic loading due to increased diameters and surface roughness of members caused by marine growth as well as to the added weight and increased inertial mass of submerged structural members, mooring components, and export electrical cables. The types of marine growth likely to occur and their effects on corrosion protection coatings are also to be considered.

7.5 Seismicity and Earthquake Related Phenomena

The effects of earthquakes on the Floating Offshore Wind Turbine with the TLP-type floating substructure to be installed in areas known to be seismically active are to be considered.

7.5.1 Levels of Earthquake Conditions

The magnitudes of the parameters characterising the earthquakes with return periods appropriate to the design life of the Floating Offshore Wind Turbine are to be determined. Two levels of earthquake conditions are to be considered to address the risk of damage and survivability:

(1) ***Strength Level***. Ground motion which has a reasonable likelihood of not being exceeded at the site during the design life of the Floating Offshore Wind Turbine.
(2) ***Ductility Level***. Ground motion for a rare, intense earthquake to be applied to evaluate the risk of structural collapse.

7.5.2 Regional and Site-Specific Data

The anticipated seismicity of an area is, to the extent practicable, to be established based on regional and site-specific data including, as appropriate, the following:

(1) Magnitudes and recurrence intervals of seismic events.
(2) Proximity to active faults.
(3) Type of faulting.
(4) Attenuation of ground motion between the faults and the site.
(5) Subsurface soil conditions.
(6) Records from past seismic events at the site where available, or from analogous sites.

7.5.3 Other Earthquake Related Phenomena

Seismic data are to be used to establish quantitative Strength Level and Ductility Level earthquake criteria describing the earthquake induced ground motion expected during the life of the Floating Offshore Wind Turbine. Both vertical and horizontal accelerations are to be considered for the design of the TLP-type floating substructure and the tendon system. In addition to ground motion, and as applicable to the site in question, the following earthquake related phenomena are to be considered:

(1) Liquefaction of subsurface soils submarine slides.
(2) Tsunamis.
(3) Acoustic overpressure shock waves.

7.6 Sea Ice or Lake Ice

For an installation site where ice hazards may occur, the effects of sea ice or lake ice on the Floating Offshore Wind Turbine are to be considered in the design. Depending on the ice conditions at the site, the Floating Offshore Wind Turbine may encounter moving ice and fast ice cover. Statistical ice data of the site are to be used as the basis for deriving parameters such as ice thickness, ice crushing strength and pack ice concentration, etc., which are required for determining the ice loads. Impact, both centric and eccentric, is to be considered where moving ice may impact the Floating Offshore Wind Turbine. Impact of smaller ice masses, which are accelerated by storm waves, and of large masses (multi-year floes and icebergs) moving under the action of current, wind, and Coriolis effect is to be considered in the design. The interaction between ice and the Floating Offshore Wind Turbine produces mutual responses. This compliance is to be considered as appropriate.

7.7 Soil Conditions

Site investigation in general is to be in accordance with the Class Rules for the design, construction, and classification of offshore installations. Soil data are to be taken in the vicinity of the anchor or tendon foundation site. An interpretation of such data is to be submitted by a recognised geotechnical consultant. To establish the soil characteristics of the site, the foundation system borings or probing are to be taken at all foundation locations to at least the anticipated depth of any pile or anchor penetrations plus a consideration for the soil variability. As an alternative, sub-bottom profile runs may be taken and correlated with at least two borings or probing in the vicinity of anchoring locations and an interpretation may be made by a recognised geotechnical consultant to adequately establish the soil profile at all anchoring locations.

7.8 Lightning

The lightning protection is to be designed in accordance with IEC 61400-24. It is not necessary for protective measures to extend to all parts of the Floating Offshore Wind Turbine if safety is not compromised.

7.9 Electric Network Conditions

The assessment of electric network conditions is to be performed in accordance with IEC 61400-3-1 (2019).

Chapter 12
Loads

1 General

This chapter pertains to the identification, definition, and determination of the loads which the Floating Offshore Wind Turbine may experience in its pre-service (load-out, transportation, installation, and commissioning) and in-service (operation, maintenance, and repair) phases. As appropriate to the planned Floating Offshore Wind Turbine, the types of loads described in Sect. 2 below are to be accounted for in the design.

2 Types of Loads to Be Considered

Loads applied to the Floating Offshore Wind Turbine are categorised as permanent loads, variable loads, and environmental loads.

2.1 Permanent Loads

Permanent loads are loads which do not change during the mode of operation under consideration. Permanent loads include, but are not limited to, the following:

(1) Weight of rotor components (blades, hub, shaft, etc.) and equipment inside the nacelle (control and protection components, gearbox, drive train components, electrical generation components, cables, etc.).
(2) Weight of nacelle housing structure, Floating Support Structure, station keeping systems, fenders, ladders, corrosion protection system, and other permanent structures.

F. Aderouane, *Design, Construction, and Global Performance Analysis of Offshore Wind Turbines*, Green Energy and Technology, https://doi.org/10.1007/978-3-031-86010-2_12

(3) Weight of machinery, equipment and systems permanently installed on the floating substructure.
(4) Weight of permanent ballast, if applicable.
(5) Permanent deformation and loads introduced during fabrication.
(6) External hydrostatic pressure.
(7) Pre-tension in mooring lines.
(8) Static earth pressure.

2.2 Variable Loads

Variable loads associated with the normal operation of a wind turbine are loads which may change during the mode of operation considered. Variable loads acting during the in-service phase include, but are not limited to, the following:

(1) Forces exerted by lifting equipment during maintenance and repair.
(2) Forces exerted on the Floating Support Structure by vessels moored to the structure or routine impact loads from a typical supply vessel that would normally service the offshore wind turbine.
(3) Loads associated with helicopter operations, if applicable.
(4) Loads associated with operations of an adjustable ballast system, if applicable.
(5) Actuation loads generated by wind turbine operations and controls including torque control from a generator or inverter, yaw and pitch actuator loads, and mechanical braking loads. The range of actuator forces is to be considered as appropriate in the calculation of response and loading. In particular, the range of friction, spring force or pressure for mechanical brakes is influenced by temperature and ageing, which are to be accounted for when calculating the response and loading during any brake event.
(6) Deformation loads due to deformation imposed on the Floating Support Structure and the station keeping system. The deformation loads include those due to temperature variations leading to thermal stress in the structure and, where necessary, loads due to soil displacements (e.g., differential settlement or lateral displacement) or due to deformations of adjacent structures. For concrete structures, deformation loads due to prestress, creep, shrinkage, and expansion are to be considered.

Where applicable, the dynamic effects of the variable loads on the Floating Offshore Wind Turbine are to be considered. Variable loads encountered during the pre-service phase are to be determined for each specific operation involved. The dynamic effects of such loads are also to be accounted for, as necessary.

2.3 *Environmental Loads*

Environmental loads are loads due to the action of wind, wave, current, ice, snow, earthquake, and other environmental phenomena as described in Chap. 11. The characteristic parameters defining environmental loads are to be appropriate to the installation site and in accordance with the requirements specified in Chap. 11. The combination and severity of environmental conditions for the design of the Floating Offshore Wind Turbine are specified in Sect. 2. Calculations of environmental loads are described in Sect. 3. Directionality is to be accounted for in the environmental criteria. Unless site-specific assessments provide evidence in support of a less stringent requirement, environmental loads are to be applied from directions producing the most unfavourable global or local effects on the Floating Support Structure or the station keeping system.

3 Design Load Conditions

3.1 *General*

Design load conditions for the design of the Floating Offshore Wind Turbine are to be represented by:

(1) *Design Load Cases (DLCs)*, which are defined to verify the design adequacy of the Floating Offshore Wind Turbine subjected to the combination of turbine operational conditions, site-specific environmental conditions, electrical network conditions and other applicable design conditions, such as specific transportation, assembly, maintenance, or repair conditions.
(2) *Survival Load Cases (SLCs)*, which are defined to verify the survivability of the station keeping system and the adequacy of air gap when the Floating Offshore Wind Turbine is subjected to the environmental conditions that are more severe than the extreme design environmental conditions. The above design load conditions are to be assessed for the design of the floating substructure and the station keeping system. The effect of the RNA and the tower on the floating substructure and the station keeping system is to be considered. Assessment of the tower design for the design load cases defined in this chapter may be required (refer to Chap. 10, Sect. 1).

The specifications of the load cases are given in the following paragraphs.

3.2 Definition of Design Load Cases (DLCs)

3.2.1 General

All relevant DLCs with a reasonable probability of occurrence and covering the most significant conditions that the Floating Offshore Wind Turbine may experience are to be considered in the design. As a minimum, the DLCs defined in Table 1 are to be assessed in the design of the Floating Offshore Wind Turbine. Further requirements for the DLCs are specified in Sect. 3.2.3 for the Floating Support Structure and Sect. 3.2.4 for the station keeping system. The DLCs specified in Table 1 are adapted from "Table 2—Design load cases" of IEC 61400-3-1 (2019) for bottom-founded offshore wind turbines, with various modifications to address unique load and response characteristics of the Floating Offshore Wind Turbine, and "Table 2—Floating Offshore Wind Turbine specific design load cases" of IEC TS 61400-3-2 (2019) for Floating Offshore Wind Turbines. Other design load cases are to be considered, whenever they are deemed relevant to a specific design. If correlation exists between an extreme environmental condition and a fault condition of the wind turbine, a realistic combination of the two is to be considered as an additional design load case.

Consideration is also to be given to the effects of draft and trim variations as well as Vortex Induced Vibration (VIV) and Vortex Induced Motion (VIM) on the Floating Support Structure and the station keeping system. For the Floating Offshore Wind Turbine to be deployed at an ice-infested offshore site, design load cases are to be specified to account for the effects of fast ice formation and moving ice (refer to Chap. 11, Sect. 7.6) on the Floating Support Structure and the station keeping system. For the Floating Offshore Wind Turbine equipped with active control systems, such as actively controlled ballast systems, station keeping systems with active tension adjustment, and sea state monitoring systems assisting in regulating turbine operations, the fault conditions of these systems that could adversely affect the Floating Offshore Wind Turbine are to consider in the definition of the DLCs. The descriptions and analysis requirements for DLCs defined in Table 1, including the amendment to the original DLCs specified in IEC 61400-3-1 (2019) and IEC TS 61400-3-2 (2019), are presented as follows. Further reference is to be made to paragraph 7.4 and paragraph 7.5.4 of IEC 61400-3-1 (2019), and paragraph 7.4 and paragraph 7.5.4 of IEC TS 61400-3-2 (2019). The following descriptions pertain to the required design load cases in Table 1.

(1) The DLC serial numbers are in general in accordance with those specified in IEC 61400-3-1 (2019) and IEC 61400-3-2 (2019).
(2) The DLCs defined in Table 1 are for the design of the Floating Support Structure and the station keeping system of the Floating Offshore Wind Turbine. As such, DLC 1.1 required by IEC 61400-3-1 (2019) for the calculation of the ultimate loads acting on the RNA is not included in Table 1.
(3) For "Power production plus occurrence of fault," DLC 2.5 is defined to address the event of low voltage ride through (LVRT), which is considered as a normal

Table 1 Design load cases

Design condition	DLC	Wind condition	Waves	Wind and wave directionality	Sea currents	Water level	Other conditions	Type of analysis	Safety factor
(1) Power production	1.2	NTM $V_{in} \leq V_{hub} \leq V_{out}$	NSS Joint prob. distribution of H_s, T_p, V_{hub}	MIS, MUL	NCM	NWLR or $\geq$ MSL		F	FDF
	1.3	ETM $V_{in} \leq V_{hub} \leq V_{out}$	NSS $H_s = E[Hs \mid V_{hub}]$	COD, UNI	NCM	MSL		S	N
	1.4	ECD $V_{in} \leq V_{hub} \leq V_{out}$	NSS $H_s = E[H_s \mid V_{hub}]$	MIS, wind direction change	NCM	MSL		S	N
	1.5	EWS $V_{in} \leq V_{hub} \leq V_{out}$	NSS $H_s = E[H_s \mid V_{hub}]$	COD, UNI	NCM	MSL		S	N
	1.6	NTM $V_{in} \leq V_{hub} \leq V_{out}$	NSS $H_s = H_s, SSS(n - yr)$	MIS, MUL	NCM	NWLR	$n = 50$ in general; $n = 100$ for tropical cyclone-prone sites (refer to Note 1 for further guidance)	S	N

(continued)

Table 1 (continued)

Design condition	DLC	Wind condition	Waves	Wind and wave directionality	Sea currents	Water level	Other conditions	Type of analysis	Safety factor
(2) Power production plus occurrence of fault	2.1	NTM $V_{in} \le V_{hub} \le V_{out}$	NSS $H_s = E[Hs \vert V_{hub}]$	COD, UNI	NCM	MSL	Normal control system fault or loss of electrical network or primary layer control function fault	S	N
	2.2	NTM $V_{in} \le V_{hub} \le V_{out}$	NSS $H_s = E[Hs \vert V_{hub}]$	COD, UNI	NCM	MSL	Abnormal control system fault or secondary layer protection function related fault	S	A
	2.3	EOG $V_{hub} = V_r \pm 2m/s(6.6ft/s) and V_{out}$	NSS $H_s = E[Hs \vert V_{hub}]$	COD, UNI	NCM	MSL	External or internal electrical fault including loss of electrical network	S	A
	2.4	NTM $V_{in} \le V_{hub} \le V_{out}$	NSS $H_s = E[Hs \vert V_{hub}]$	COD, UNI	NCM	MSL	Control system fault, electrical fault or loss of electrical network	S	FDF

(continued)

Table 1 (continued)

Design condition	DLC	Wind condition	Waves	Wind and wave directionality	Sea currents	Water level	Other conditions	Type of analysis	Safety factor
	2.5	NTM $V_{in} \leq V_{hub} \leq V_{out}$	NSS $H_s = E[Hs\|V_{hub}]$	COD, UNI	NCM	NWLR or $\geq$ MSL	Low voltage ride through	S	N
(3) Start Up	3.1	NWP $V_{in} \leq V_{hub} \leq V_{out}$	NSS $H_s = E[Hs\|V_{hub}]$	COD, UNI	NCM	NWLR or $\geq$ MSL		F	FDF
	3.2	EOG $V_{hub} = V_r \pm 2m/s(6.6ft/s) and V_{out}$	NSS $H_s = E[Hs\|V_{hub}]$	COD, UNI	NCM	MSL		S	N
	3.3	EOG $V_{hub} = V_r \pm 2m/s(6.6ft/s) and V_{out}$	NSS $H_s = E[Hs\|V_{hub}]$	MIS, wind direction change	NCM	MSL		S	N
(4) Normal Shut Down	4.1	NWP $V_{in} \leq V_{hub} \leq V_{out}$	NSS $H_s = E[Hs\|V_{hub}]$	COD, UNI	NCM	NWLR or $\geq$ MSL		F	FDF
	4.2	EOG $V_{hub} = V_r \pm 2m/s(6.6ft/s) and V_{out}$	NSS $H_s = E[Hs\|V_{hub}]$	COD, UNI	NCM	MSL		S	N
(5) Emergency Stop	5.1	NTM $V_{hub} = V_r \pm 2m/s(6.6ft/s) and V_{out}$	NSS $H_s = E[Hs\|V_{hub}]$	COD, UNI	NCM	MSL		S	N

(continued)

Table 1 (continued)

Design condition	DLC	Wind condition	Waves	Wind and wave directionality	Sea currents	Water level	Other conditions	Type of analysis	Safety factor
(6) Parked (standing still or idling)	6.1	EWM Turbulent Wind Model $V_{hub} = V_{10min,n-yr}$	ESS $H_s = H_{s,n-yr}$	MIS, MUL	ECM *n*-yr Current	EWLR *n*-yr Water Level	$n = 50$ in general; $n = 100$ for tropical cyclone-prone sites (refer to Note 1 for further guidance)	S	N
	6.2	EWM Turbulent Wind Model $V_{hub} = V_{10min,n-yr}$	ESS $H_s = H_{s,n-yr}$	MIS, MUL	ECM *n*-yr Current	EWLR *n*-yr Water Level	Loss of electrical network $n = 50$ in general; $n = 100$ for tropical cyclone-prone sites (refer to Note 1 for further guidance)	S	A
	6.3	EWM Turbulent Wind Model $V_{hub} = V_{10min,n-yr}$	ESS $H_s = H_{s,n-yr}$	MIS, MUL	ECM *n*-yr Current	NWLR	Extreme yaw misalignment $n = 1$	S	N
	6.4	NTM $V_{hub} = V_{10min,n-yr}$	NSS Joint prob. distribution of H_s, T_p, V_{hub}	MIS, MUL	NCM	NWLR or $\geq$ MSL	$n = 1$	F	FDF
(7) Parked and fault conditions	7.1	EWM Turbulent Wind Model $V_{hub} = V_{10min,n-yr}$	ESS $H_s = H_{s,n-yr}$	MIS, MUL	ECM *n*-yr Current	NWLR	$n = 1$	S	A

(continued)

Table 1 (continued)

Design condition	DLC	Wind condition	Waves	Wind and wave directionality	Sea currents	Water level	Other conditions	Type of analysis	Safety factor
	7.2	NTM $V_{hub} = V_{10min,n-yr}$	NSS Joint prob. distribution of H_s, T_p, V_{hub}	MIS, MUL	No current	NWLR or $\geq$ MSL	$n = 1$	F	FDF
(8) Transport, assembly, maintenance and repair	8.1	To be defined by the Fabricator and/or Owner						S	T
	8.2	EWM Turbulent Wind Model $V_{hub} = V_{10min,n-yr}$	ESS $H_s = H_{s,n-yr}$	MIS, MUL	ECM *n*-yr Current	NWLR	$n = 1$	S	A
	8.3	NTM $V_{hub} = V_{10min,n-yr}$	NSS Joint prob. distribution of H_s, T_p, V_{hub}	MIS, MUL	NCM	NWLR or $\geq$ MSL	No grid during installation period $n = 1$	F	FDF
	8.4	To be defined by the Fabricator and/or Owner						F	FDF
(9) Power production	9.1	NTM $V_{in} \leq V_{hub} \leq V_{out}$	SSS $H_s = H_{s,SSS}$	MIS, MUL	NCM	NWLR	Transient condition between intact and redundancy check condition	S	A
	9.2	NTM $V_{in} \leq V_{hub} \leq V_{out}$	SSS $H_s = H_{s,SSS}$	MIS, MUL	NCM	NWLR	Redundancy check condition	S	A
	9.3	NTM $V_{in} \leq V_{hub} \leq V_{out}$	NSS $H_s = E[Hs \mid V_{hub}]$	MIS, MUL	NCM	MSL	Leakage (damage stability)	S	A

(continued)

Table 1 (continued)

Design condition	DLC	Wind condition	Waves	Wind and wave directionality	Sea currents	Water level	Other conditions	Type of analysis	Safety factor
10) Parked (standing still or idling)	10.1	EWM $V_{hub} = V_{10min,50-yr}$	ESS $H_S = H_{S,50-yr}$	MIS, MUL	ECM 50-yr Currents	EWLR 50-yr Water Level	Transient condition between intact and redundancy check condition	S	A
	10.2	EWM $V_{hub} = V_{10min,50-yr}$	ESS $H_S = H_{S,50-yr}$	MIS, MUL	ECM 50-yr Currents	EWLR 50-yr Water Level	Redundancy check condition	S	A
	10.3	EWM $V_{hub} = V_{10min,1-yr}$	ESS $H_S = H_{S,1-yr}$	MIS, MUL	ECM 50-yr Currents	NWLR	Leakage (damage stability)	S	A

Notes (1) The reduction of the return period of 100 years to 50 years in tropical cyclone prone areas may be acceptable provided additional safety factors are applied or robustness assessments for the Survival Load Cases are performed to justify such reduction. (2) The symbols and abbreviations used in the table are summarised as follows: A—abnormal; COD—co-directional (aligned) wind and wave direction; DLC—design load case; ECD—extreme coherent gust with direction change; ECM—extreme current model; EDC—extreme direction change; EOG—extreme operating gust; ESS—extreme sea state; ETM—extreme turbulence model; EWLR—extreme water level range; EWM—extreme wind speed model; EWS—extreme wind shear; F—fatigue; FDF—fatigue design factor; $H_{n\text{-}yr}$—extreme wave height with a return period of n years; H_s—significant wave height; $H_{s,n\text{-}yr}$—significant wave height of the extreme sea state with a return period of n years; $H_{s,SSS}$—significant wave height of the severe sea state with a return period of n years; MIS—misaligned wind and wave directions; MSL—mean sea level; MUL—multi-directional wind and wave; N—normal; n—return period of an extreme environmental condition; NCM—normal current model; NSS—normal sea state; NTM—normal turbulence model; NWLR—normal water level range; NWP—normal wind profile model; S—strength; SSS—severe sea state; T—transport, assembly, maintenance and repair; Tp—peak period of wave spectrum; UNI—uni-directional wind and wave directions; $V_{10min,1\text{-}yr}$—10-min mean wind speed at hub height with a return period of n years; $V_{10min,50\text{-}yr}$—10-min mean wind speed at hub height with a return period of n years; $V_{10min,n\text{-}yr}$—10-min mean wind speed at hub height with a return period of n years; $V_{3sec,n\text{-}yr}$—3-s mean wind speed at hub height with a return period of n years; $V_{(zhub)}$—steady wind speed at hub height, z_{hub}; $V_{1min,n\text{-}yr}$—1-min mean wind speed at hub height with a return period of n years; V_{hub}—10-min mean wind speed at hub height; V_{in}—cut-in wind speed; V_{out}—cut-out wind speed; V_r—rated wind speed; $V_{r\pm 2\ m/s}$—(66 ft/s) sensitivity to the wind speeds in the given range condition. The design low voltage ride through event is to be specified by voltage drop and duration. In addition, DLC 2.6 is defined to address fault of sea state limit protection system which is considered as an abnormal condition.

(4) As a new addition, DLC 4.3 is defined to address the normal shutdown process when the sea state exceeds the maximum operational limit as defined in the Operating Manual. Depending upon the requirement of the Operating Manual, the emergency stop may need to be considered instead of the normal shutdown.

(5) The design environmental conditions referred in Table 1 for wind, wave, current, and water level range are in accordance with the definitions specified in Chap. 11 of this book. Detailed references are listed in the table notes.

(6) Site-specific extreme wind speeds with various return periods are used to define the environmental conditions in DLC 6.1 to 6.4, DLC 7.1 and 7.2, and DLC 8.2 and 8.3 in Table 1. This differs from IEC 61400-3-1 (2019) where definitions of these DLCs are related to RNA's Reference Wind Speed (V_{ref}) and the conversion factors are prescribed for different return periods.

(7) Currents are to be considered in the fatigue design load cases.

(8) The return period chosen for the extreme environmental conditions of DLC 6.1 and DLC 6.2 and for the severe wave conditions of DLC 1.6 is not to be less than 50 years.

(9) DLC 6.2 assumes a loss of connection to electrical power network at an early stage of the storm containing the extreme wind conditions. A nacelle yaw misalignment ranging between –180° and + 180° is required to be considered in DLC 6.2. Load calculations are to be based on the misalignment angle that results in the most unfavourable responses in the Floating Offshore Wind Turbine. The range of yaw misalignment may be reduced to account for the contribution from an active or passive yaw control system, provided that the designer can justify that:

- Such a system can achieve the assumed reduction of yaw misalignment under site-specific conditions.
- An appropriate monitoring and maintenance programme is implemented to maintain the effectiveness of yawing control during the design life of the Floating Offshore Wind Turbine.

(10) For those load cases, including DLC 1.6, DLC 6.1, 6.2, 6.3, 7.1 and 8.2, which in general require time domain dynamic simulations for the combined turbulent wind and stochastic storm waves, the simulation time duration may differ from the reference periods of wind speed and significant wave height. The extreme responses determined for the wave condition used in the simulation are to be equivalent to those obtained from 3-h simulations. IEC 61400-3-1 (2019), paragraph 7.5.6 recommends at least six 1-h constrained wave simulations be performed. Other approaches may be taken if the estimated extreme response is shown not less than that obtained with 1-h realisations. Depending upon the type of the Floating Support Structure and the station keeping system of a specific design as well as site conditions, other simulation time durations may need to be used along with an appropriate adjustment to the wind model and/

or wave model such that the extreme responses can be adequately estimated. Additional requirements of time domain analyses are provided in Chap. 13, Sect. 6.

(11) Where a wind speed range is indicated in Table 1, wind speeds leading to the most unfavourable responses are to be considered for the design of the Floating Offshore Wind Turbine. When the range of wind speeds is represented by a set of discrete values, the interval between two adjacent discrete wind speeds is not to be greater than 2 m/s (6.6 ft/s), In addition, the turbine Rated Wind Speed (V_r), where applicable, is to be included as one of the discrete wind speeds used in the load calculation.

(12) If site-specific directional data are not available, the direction of applied environmental conditions is to be determined to produce the most unfavourable effect on the Floating Offshore Wind Turbine. For DLC 6.1, 6.2, 6.3, 7.1 and 8.2, the misalignment between wind and wave directions is to be considered up to 90° for extreme environmental conditions governed by tropic cyclones. For single point mooring systems such as a turret mooring system, a heading analysis can be performed with consideration of directionality of site-specific wave, wind, and current conditions. When site-specific directional data are not available, the directional combination in the Class guidance relating to position mooring systems is to be considered.

(13) Extreme metocean conditions in a specific load case (e.g., DLC 6.1, 6.2, 6.3, 7.1 and 8.2) are formed by combining the extreme wind (EWM), the extreme wave (ESS), the extreme current (ECM), and the extreme water level range (EWLR). The probability of joint occurrence of these environmental parameters is to be considered when establishing extreme metocean conditions. Consideration is to be given to the peak wind, peak wave, and peak current condition (refer to, e.g., API RP 2MET, 2019, as appropriate to site conditions and a specific design of the Floating Offshore Wind Turbine. Combining all individual extremes with the same return period is normally a conservative approach.

(14) For those DLCs denoted by 'S' in the 'Type of Analysis' column in Table 1 for the structural strength design, the effect of environmental loads is to be combined with the effect of permanent loads and variable loads (refer to Sect. 2). Combinations of the load effects that produce the most severe local and global effects on the Floating Support Structure and/or the station keeping system as determined by the requirements of the pre-service and in-service phases, as well as different nature of structures, are to be used.

(15) The description for DLC 8.x is revised to 'Temporary (Load-out, transportation, installation, maintenance and repair).' Design environmental conditions for transportation (dry-tow) are to be based on a 10-year return period for conditions along the selected transit route unless a weather routing plan is implemented for the voyage. For field transit (wet-tow), environmental conditions specified by the Owner, normally with a return period of 1 year, may be applied. For other 'Temporary' operations, the return period of design environmental conditions may be as specified by the Owner. The Owner is responsible for assuring that operational plans and environmental monitoring for these

temporary phases are compatible with the environmental conditions used in the design.

(16) A new DLC 8.2 is added to account for the fatigue damage occurred during the temporary phase. The serial numbers of the subsequent DLCs are revised accordingly.

(17) Wind and wave directionality for DLC 1.3, 1.5, 1.6 and 8.3 is to consider the misalignment of wind and wave directions. If site-specific directional data are not available, the direction of applied environmental conditions is to produce the most unfavourable effect on the Floating Support Structure.

(18) For power production, additional design load cases DLC 9.1, 9.2 and 9.3 are adapted from IEC TS 61400-3-2 with some modifications. For the parked (standing still or idling) condition, additional design load cases DLC 10.1, 10.2 and 10.3 are adapted from IEC TS 61400-3-2 with some modifications. The NTM wind model is used in combination with severe sea state (SSS) for DLC 9.1 and DLC 9.2. The extreme wind model (EWM) is used in combination with extreme sea state (ESS) with 1-year return period for DLC 10.3 according to damage stability requirements. DLC 9.1 and 10.1 address a transient situation between the intact condition (all mooring lines or tendons are intact) and the condition after the loss of one mooring line or tendon. DLC 9.2 and 10.2 address the situation after one mooring line or tendon breaks and the structure has reached a new mean position. The damaged condition with one broken line is to be assessed in accordance with DLC 9.1, 9.2, 10.1 and 10.2 for the redundant mooring system (refer to 15/2.7). For the TLP tendon made up of steel tubulars, DLC 9.1, 9.2, 10.1 and 10.2 are not applicable. DLC 9.3 and 10.3 address the situation for floating substructure with more than one compartment to evaluate flooding conditions according to damage stability requirements.

(19) For each DLC, if a wind, wave, swell, and current misalignment can lead to higher loading for the Floating Offshore Wind Turbine, this misalignment is to be considered.

(20) For guidance on the suggested number and length of simulations, refer to Part 3.

3.2.2 DLCs for Floating Support Structures

Section 3.2 is to be consulted for the requirements of the DLCs. For each DLC defined in Table 1, the 'Type of Analysis' is denoted 'S' for the strength analysis or 'F' for the fatigue analysis. Results of the strength analysis are used in the structural assessment against acceptance criteria pertaining to the yielding and buckling. Results of the fatigue analysis are used in the structural assessment against criteria pertaining to fatigue performance. Those DLCs indicated with 'S,' are further classified as:

(1) 'N' representing normal design conditions, which are expected to occur frequently during the design life of the Floating Offshore Wind Turbine. The corresponding operational mode of the turbine is in a normal state or with minor faults or abnormalities.

(2) 'A' representing abnormal design conditions, which are less likely to occur than normal design conditions. They usually correspond to design conditions with severe faults that result in activation of system protection functions.
(3) 'T' representing the design conditions relevant to temporary operations including load-out, transportation, installation, maintenance, and repair of the Floating Offshore Wind Turbine.

The type of design conditions, 'N,' 'A,' or 'T' determines the safety factors in the strength design criteria to be applied to the Floating Support Structure. The safety factors referred in Table 1 are specified in Chap. 14, Sect. 3.

3.2.3 DLCs for Station Keeping Systems

Section 3.2 is to be consulted for the requirements of the DLCs. The DLCs specified in Table 1 are to be applied, as a minimum, to the design of the station keeping system. Additional load cases are to be considered, whenever they are deemed relevant to the integrity of the station keeping system. These additional load cases are to include, but are not limited to, consideration of the following conditions, where applicable:

(1) Global yawing moment exerted on the Floating Support Structure due to unbalanced rotor aerodynamic loads caused by the shade effect or the wake effect of neighbouring floating offshore wind turbines.
(2) Vortex Induced Motion (VIM) of the floating substructure due to the site current conditions established in accordance with Chap. 11, Sect. 5.1.
(3) Vortex Induced Vibration (VIV) fatigue due to the site-specific current condition established in accordance with Chap. 11, Sect. 5.1.
(4) Earthquake-induced foundation movements on the design of the tendon system for the TLP-type floating substructure located in seismically active areas (refer to Chap. 11, Sect. 7.5).
(5) Where the tendon is designed to be subject to planned replacement, the temporary removal of one tendon line is to be analysed using DLC 6.1 along with the 5-year return extreme environmental conditions.

The safety factors for the station keeping system design are specified in Chap. 15 of this book. The association between safety factors and the designation 'N,' 'A' and 'T' representing turbine operating conditions, as described in Sect. 3.2.3 for the design of the Floating Support Structures, are not applicable to the station keeping system design.

3.3 Definition of Survival Load Cases (SLCs)

The Survival Load Cases (SLCs) are used for the robustness check of the station keeping system and the air gap (also known as deck clearance or freeboard). As a minimum, the SLCs specified in Table 2 are to be assessed in the robustness check of the Floating Offshore Wind Turbine. The probability of joint occurrence of environmental parameters are to be considered when establishing survival metocean conditions, with the consideration of the peak wind, peak wave, and peak current condition (e.g., API RP 2MET). The effect of environmental loads is to be combined with the effect of permanent loads and variable loads. Combinations of the load effects that produce the most unfavourable effects on the station keeping system or the air gap are to be used to assess the design adequacy. The safety factors applicable for the survival load case are described in Chap. 15. The differentiation of 'N,' 'A' and 'T' turbine operating conditions is not applicable to the survival load cases. For the TLP-type floating substructure, additional robustness checks of its station keeping system are to include.

(1) The assessment of the minimum tendon tension for the Floating Offshore Wind Turbine subjected to the SLCs defined in Table 2.
(2) The strength assessment of the 'one-tendon removed' case, if applicable, for the Floating Offshore Wind Turbine carrying the parked RNA and subjected to a 50-year return extreme environmental condition.

Table 2 Survival load cases

Design condition	Wind condition	Waves	Wind and wave directionality	Sea currents	Water level
Parked RNA; Intact Blades; Intact Hull, and Station Keeping System	SurWM $V_{hub} = V_{10min,n-yr}$	SurSS $H_s = H_{s,n-yr}$	MIS, MUL	SurCM *n*-yr Currents	SurWLR *n*-yr Water Level
Parked RNA; Damaged Blade(s); Intact Hull and Station Keeping System	SurWM $V_{hub} = V_{10min,500-yr}$	SurSS $H_s = H_{s,500-yr}$	MIS, MUL	SurCM 500-yr Currents	SurWLR 500-yr Water Level

Notes (1) 'Parked RNA, Damaged Blade(s); Intact Hull and Station Keeping System' case is to be assessed if one turbine blade or multiple turbine blades cannot remain intact under the storm wind condition with a return period of 500 years (i.e. n < 500 years). (2) The symbols and abbreviations used in the table are summarised as follows: $n - yr$—maximum return period (n years) of the storm wind condition that turbine blades can sustain and remain intact or 500 years, whichever is less; $H_{s,n-yr}$—significant wave height with a return period of n years; $H_{s,500-yr}$—significant wave height with a return period of 500 years; $V_{10min,n-yr}$—10-min mean wind speed at hub height with a return period of n years; $V_{10min,500-yr}$—10-min mean wind speed at hub height with a return period of 500 years; SurWM—survival wind model (Chap. 11, Sect. 3.6.1); SurSS—survival sea state (Chap. 11, Sect. 4.5); SurCM—survival current model (Chap. 11, Sect. 5.4); SurWLR: survival water level range (Chap. 11, Sect. 6.4); (3) Other symbols and abbreviations used in the table are defined in the Notes of Table 11.1 of Chap. 11

(3) The fatigue life assessment of the tendon system, in accordance with API RP 2T, for the Floating Offshore Wind Turbine carrying the parked RNA and subjected to a single extreme environmental event with a return period of 50 years.

4 Determinations of Environmental Loads

4.1 General

Environmental loads are to be determined using analytical methods compatible with the environmental condition models established in compliance with Chap. 11. Recognised load calculation methods are to be employed that are proven sufficiently accurate in practice and are shown to be appropriate to the characteristics of the Floating Offshore Wind Turbine and site conditions. Model or onsite test data can also be employed to establish environmental loads.

4.2 Wind Loads

Wind loads and local wind pressures are to be determined based on analytical methods or wind tunnel tests using a representative model of the Floating Offshore Wind Turbine. Static and dynamic wind load effects generated directly by inflowing wind and indirectly by the wind generated motions of the Floating Offshore Wind Turbine and turbine operations are to be considered.

4.2.1 Aerodynamic Loads Generated by the Rotor

Aerodynamic loads induced by airflow passing through the rotor are determined by the mean wind speed and turbulence across the rotor plane, rotor rotational speed, air density, and aerodynamic shapes of wind turbine components, as well as interactive effects such as aero-elasticity and rotational sampling. Aerodynamic loads due to these effects are to be calculated using recognised methods and computer programmes. The 10-min mean wind speed at hub height, (i.e., V_{hub} as defined in Chap. 11, Sect. 3.2.1), and the wind models defined in Chap. 11, Sect. 3 are used in the definition of design load conditions in Sect. 3. For floating offshore wind turbines installed in a wind farm, the potential shadow effect and wake effect on the loads are to be considered for both the strength and fatigue analyses. For large wind farms, an increase in the turbulence intensity or terrain roughness is to be considered. The mutual influence of wind turbines through the wake interaction behind the rotor is to be considered up to 10 times of rotor diameter. Refer to IEC 61400-1 for the guidance on the wake effects from neighbouring wind turbines.

4.2.2 Rotor Thrust Force

For calculating the overturning moment in the stability analysis (refer to Chap. 16, Sect. 2), the overturning moment due to quasi-static wind load on turbine RNA may be calculated using the thrust coefficient, C_T, determined by the rotor properties, control algorithm and turbine operating conditions. The thrust force generated by wind perpendicular to the swept area of the blades may be estimated by the following equation:

$$F_{thrust} = \left(\frac{\rho_a}{2}\right) C_T A_{thrust} V_{hub}^2$$

where:

F_{thrust} = wind drag load, in N (lbf).
ρ_a = mass density of air, in kg/m^3 (Slug /ft^3).
C_T = thrust coefficient (dimensionless).
A_{thrust} = swept area of the blades, in m^2 (ft^2).
V_{hub} = 10-min mean wind speed at hub height, in m/s (ft/s).

4.2.3 Wind Forces on Exposed Structural Components

For wind pressure normal to flat surfaces, such as nacelle and exposed above-water components of the Floating Support Structure, or normal to the axis of members not having flat surfaces, such as the tower, the wind loading can be considered either as a steady wind force or as a combination of steady and time-varying force, as described below:

(1) When wind force is considered as a constant (steady) force, the wind velocity based on the 1-min mean wind speed is to be used for calculating the wind load.
(2) Effect of the wind spectrum can be considered by considering wind loading as a combination of steady load and a time-varying component calculated from a suitable wind spectrum. API RP 2T provides further guidance on calculating the dynamic component of wind force on exposed structures. For this approach, the wind velocity based on 1-h mean wind speed is used for calculating the steady wind force. The first approach is preferred to this approach when the wind energy spectrum cannot be derived with confidence.

The following relation is to be used for calculating the steady wind force when using the first approach described above:

$$F_{wind} \left(\frac{\rho_a}{2}\right) C_s A_{wind} V_{wind}^2$$

where:

F_{wind} = steady wind force, in N (lbf).

Table 3 Shape coefficients C_s

Shape	***Values of*** C_s
Sphere	0.40
Cylindrical shape	0.50
Major flat surfaces and overall projected area of the Support Structure	1.00
Isolated structural shapes (cranes, angles, beams, channels, etc.)	1.50
Under deck areas (smooth)	1.00
Under-deck areas (exposed beams and girders)	1.30
Derricks or truss cranes (each face)	1.25
Sides of buildings	1.50

ρ_a = mass density of air, in kg/m^3 (Slug/ft^3).
C_s = shape coefficient (dimensionless).
A_{wind} = projected area of windage on a plane normal to the direction of the considered force, in m^2 (ft^2).
V_{wind} = 1-min mean wind speed at a given elevation above the SWL, in m/s (ft/s).

In the absence of experimental data, values in Table 3 for the shape coefficient (C_s) are to be applied. The conversion of mean wind speeds with different averaging time durations and the applicable wind profile are to follow Chap. 11, Sect. 3.2. For any direction of wind approaching to the structure, the wind force on flat surfaces is to be assumed to act normal to the surface. The wind force on cylindrical objects is to be assumed to act in the direction of the wind. The area of open truss works commonly used for derricks and crane booms may be approximated by taking 30% of the projected area of both the windward and leeward sides with the shape coefficient taken in accordance with Table 3. Where one structural member shields another from direct exposure to the wind, shielding may be considered. The two structural components are to be separated by not more than seven times the width of the windward component for a reduction to be taken in the wind load on the leeward member. Cyclic loads due to Vortex Induced Vibration of structural members are to be investigated, where applicable. Both drag and lift components of load due to vortex induced vibration are to be considered. The effects of wind loading on structural members or components that are not normally exposed to wind loads after installation are to be considered, especially during load-out or transportation operations.

4.3 Wave Loads

The wave forces acting on the Floating Offshore Wind Turbine consist of three primary components, [i.e., first order forces at wave frequencies, second order forces at frequencies lower than the wave frequencies and steady components of the second order forces (also known as mean drift forces)]. The mean and oscillatory low frequency drift forces may be determined by model tests or using hydrodynamic computer programmes benchmarked against model test results or other data. For the TLP-type floating substructure, high-frequency wave loads also need to be considered as they may excite the floating substructure at its natural periods in heave, roll and pitch with typical natural periods ranging from 1 to 5 s. For slender members that do not significantly alter the incident wave field, semi-empirical formulations, such as Morison's equation, may be used. For calculation of wave forces on structural configurations that significantly alter the incident wave field, appropriate methods which account for both the incident wave force (e.g., Froude-Krylov force) and the forces resulting from wave diffraction are to be used. In general, Morison's equation is applicable to slender members with diameters (or equivalent diameters giving the same cross-sectional areas parallel to the flow) less than 20% of the wave lengths.

For the floating substructure consisting of both large (columns and pontoons) members and slender members (bracing members), a combination of diffraction and Morison's equation can be used to calculate hydrodynamic characteristics and hydrodynamic loading. Alternatively, suitable model test results or full-scale measurements can be used. For installation sites where the ratio of water depth to wavelength is less than 0.25, nonlinear effects of wave action are to be considered. This may be fulfilled by modifying linear diffraction theory to account for nonlinear effects or by performing model tests. Wave force calculations are to consider shallow water effects which increase the current due to blockage effects, change the system natural frequency due to nonlinear behaviour of moorings, and alter wave kinematics. Wave slamming loads are to be considered for structural members such as pontoons, columns, braces, and members forming the underside of the topside deck structure that are subjected to wave slamming during transport and operation. Breaking wave slamming loads are also to be considered, if applicable. Guidance on the breaking wave hydrodynamics and the slamming loads exerted by a breaking wave on a cylindrical member is given in IEC 61400-3-1 (2019), Annex C. The FPI Rules provides guidance on the calculation of local slamming pressure for scantling design. The effect of wave slamming loads, if relevant, is to be considered in the design of the station keeping system. Green water effects are to be considered, as appropriate, for the strength of affected structures and for the stability analysis as described in Chap. 16.

4.4 Current Loads

Current induced loads on submerged hull, mooring or tendon lines, export electrical cables or any other submerged objects associated with the Floating Offshore Wind Turbine are to be determined based on analytical methods, model test data or full-scale measurements. When currents and waves are superimposed, the current velocity is to be added vectorially to the wave induced water particle velocity prior to computation of the total force. Current profiles used in the design are to be representative of the expected conditions at the installation site and determined in accordance with Chap. 11, Sect. 5. Current force, $F_{current}$, on the submerged part of any structure is calculated as the drag force by the following equation:

$$F_{current} = \frac{1}{2}\rho_{water} C_D A_{current} u_c |u_c|$$

where:

$F_{current}$ = current force, in N (lbf).
ρ_{water} = mass density of water, kg/m^3 (Slug/ft^3).
C_D = drag coefficient in steady flow (dimensionless).
$A_{current}$ = projected area exposed to current, in m^2 (ft^2).
u_c = current velocity vector normal to the plane of projected area, in m/s (ft/s).

For long cylindrical members with large length-diameter ratios, lift force may become significant and is to be considered in the design. Vortex Induced Vibration (VIV) is a resonant response caused by vortex shedding at resonant frequencies in current. Effects of VIV on fatigue and increased drag loads are to be assessed for structural members and for the station keeping system, where applicable. Dynamic amplification due to vortex shedding is to be considered if deemed necessary. Vortex shedding may excite large resonant hull motions, particularly for a single column hull structure such as a Spar and a single column TLP, as well as a multicolumn hull structure with deep draft. These motions are commonly termed Vortex Induced Motions (VIM). Hull VIM, where applicable, and its effects on the strength and fatigue of the station keeping system and the export electrical cable are to be assessed.

4.5 Ice and Snow Accumulation Induced Loads

At locations where the Floating Offshore Wind Turbine is subjected to ice and snow accumulation, increased weight and change in effective area of structural members due to accumulated ice and snow are to be considered. Particular attention is to be paid to increases in aerodynamic and hydrodynamic loading due to the change in size and surface roughness of both non-rotating and rotating parts caused by ice and snow accumulation. Effect of ice and snow accumulation is also to be considered in the stability analysis if deemed necessary.

4.6 *Earthquake Loads*

For the Floating Offshore Wind Turbine supported by the tendon system and located in seismically active areas, the Strength Level and Ductility Level earthquake induced ground motions (refer to Chap. 11, Sect. 7.5) are to be determined based on seismic data applicable to the installation site. Reference is made to API RP 2T for the guidance on designing the TLP-type Floating Offshore Wind Turbine for earthquake loading. Earthquake ground motions are to be described by either applicable ground motion records or response spectra consistent with the return period appropriate to the design life of the Floating Offshore Wind Turbine. Available standardised spectra applicable to the region of the installation site are acceptable provided such spectra reflect site-specific conditions affecting frequency content, energy distribution, and duration. These conditions include:

- The type of active faults in the region.
- The proximity of the site to the potential source faults.
- The attenuation or amplification of ground motion between the faults and the site.
- The soil conditions at the site.

As appropriate, effects of soil liquefaction, shear failure of soft mud and loads due to acceleration of the hydrodynamic added mass by the earthquake, submarine slide, tsunamis and earthquake generated acoustic shock waves are to be considered.

4.7 *Marine Growth*

The following effects of anticipated marine growth are to be accounted for in the design.

- Increase in hydrodynamic diameter.
- Increase in surface roughness used in the determination of hydrodynamic coefficients (e.g., lift, drag and inertia coefficients).
- Increase in permanent load and inertial mass.

The amount of accumulation assumed for a design is to reflect the extent of and interval between cleaning of submerged structural parts.

4.8 *Ice Loads*

Ice loads acting on the Floating Offshore Wind Turbine may include both static and dynamic loads. Static loads can be generated by ice features lodged against the structure or by temperature fluctuations. Dynamic loads are normally caused by moving ice interactions with the Floating Support Structure and the station keeping system. The global forces exerted by ice on the Floating Offshore Wind Turbine as a whole and local concentrated load on structural elements are to be considered. The effects of rubble piles on the development of larger areas and their forces on the Floating Support Structure are to be considered. Further design considerations can be found in ISO 19906 and API RP 2N.

Chapter 13
Global Performance Analysis

1 General

Global performance analyses are to determine the global effects of environmental conditions and other loads on the Floating Offshore Wind Turbine and its components including the tower, floating substructure, mooring lines or tendons, anchors, export electrical cable, etc. Global performance analyses are to be carried out for all critical conditions in the pre-service and in-service phases, represented by the design load conditions specified in Chap. 12, Sect. 3. Global performance analyses are intended to determine the following parameters:

(1) Motions of the floating substructure in six degrees of freedom.
(2) Mooring line (or tendon) and export electrical cable tensions, including the maximum and minimum tensions, and fatigue loads for the component design.
(3) Tower base loads for the floating substructure analysis.
(4) Tower top accelerations for the RNA design or selection.
(5) Critical global forces and moments, or equivalent design wave heights and periods as appropriate, for the hull structural analysis.
(6) Hull hydrodynamic pressure loads for hull structural analysis.
(7) Accelerations for the determination of inertia loads.
(8) Air gap (also known as deck clearance or freeboard).
(9) Separation of resonance peaks, if required.

Detailed guidance on global performance analyses for Floating Offshore Wind Turbines is provided in Part 3. Common practice in global performance analyses for floating offshore oil and gas production installations, as summarised in the FPI Rules as well as applicable recognised standards such as those published by API and ISO, may be adapted for application to the Floating Offshore Wind Turbine, provided that the distinctive load and response characteristics of the Floating Offshore Wind Turbine are account for. The subjects that calling for special considerations include, but are not limited to, the following:

F. Aderouane, *Design, Construction, and Global Performance Analysis of Offshore Wind Turbines*, Green Energy and Technology, https://doi.org/10.1007/978-3-031-86010-2_13

(1) Metocean data that have a sufficient coverage on joint occurrence of wind, wave, current and water level conditions, with a special attention to wind conditions required by turbine load analyses.
(2) Dynamic interactions among the RNA, the tower, the floating substructure, the station keeping system and the power cable.
(3) Actions of turbine's safety and control systems.
(4) Time scale difference between wind speeds (normally 10 min or 1 h) and storm waves (normally 3 h).
(5) Simulation time duration that is sufficient to capture statistics of responses.
(6) Number of realisations (random seeds) that can achieve statistical convergence for the global responses of the Floating Offshore Wind Turbine subjected to turbulent wind and irregular wave loading simultaneously.

2 Wave-Induced Motion Responses

The wave-induced response of the Floating Offshore Wind Turbine normally consists of three categories of response, i.e., first order (wave frequency) motions, low frequency or slowly varying motions and steady drift. For the TLP-type floating substructure, high-frequency heave, roll and pitch motions could also be excited.

(1) ***First Order Motions***. These motions have six degrees of freedom (surge, sway, heave, roll, pitch, and yaw) and are at wave frequencies. They can be obtained from model tests in regular or random waves or by numerical analysis in the frequency or time domain.
(2) ***Low Frequency Motions***. These motions are induced by low frequency components of second order wave forces. The low frequency motions can be substantial, particularly at frequencies near the natural frequency of the Floating Offshore Wind Turbine. The low frequency motion-induced load in the mooring lines or tendons may act as a dominating design load for the station keeping system. The low frequency motions are to be calculated using appropriate motion analysis software or by model tests.
(3) ***Steady (Mean) Drift***. The Floating Offshore Wind Turbine in waves experiences a steady drift along with the first and second order motions. The mean wave drift force and yawing moment are induced by the steady component of the second order wave forces. Mean drift forces and yawing moments are to be calculated using appropriate motion analysis programmes or model tests.
(4) ***High-Frequency Responses***. High-frequency responses of the TLP-type floating substructure are significantly affected by various nonlinear excitation mechanisms. The relevant design guidance can be found in API RP 2T.

3 Critical Global Loads and Responses

Stochastic methods for global performance analyses of the Floating Offshore Wind Turbine provide a rational approach to calculate the global responses to irregular waves and turbulent winds. The results of global performance analyses are either in the format of response spectra in the frequency domain or the response time series in the time domain. For the structural design, the complexity of the Floating Support Structure makes either the spectral-based or the time domain stress analyses an incredibly challenging process. Furthermore, spectral-based stochastic stress analyses are not able to establish the correlation between external loads and internal forces/moments and stress distributions, making it difficult to optimise the structural design. To address these issues and provide an engineering viable design method, the "design wave approach" is commonly adopted by the offshore oil and gas industry for the strength analysis of floating offshore structures. The merits of stochastic methods can be retained when the design waves are derived using the extreme stochastic values of pre-determined critical response parameters, which are compatible with a specific structural configuration. The procedures of development of the stochastic equivalent design wave cases for structural analysis are described in the MOU Rules as well as API RP 2T.

4 Air Gap (Deck Clearance or Freeboard)

Unless topside deck structures, equipment on deck, and the tower, whenever relevant, are satisfactorily considered for direct passage of waves and wave impact, reasonable clearance between the wave crest and the structures for which wave forces are not considered in the design is to be established for all afloat modes of operation. A minimum air gap of 1.5 m (5 ft) is to be provided between the 50-year return maximum wave crest elevation above the highest still water level (HSWL) (refer to Chap. 11, Sect. 5.1) and the lowest edge of the floating substructure for which wave forces are not included in the design. Consideration is to be given to the effect of wave run-up and motions of the floating substructure. Local wave crest elevation is to be account for as appropriate. The requirement of air gap is to be checked for the design load case DLC 1.6, DLC 6.1, and DLC 6.2, as well as DLC 10.1, DLC 10.2 and DLC 10.3 specified in Chap. 12, Sect. 3.2. Under the survival load cases, as specified in Chap. 12, Sect. 3.3, the air gap is not to be smaller than zero. The air gap criterion is also to be checked at various locations on the underside of topside deck. If the air gap criterion for the survival load cases is not satisfied, the anticipated local and global wave forces (including slamming) are to be suitably considered in the design of the station keeping system. The air gap is normally determined by an appropriate model test. Alternatively, the air gap can also be calculated using a detailed global performance analysis that accounts for relative motions between the floating substructure and waves. The following items are to be considered in the determination of the air gap:

(1) Motions of the floating substructure in six degrees of freedom.
(2) Restraints provided by the station keeping system.
(3) Nonlinearity of wave profile.
(4) Wave diffraction and run-up.
(5) Tide and water level effects, if applicable.
(6) Various environmental headings.
(7) Draft of the floating substructure.

A general description of air gap analysis can be found in Part 3. Relevant guidance on air gap and wave impact analysis is provided in the applicable Class guidance on air gap and wave impact analysis for semi-submersibles.

5 Model Testing

Model testing provides an independent check of system responses under simulated environmental conditions. It is also used for deriving some of the design parameters, such as the air gap and nonlinear effects, particularly for an innovative design. Model testing and numerical analyses are not to replace but are to complement each other. The primary objectives of model testing are listed below:

- To determine the responses of a particular design, such as to calibrate low-frequency and high-frequency damping coefficients.
- To verify analysis tools for prediction of system responses or to correlate the analysis results.
- To derive some design information as a substitute for numerical analysis.

Appropriate environmental conditions are to be selected in the model testing. Consideration is to be given to the model scaling for the Floating Offshore Wind Turbine where both hydrodynamic and aerodynamic load effects need to be accounted for. Additional guidance on hydrodynamic model tests for floating offshore structures can be found in API RP 2T and ISO 19904-1.

6 Analysis Methodologies

6.1 General

Because significant interactions could occur among the turbine RNA, the tower, the floating substructure, the station keeping system and, if relevant, the power cable, an integrated model (also termed "a coupled model") including these components is recommended to be used for global performance analyses. An alternative method, where the dynamic analyses of the station keeping system or the export electrical cable are performed separately by using the motions of the floating substructure as boundary conditions, may also be acceptable, provided the coupling effects are adequately accounted for. Refer to the Part 3 for further guidance. It is to be demonstrated that the global performance analysis software used in the design has the capability of appropriately considering interactions among aerodynamic loads, hydrodynamic loads, actions of turbine safety and control systems and structural dynamic responses of the Floating Offshore Wind Turbine. The analysis procedure is to be developed with consideration of the application limit of the selected software. Both industry-recognised software and in-house software may be used for the analyses. However, in-house software is to be adequately validated against model tests or industry-recognised software.

6.2 Frequency Domain Analyses

Frequency domain analyses solve the equations of motion using methods of harmonic analysis or methods of Laplace and Fourier transformations. To evaluate the wave-frequency responses of the Floating Offshore Wind Turbine, linear wave theory is usually employed in the wave frequency analysis. Alternative methods may be applied to evaluate the effects of finite amplitude waves. The low frequency motion analysis is to be carried out to evaluate the effects caused by wind dynamics and wave drift forces. The damping levels used in the analyses are to be properly determined and documented. For the TLP-type floating substructure, where second-order sum-frequency effects are determined to be significant, the high frequency springing responses of the floating substructure and tendons are to be evaluated. Frequency domain analyses for aerodynamic responses of turbine RNA and effects of turbine control systems are to be properly formulated. Preferably, combined aerodynamic, hydrodynamic and control system actions in the frequency domain are used in the calculation of the dynamic responses of the Floating Offshore Wind Turbine. Frequency-domain analyses, by nature, cannot capture nonlinear dynamic interactions among the components of the Floating Offshore Wind Turbine. They are also unable to account for transient responses as well as nonlinear aerodynamic and hydrodynamic load effects. Because of these limitations, most of the currently available simulation software for floating offshore wind turbines is based on the time domain analysis approach as described in Sect. 6.3. Frequency domain analyses are normally performed to calculate the hydrodynamic coefficients which are used as input to time domain analyses.

6.3 Time Domain Analyses

The time domain analysis procedure consists of a numerical solution of the equations of motion for the Floating Offshore Wind Turbine subjected to external forces exerted by environmental conditions and the operation of the wind turbine. Time series of wind and wave conditions are generated for simulating turbulent wind conditions and stochastic wave elevations and kinematics. Time domain analyses are the preferable approach for global performance analysis of the Floating Offshore Wind Turbine, primarily because they can provide a rational means of modelling the nonlinear and transient effects in global responses of a floating offshore wind turbine. These nonlinear effects include hydrodynamic drag forces, finite wave amplitude effects, nonlinear restoring forces from moorings, and effects of motion suppression devices or components (e.g., heave plates). When an integrated (coupled) model is used for global performance analyses, coupling effects among responses of the turbine RNA, the tower, the floating substructure, the station keeping system, and the export electrical cable can be accounted for at each incremental analysis time step. A more realistic simulation of the effects of a turbine control system and turbine's operating conditions can also be achieved using this approach. In time domain analyses, the most probable maximum responses are to be predicted using appropriate distribution curves fitted to the simulation results or other recognised statistical techniques. Time domain analyses are to be carried out for a sufficiently long time to achieve stationary statistics, particularly for low frequency responses. Multiple realisations of the same conditions may be necessary to generate adequate data for statistical analysis and to verify consistency of the simulation. The designer is to demonstrate the adequacy of the selected simulation time duration and the number of realisations. For the TLP-type floating substructure, the ringing (the high frequency vertical vibration excited by impulsive loading) and springing (the high frequency vertical vibration excited by cyclic loading at or near the resonant periods) responses of the floating substructure and the tendon are to be considered as appropriate. Further guidance on high frequency ringing and spring analyses can be found in API RP 2T. The effect of Vortex Induced Motions (VIM, refer to Chap. 12, Sect. 4.3) for the floating substructure having Spar or single column TLP floating substructures or other deep-draft floating substructures are to be accounted for as appropriate.

Chapter 14
Design of Floating Substructures

1 General

This chapter provides general requirements to be applied in the design of the Floating Support Structure. The criteria in 14/3 and 14/4 dealing specifically with the steel and concrete structural design are to be applied along with the requirements specified in this chapter. Design documentation of structures to be submitted for review is specified in 8/4. General design considerations for the structures are provided in Chap. 10. The design criteria in this chapter are to be applied to the design of the floating substructure. Compliance of the tower design with the requirements in this chapter may be required (refer to Chap. 10).

2 Format of Design Approach

The Floating Support Structure can be considered as having two parts. The first is the floating substructure, which in this book can be one of four types (i.e. Column-Stabilised, TLP, Spar-type, or Barge. There may also be a topside structure atop the hull. If the topside structure is integrated with the hull, the hull and the topside structure are considered together. If the topside is non-integral with the hull, it can be designed independent of the hull although any significant interaction between the hull and the topside is to be suitably considered. The second part of the Floating Support Structure is the tower. The topside structure atop the hull, the hull interface structure to the power cable, and structural components supporting the tower on the hull are considered part of the floating substructure in this book. The design requirements for the steel hull and its steel topside structure are given in 14/3.4.2, which is based on the Working Stress Design (WSD) formatted criteria, or 14/3.4.6, which provides the Load and Resistance Factor Design (LRFD) formatted criteria. The design requirements for the concrete components of the floating substructure

F. Aderouane, *Design, Construction, and Global Performance Analysis of Offshore Wind Turbines*, Green Energy and Technology, https://doi.org/10.1007/978-3-031-86010-2_14

are specified in 14/4, which is based on the LRFD formatted criteria. In addition to the design approaches outlined in this book, suitable considerations are to be given to the serviceability of the structure relative to excessive deflection, vibration and, in the case of concrete, cracking.

2.1 Design Load Conditions

Load combinations are to reflect the design load conditions as specified in Chap. 12. With reference to Chap. 12, suitable stochastic methods are to be used to establish design loads. The influence of the less severe environmental loads in combination with the turbine operational loads is to be investigated for their potential to produce maximum peak load effects in the Floating Support Structure. Dynamic analyses are to be performed to assess effects of environmental or other types of loads. When a fatigue analysis is performed, a long-term distribution of stress range, with proper consideration of dynamic effects, is to be determined for relevant loadings anticipated during the pre-service and in-service phases of the Floating Offshore Wind Turbine.

2.2 Design Life

The design life of the Floating Support Structure, which constitutes part of the Floating Offshore Wind Turbine, is to be in accordance with 10/3.

2.3 Long-term and Secondary Effects

Consideration is to be given to the following effects, as appropriate to the planned Floating Support Structure:

(1) Local vibration due to machinery, equipment, and vortex shedding.
(2) Stress concentrations at critical joints.
(3) Secondary stresses induced by large deflection.
(4) Cumulative fatigue.
(5) Corrosion.
(6) Abrasion due to ice.
(7) Freeze–thaw action on coatings and the concrete.

2.4 *Zones of Exposure*

Measures taken to mitigate the effects of corrosion are to be specified and described by the following definitions for corrosion protection zones.

(1) ***Submerged Zone***. That part of the Floating Support Structure below the Splash Zone.
(2) ***Splash Zone***. That part of the Floating Support Structure. Characteristically, the Splash Zone is not easily accessible for field painting, nor protected by cathodic protection.
(3) ***Atmospheric Zone***. That part of the Floating Support Structure above the Splash Zone.

Additionally, for the Floating Offshore Wind Turbine located in areas subjected to floating or submerged ice, the portion of the Floating Support Structure expected to encounter floating or submerged ice is to be designed with consideration for such contact.

3 Steel Structures

3.1 *General*

3.1.1 Materials

The requirements of this chapter are specified for structures constructed of steel, which is manufactured and has properties in accordance with Chaps. 9 and 10/10. Where it is intended to use steel or other materials having properties differing from those of Chap. 9, their applicability will be considered by Class upon reviewing the specifications of alternative materials and the proposed methods of fabrication.

3.1.2 Corrosion Protection

Materials are to be protected from corrosion using a corrosion protection system such as coatings. The system is to be effective from the time the floating substructure is initially placed on site. Where the sea environment contains unusual contaminants, any special corrosive effects of such contaminants are to be considered. For the design of protection systems, reference is to be made to the publications from NACE International: SP0176 and SP0108, or other recognised standards.

3.1.3 Access for Inspection

In the design of the floating substructure, consideration is to be given to providing access for inspection during construction and, to the extent practicable, for survey after construction. Any openings on the floating substructure for the purpose of providing access to the structure are to be evaluated to verify there is no adverse effect on the integrity and buoyancy of the structure.

3.2 Scantling Design of the Hull Structures

The initial scantling design of hull structural members, including plating, stiffeners, girders, brackets, etc., is to be in accordance with the FPI Rules. The aspects that are not covered by the FPI Rules are to be based on recognised standards. The obtained initial scantlings are to be considered as minimum values. The application of the design criteria specified in the remainder of this chapter cannot be used to justify reductions in these scantlings. However, the scantlings are to be suitably increased where required by the criteria specified in the remainder of this chapter. Pontoons, columns, tanks, and bracing members may be considered either as framed or unframed shells. Ring girders, bulkheads, or other suitable diaphragms are to be adequate to maintain shape and stiffness under all anticipated loadings in association with established analysis methods.

3.3 Structural Strength and Fatigue Analysis

The steel structure of the floating substructure is to be designed and analysed for the loads to which it is likely to be exposed during the pre-service and in-service phases. Loads to be investigated are to include at least those relating to both realistic operating and environmental conditions combined with permanent and variable loads (refer to Chap. 12) that are appropriate to the functions and operations of the Floating Offshore Wind Turbine. Loading conditions are to be in accordance with 12/3 for the strength and fatigue analysis of the floating substructure. The general guidance on choosing appropriate approaches to perform structural analyses for the floating substructure is provided in the FPI Rules. The designer is to verify that the adopted structural analysis method is suitable for specific structural behaviours and can lead to accurate analysis results. Structural responses obtained from the structural analysis are to be checked against the design criteria specified in 14/3.7 for the strength (yielding and buckling) assessment or 14/3.7.1.4 for the fatigue assessment.

3.4 *Strength Design Criteria*

3.4.1 General

The strength design criteria specified in this chapter are to be applied to those design load cases relevant to the strength assessment. A minimum set of Design Load Cases (DLCs) for strength assessment is defined in Table 12.1, where the 'Type of Analysis' column is denoted 'S.'

3.4.2 Working Stress Design (WSD) Approach

When the strength design of the floating substructure is based on the WSD approach, the design acceptance criteria are expressed in terms of appropriate allowable stresses. Linear, elastic methods can be employed in structural analyses provided proper measures are taken to prevent general and local buckling failure. The loading conditions for the strength analysis are to be in accordance with 14/3.3. A factor of 1.0 applies to all categories of load. When a variable load is considered a favourable load that relieves total load responses, the minimum value of this variable load is to be used in the load combination. The safety factors specified in this chapter are to be applied in conjunction with the normal (N) and abnormal (A) design conditions as well as the temporary design conditions (T) for load-out, transportation, installation, maintenance, and repair of the Floating Offshore Wind Turbine, as defined in 12/3.2 and Table 12.1.

3.4.3 Individual Stresses in Structural Members

Individual stress components or direct combinations of such stresses in a structural member are not to exceed the allowable stress as obtained from the following equation:

$$F_{allowable} = \frac{F_y}{C_{SF}}$$

where

$F_{allowable}$ = allowable stress
F_y = specified minimum yield strength
C_{SF} = safety factor

- For the normal design conditions (designated 'N' in the column entitled ' Safety Factor' in Table 12.1)

= 1.5 for axial or bending stress
= 2.5 for shear stress

- For the abnormal design conditions (designated 'A' in the column entitled 'Safety Factor' in Table 12.1)

= 1.25 for axial or bending stress
= 2.0 for shear stress

- For the temporary design conditions (designated 'T' in the column entitled 'Safety Factor' in Table 12.1)

= 1.67 for axial or bending stress
= 2.75 for shear stress

3.4.4 Buckling Strength of Structural Members Subjected to a Single Action

Buckling is to be considered for a structural element subjected to compressive axial load or bending moment. The computed compressive or bending stress is not to exceed the allowable stress as obtained from the following equation:

$$F_{allowable} = \frac{F_y}{C_{SF}}$$

where

$F_{allowable}$ = allowable stress
F_{CR} = critical buckling strength of a structural member subjected to axial compression or critical bending strength of a structural member subjected to bending moment, as defined in the applicable Class guidance for performing buckling and ultimate strength assessment for offshore structures
C_{SF} = safety factor

- For the normal design conditions (designated 'N' in the column entitled 'Safety Factor' in Table 12.1)

=1.5/Ψ

- For the abnormal design conditions (designated 'A' in the column entitled 'Safety Factor' in Table 12.1)

=1.25/Ψ

- For the temporary design conditions (designated 'T' in the column entitled 'Safety Factor' in Table 12.1)

=1.67/Ψ

Ψ = adjustment factor, as defined in the Class guidance for performing buckling and ultimate strength assessment for offshore structures.

3.4.5 Structural Members Subjected to Combined Axial Load and Bending

Structural members subjected to axial tension or compression in combination with bending are to be designed according to the Class guidance for performing buckling and ultimate strength assessment for offshore structures in conjunction with the utilisation factors, which are the reciprocals of the corresponding safety factors specified in 14/3.4.2.

Allowable Stress of Plated Structures

For plated structures where the equivalent stress is determined using the von Mises equivalent stress criterion, the equivalent stress is not to exceed the allowable stress as obtained from the following equation:

$$F_{allowable} = \frac{F_y}{C_{SF}}$$

where

$F_{allowable}$ = allowable stress
F_y = specified minimum yield strength
C_{SF} = safety factor

- For the normal design conditions (designated 'N' in the column entitled 'Safety Factor' in Table 12.1) = 1.33
- For the abnormal design conditions (designated 'A' in the column entitled 'Safety Factor' in Table 12.1) = 1.11
- For the temporary design conditions (designated 'T' in the column entitled 'Safety Factor' in Table 12.1) = 1.5

Buckling Strength of Plated Structures

The buckling strength of plated structures is to be designed according to the Class guidance for performing buckling and ultimate strength assessment for offshore structures in conjunction with the utilisation factors, which are the reciprocals of the corresponding safety factors specified in 14/3.2.

3.4.6 Load and Resistance Factor Design (LRFD) Approach

In lieu of the WSD approach described in 14/3.4.2, the design of steel structures can also be based on the LRFD approach. The loading conditions for the strength analysis are to be in accordance with 14/2.1. The partial safety factors (γf) for loads,

Table 1 Partial safety factors (γf) for environmental loads

Normal (N)	Abnormal (A)	Transport, assembly, maintenance and repair (T)
1.35	1.1	1.5

Note
The design conditions represented by '*N*,' '*A*' and '*T*' are defined in 12/3.2 and Table 12.1

as specified in Table 1, are to be applied to the environmental loads. A partial safety factor of 1.0 is to be applied to permanent loads and variable loads, where they are combined with the design environmental loads. However, where a permanent load or a variable load is considered as a favourable load that relieves total load responses, a partial safety factor of 0.9 is to be applied to this load. Where a variable load is considered a favourable load, the minimum value of this variable load is to be used in the load combination.

Alternatively, the partial load factors specified in IEC 61400-3-2 (2019) can be applied to the load effects determined by the dynamic analysis of the floating offshore wind turbine using unfactored loads. The resistance factors are to be determined in accordance with ISO 19904-1 and the strength capacities are to be determined in accordance with the Class guidance for load and resistance factor design (LRFD) criteria for offshore structures or equivalent. Where the resistance concerns bolted connections and fillet and partial penetration welds, the minimum resistance factor of 1.30 is to be applied.

3.5 Fatigue Assessment

The fatigue assessment is to be performed to verify adequate capacity against fatigue failure within the design life of the Floating Offshore Wind Turbine. The fatigue assessment is to be performed for structural members and connections where fatigue is a probable mode of failure, or for which experience is insufficient to justify safety from possible cumulative fatigue damage. Emphasis is to be given to structural members and connections that are difficult to inspect and repair once the Floating Offshore Wind Turbine is in service and to those areas susceptible to corrosion-accelerated fatigue. Guidance on the fatigue analysis of the floating substructures is provided in Appendix 6. For structural members and connections that require a detailed assessment of cumulative fatigue damage, the calculated fatigue life is not to be less than the design life of the Floating Offshore Wind Turbine times the safety factors for fatigue life (i.e., fatigue design factors (FDFs), as listed in Table 2). The fatigue resistance of structural details is to be evaluated in accordance with the Class Rules relating to fatigue assessment of offshore structures. Suitable S–N curves are to be selected, with special attention being given to the application and limitations of those curves. The loading conditions for the fatigue assessment are to be in accordance with 14/3.3. A minimum set of Design Load Cases (DLCs) for fatigue

Table 2 Safety factors for fatigue life of structures (fatigue design factors [FDF])

Importance	Inspectable and repairable	
	Yes	No
Non-Critical	2	3
Critical	3	5

Notes

(1) 'Critical' indicates that failure of these structural items would result in the rapid loss of structural integrity and produce an event of unacceptable consequence

(2) FDF = 1.0 may be applicable to Inspectable and repairable non-critical structural members above the splash zone

(3) For critical structures of the topside structure that is non-integral with the hull structure and the turbine tower above the splash zone, FDF = 2.0 may be applied, provided these structures can be inspected during the periodical survey or when structural damage is suspected such that critical crack development can be detected and repaired

assessment is specified in Table 12.1, where 'F' in the column titled 'Type of Analysis' designates the fatigue assessment. Loading history of the floating substructure during pre-service and in-service phases is to be accounted for. Fatigue analyses are to be carried out using an appropriate loading spectrum or time series in accordance with the accepted theories in calculating accumulated damage. In the case that the design is based on the LRFD approach, the load factors for all load categories are to be taken as 1.0 in the fatigue assessment.

3.6 Analysis and Design of Structures

3.6.1 Primary Structures

General

Structural analyses are to be performed to assess yielding, buckling and fatigue of the hull, the integrated topside deck structure, the turbine tower if classed, and main intersections of hull primary structural components to the topside deck structure and the turbine tower. The analyses are to verify the adequacy of scantlings established in the basic design as described in 14/3.2, but they are not to be used to reduce these scantlings. Reference is made to the FPI Rules for general procedures and methodologies of structural analysis and design for the floating substructure.

Global Strength Analysis

The primary structural components of the hull integrated topside deck structure and turbine tower if classed are to be analysed using the design load conditions specified in 14/2.1. The analyses are to be performed using recognised calculation methods which are to be fully documented and referenced. The global strength of the floating substructure is to be designed to resist maximum global effects induced by the critical global loads/responses described in 13/3. The type of critical global loads/responses depends on the specific configuration of the floating substructure.

Major Joint Analysis—Analysis for Main Intersections of Primary Structures

When details of main intersections are not adequately captured in a global strength model, local FEM analyses are to be used to design these areas. For the Column-Stabilised or TLP-type floating substructure, the main intersections include connections of pontoon to pontoon, column to pontoon, column to topside deck structure, and the tower to hull structure. For twin-pontoon Column-Stabilised floating substructures, special attention is to be given to brace connections to braces, columns, pontoons, and topside deck structure. For the Spar-type floating substructure with truss connections, the main intersections include connections of top section to topside deck structure, truss to hull structure, and the tower to hull structure. Connections of tubular structural members are to be designed to provide effective load transmission between joined members, to minimise stress concentration, and to prevent excessive punching shear. Connection details are also to be designed to minimise undue constraints against overall ductile behaviour and to minimise the effects of post-weld shrinkage. Undue concentration of welding is to be avoided. For Barge-type floating substructures, the main intersections include connections of the ring pontoon and the tower to hull structure. For all types of floating substructure, when the tower and its connections to the RNA and/or the floating substructure are within the scope of classification, the interface structures of the tower to the RNA and/or the floating substructure are to be considered in the analysis. Structural analyses are to be performed to assess yielding, buckling, and fatigue of the tower pedestal and foundation (hull structure supporting the pedestal) subjected to static and dynamic loads. Dynamic response of the tower is to be determined in accordance with the requirements of Chap. 13 for global performance analyses and with consideration of the Design Load Cases specified in 12/3.2 for the floating substructure.

Fatigue Assessment

Fatigue assessment of the primary structures mentioned in 14/3.7.1 is to follow 14/3.7.14. Special attention is to be given to the major joints mentioned in 14/3.7.2. Attention is also to be given to the designs of structural notches, cutouts, brackets, toes, and abrupt changes of structural sections where they are prone to fatigue damages.

Non-integrated Deck

The design of non-integrated decks is to adhere to the FPI Rules.

3.6.2 Other Major Structures

General

Depending on specific features of the floating substructure, additional analyses may be required to verify the design of other structural components, such as:

(1) Hull structural interface with the station keeping system (fairlead, chain stopper, and winch foundations).
(2) Hull structural interface with the power cable system (I-tube and J-tube support structure and foundation).
(3) Equipment/machinery support structures and their interface to the topside deck structure or the hull structure.
(4) Topside deck structural interface with deck modules, if relevant.

Structural Interface with Equipment/Machinery Foundations or Deck Modules

The topside deck structure or the hull structure may require reinforcements to resist reaction forces from foundations of equipment/machinery, such as crane, power cable, and deck modules. The reinforcements of the topside deck structure or the hull structure are referred to as backup structures. The forces to be resisted by the backup structures are to be designed for maximum anticipated gravity, functional, and environmental loads together with inertia loads induced by motions of the floating substructure and verified against the design criteria and requirements specified in 14/3.4. If deemed necessary, the fatigue strength is to meet the requirements of 14/3.7.1.4.

Hull Structural Interface with the Station Keeping System

Each individual foundation and backup structure of the fairlead, chain jack, winch, or tendon porch is to be designed for the breaking strength of the mooring line or tendon with a safety factor of 1.25. The foundation and backup structure for multiple fairleads, chain jacks, winches or tendon porches are to be designed for the maximum anticipated mooring or tendon loads and verified against the design criteria and requirements specified in 14/3.4.2. The fatigue life is to satisfy the requirements in 14/3.7.1.4 with consideration of the effects of both local drag and inertia loads on the mooring lines or tendons and the global motions of the floating substructure.

Hull Structural Interface with the Power Cable System

I-tube and J-tube support structure and foundation for the power cable system are to be designed for the maximum anticipated static and dynamic cable loads based on cable dynamic analysis and verified against the design criteria and requirements specified in 14/3.4.2 or 14/3.4.6. The fatigue life is to satisfy the requirements in 14/3.7.1.4 with consideration of the effects of both local drag and inertia loads on I-tube and J-tube support structure, dynamic cable loads, and the global motions of the floating substructure.

Helicopter Deck

The design of the helicopter deck is to comply with the requirements of the MOU Rules.

3.6.3 Local Structures

Structures that do not directly contribute to the overall strength of the floating substructure (i.e., their loss or damage does not impair the structural integrity of the floating substructure) are local structures. Local structures are to be adequate for the nature and magnitude of applied loads. The criteria specified in 14/3.4 are in general applicable to the design of local structural components, except for those structural parts whose primary function is to absorb energy, in which case sufficient ductility is to be demonstrated.

3.6.4 Guards and Rails

Guards and rails are to comply with the requirements of the MOU Rules, for application to the perimeters of the floating substructure. Alternative arrangements, such as a minimum 1.07 m (42 in.) high and two-tier evenly spaced handrail with a kick-board, may be considered by Class, provided they are also acceptable to the Coastal State or other governmental authorities having jurisdiction.

3.6.5 Vortex Shedding Strakes

Vortex shedding strakes may be installed to reduce the VIM effects on the deep-draft hull. Yield, buckling and fatigue strengths of vortex shedding strakes are to be checked with consideration of the effects of local drag and inertia loads as well as the effects of global motions of the hull.

3.6.6 Appurtenances

Main appurtenances attached to the exterior of the hull are to be evaluated with consideration of the effects of local drag and inertia loads together with any appropriate consideration of global action of the floating substructure. The backup structures are also to be designed for the same loads and safety factors, as a minimum.

3.6.7 Temporary Structures

Structures built for temporary use in the pre-service conditions are not subject to Class review. However, the arrangements and details of these structures are to be submitted for reference to verify the adequacy of the local and global strength of the hull and topside deck to support these temporary structures during operation in the pre-service condition. The backup structures are to be designed for the safety factors outlined in 14/3.4 for temporary design conditions.

3.6.8 Protection of Deck Openings

All openings on the deck are to comply with the appropriate Class Rules for marine vessels.

3.7 Structural Connections

Structure connections are to be accomplished by positive, controlled means such as welding, grouting, or other mechanical connectors. Such attachments are to be capable of withstanding the anticipated static and long-term cyclic loadings. Details of mechanical connectors are to be submitted for review. The welding of steel for the floating substructure and the design of welded connections are to follow the requirements of Chap. 9. Requirements for the connection of welded tubular members are described in 14/3.7.2. In the design of grouted connections, consideration is to be given to the use of mechanical shear connectors as their presence increases the strength of the connection and alleviates the effect of long-term grout shrinkage. Adequate clearance in the annulus between an inner and an outer member is to be provided for proper placement of the grout. Reliable means for the introduction of the grout to the annulus are to be provided to achieve complete filling of the annulus and to minimise the possibility of dilution of the grout and the creation of voids in the grout. The particulars of grouting mixtures used in the grouted connections are to be submitted for review. For grout material, reference can be made to Part 1.

For grouted connections subjected to axial loads, general references are to be made to API RP 2A. Special attention is to be paid to the limitation of geometric configuration required by the design criteria in API RP 2A. For grouted connections whose geometries are not covered by the existing design criteria, special consideration is to be given to the effects of reduced confinement on allowable bond stress, and suitable analyses or tests are to be submitted for review. For those grouted connections expected to resist bending moments, their strength is to be assessed by suitable analysis methods or by tests. The assessment results are to be submitted for review. For bolted flange connections, special care is to be taken to verify evenness of contact surface to avoid overstressing of bolts. The design and installation are to be in accordance with recognised standards such as the AISC Steel Construction Manual. Consideration is to be given to friction factors, relaxation, stress corrosion cracking, bolt fatigue, brittle failure, and other factors or combinations that may be present.

4 Concrete Structures

4.1 General

The requirements of this chapter are to be applied to the floating substructure constructed of reinforced and prestressed concrete.

4.1.1 Materials

Unless otherwise specified, the requirements of this chapter are intended for the floating substructure constructed of materials manufactured and having properties as specified in 9/3. Use of materials having properties differing from those specified in 9/3 will be specially considered by Class. Specifications for alternative materials, details of the proposed methods of manufacture and, where available, evidence of satisfactory previous performance, are to be submitted for review. For structural lightweight concrete, the reference is made to ACI 213R, and lightweight aggregates are to conform to the requirements of ASTM C330.

4.1.2 Durability

Materials, concrete mix proportions, construction procedures and quality control are to be chosen to produce satisfactory durability for structures located in a marine environment. Issues to be specifically addressed include chemical deterioration of concrete, corrosion of the reinforcement and hardware, abrasion of concrete, freeze–thaw durability, and fire hazards as they pertain to the zones of exposure defined in 14/2.4. Test mixes are to be prepared and tested early in the design phase to verify that proper values of strength, creep, alkali resistance, etc. can be achieved.

4.1.3 Access for Inspection

The components of the structure are to be designed to enable their inspection during construction and, to the extent practicable, periodic survey after installation.

4.2 Design Method

The criteria specified in this chapter for concrete structures are based on the Load and Resistance Factor Design (LRFD) approach.

4.2.1 Load Magnitude

The magnitude of a design load for a given type of loading k is obtained by multiplying the load, F_k, by the appropriate load factor, c_k (i.e., design load $= c_k F_k$).

4.2.2 Design Strength

In the analysis of sections, the design strength of a given material is obtained by multiplying the material strength, f_k, by the appropriate strength reduction factor, φ (i.e., design strength $= \varphi f_k$). The material strength, f_k, for concrete is the specified compression strength of concrete ($f_c\prime$) after 28 days and for steel is the minimum specified yield strength (f_y). Refer also to 14/4.6.2.

4.2.3 Design Reference

Design considerations for concrete floating substructure not directly addressed in this book are to follow the requirements of the ACI 318 and ACI 357, or equivalent.

4.3 Design Requirements

4.3.1 General

The strength of the concrete floating substructure of a Floating Offshore Wind Turbine is to be such that adequate safety exists against failure of the structure or its components. Among the modes of failure to be considered are the following:

(1) Loss of overall equilibrium.
(2) Failure of critical section.
(3) Instability resulting from large deformation.
(4) Excessive plastic or creep deformation.

The serviceability of the floating substructure is to be assessed. The following items are to be considered in relation to their potential influences on the serviceability of the structure:

(1) Cracking and spalling.
(2) Deformation.
(3) Corrosion of reinforcement or deterioration of concrete.
(4) Vibration.
(5) Leakage.

4.3.2 Load Combinations

Loads that produce the most unfavourable effects on the concrete structure during the pre-service and in-service phases are to be considered. Loads to be investigated are to include at least those relating to both realistic operating and environmental conditions combined with permanent and variable loads, as defined in 12/1, that are appropriate to the functions and operations of the Floating Offshore Wind Turbine. Load combinations are to reflect the design load conditions as specified in 12/3 for the strength and fatigue analysis of the floating substructure. The load combination is to be in accordance with 14/3.4.6 in general. The load categories referred in this chapter (i.e., permanent loads, variable loads, and environmental loads) are defined in 12/2. The partial safety factors for loads are defined in 14/3.4.6. The Design Load Cases (DLCs) in 12/3.2 are to be assessed as a minimum requirement of design load conditions. DLCs for ice conditions are to be considered in accordance with 12/3.2 for the installation site where ice is expected to occur.

4.3.3 Strength Reduction Factors

The strength of a member or a cross-section is to be calculated in accordance with 14/4.6 and it is to be multiplied by the following strength reduction factor, φ:

(1) In the case of bending without axial tension, $\varphi = 0.90$.

(2) In the case of axial compression or axial compression combined with bending:

- For reinforced members with spiral reinforcement, $\varphi = 0.70$.
- For other reinforced members (excluding slabs and shells), $\varphi = 0.65$.

The values given in the above for two types of members may be increased linearly to 0.9 as φP_u decreases from $0.1 f_{c'} A_g$ or φP_b, whichever is smaller, to zero, where:

$f_{c'}$ = specified compression strength of concrete
A_g = gross area of section
P_u = axial design load in compression member
P_b = axial load capacity assuming simultaneous occurrence of the ultimate strain of concrete and yielding of tension steel

- For slabs and shells, $\varphi = 0.70$

(3) In the case of shear and torsion, $\varphi = 0.75$
(4) In the case of bearing on concrete, $\varphi = 0.65$, except for post-tensioning anchorage bearing. For bearing on concrete in post-tension anchorage, $\varphi = 0.85$. Alternatively, the expected strength of concrete members can be determined by using idealised stress–strain curves and material factors (c_M) given in ACI 357R. The material factors applied to the stress–strain curves limit the maximum stress to achieve the desired reliability like using the strength reduction factors given above. The strength reduction factors (φ) and the material factors (c_M) are not to be used simultaneously.

4.3.4 Fatigue

The fatigue strength of the concrete floating substructure of an offshore wind turbine is considered satisfactory if under the unfactored fatigue loads (i.e., $c_k = 1$) the following conditions are satisfied. The fatigue analysis is based on the stress in a critical section. The stress range is the stress range derived from all fatigue load cycles with a 1-year return probability level:

(1) The stress range in reinforcing or prestressing steel does not exceed 138 MPa (20 ksi), or where reinforcement is bent, welded, or spliced, 69 MPa (10 ksi).
(2) There is no membrane tensile stress in concrete and not more than 1.4 MPa (200 psi) flexural tensile stress in concrete.
(3) The stress range in compression in concrete does not exceed $0.5 f_{c'}$ where $f_{c'}$ is the specified compressive strength of concrete.
(4) Where maximum shear exceeds the allowable shear of the concrete alone, and where the cyclic range is more than half the maximum allowable shear in the concrete alone, all shear is taken by reinforcement. In determining the allowable shear of the concrete alone, the influence of permanent compressive stress may be accounted for.

(5) In situations where fatigue stress ranges allow greater latitude than those under the serviceability requirements given in Table 1, the latter condition is to assume precedence.
(6) Bond stress does not exceed 50% of that permitted for static loads. If lap splices of reinforcement or pretensioning anchorage development are subjected to cyclic tensile stresses greater than 50% of the allowable static stress, the lap length or prestressing development length is to be increased by 50%.

Where the above nominal values are exceeded, an in-depth fatigue analysis is to be performed. In such an analysis, the possible reduction of material strength is to be accounted for based on appropriate data (S–N curves) corresponding to the 95th percentile of specimen survival. In this regard, consideration is to be given not only to the effects of fatigue induced by normal stresses, but also to fatigue effects due to shear and bond stresses under unfactored load combinations. Particular attention is to be given to submerged areas subjected to the low-cycle, high-stress components of the loading history. In prestressed members containing unbonded reinforcement, special attention is to be given to the possibility of fatigue in the anchorages or couplers that may be subjected to corrosive action. Fatigue analyses are to be carried out using an appropriate loading spectrum or time series in accordance with the accepted theories in calculating accumulated damage. The fatigue design factors (FDFs) are defined in Table 2, except that the calculated fatigue life is to be at least twice the design life of the Floating Offshore Wind Turbine.

4.4 Serviceability Requirements

4.4.1 Serviceability

The serviceability of the concrete floating substructure is to be checked using stress–strain diagrams, as depicted in Figs. 1 and 2. The strength reduction factor, φ, and partial safety factors for loads, c_k, are to be taken as 1.0. The unfactored ($c_k = 1.0$) load combination of most unfavourable permanent loads, variable loads and the environmental loads is to be applied. Using this method, the reinforcing stresses are to be limited in compliance with Table 3. Additionally, for hollow structural cross-sections, the maximum permissible membrane strain across the walls is not to cause cracking under any combination of unfactored loads. For structures prestressed in one direction only, tensile stresses in reinforcement transverse to the prestressing steel are to be limited so that the strains at the plane of the prestressing steel do not exceed $\frac{D_{ps}}{E_S}$, where D_{ps} is as defined in 7–3/Table 3 and E_S is the modulus of elasticity of reinforcement (refer to 14/4.6.2). Alternative criteria such as those which directly limit crack width may also be considered.

Fig. 1 Idealised stress–strain relation for concrete in uniaxial compression E_c is defined in 12/4.4

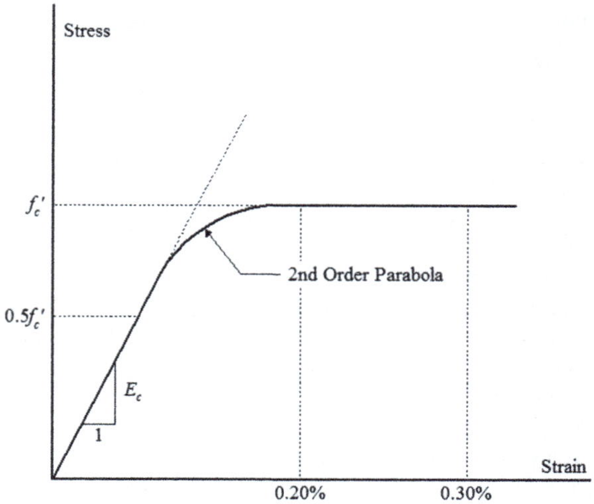

Fig. 2 Idealised stress–strain relation for non-prestressed steel in uniaxial tension $E_s = 200 \times 10^3$ MPa (29×10^3ksi)

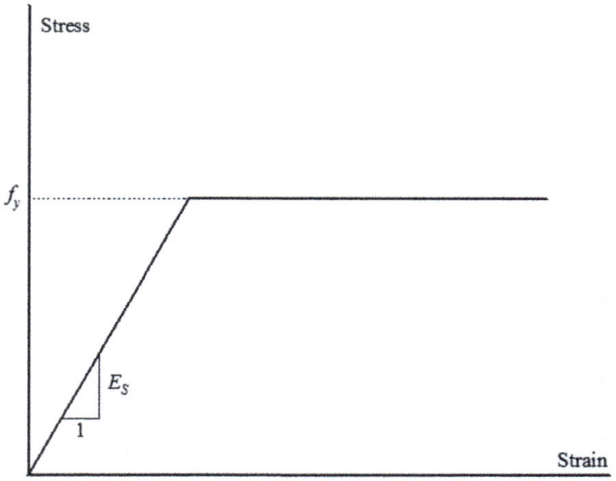

4.4.2 Liquid-Containing Structures

The following criteria are to be satisfied for liquid-containing structures to verify adequate design against leakage:

(1) The reinforcing steel stresses are to be in accordance with 14/4.5.1.
(2) The compression zone is to extend over 25% of the wall thickness or 205 mm (8 in.), whichever is less.
(3) There is no membrane tensile stress unless other construction arrangements are made, such as the use of special barriers to prevent leakage.

Table 3 Allowable tensile stresses for prestress and reinforcing steel to control cracking

Stage	Loading	Allowable stress, MPa (ksi)	
		Reinforcing Steel, f_s	Prestressing Tendons, D_{ps}
Construction: where cracking during construction would be detrimental to the completed structure	All loads on the structure during construction	160 (23.0)	130 (18.5)
Construction: where cracking during construction is not detrimental to the completed structure	All loads on the structure during construction	210 (30.0) or 0.6 f_y, whichever is less	130 (18.5)
Transportation and installation	All loads on the structure during transportation and installation	160 (23.0)	130 (18.5)
At offshore site	Dead and live loads plus design environmental loads	0.8 f_y	

f_y—yield stress of the reinforcing steel

f_s—allowable stress in the reinforcing steel

D_{ps}—increase in tensile stress in prestressing steel with reference to the stress at zero strain in the concrete

4.5 Analysis and Design

4.5.1 General

The analysis of a concrete structure may assume the use of linearly elastic materials and linearly elastic structural behaviour in accordance with the requirements of ACI 318 and the additional requirements of this chapter. The material properties to be used in analysis are to conform to 14/3. However, the inelastic behaviour of concrete based on the true variation of the modulus of elasticity with stress and the geometric nonlinearities, including the effects of initial deviation of the structure from the design geometry, are to be considered whenever their effects reduce the strength of the floating substructure. The beneficial effects of the concrete's nonlinear behaviour may be accounted for in the analysis and design of the structure to resist dynamic loadings.

4.5.2 Material Properties for Structural Analysis

Specified Compressive Strength

The specified compressive strength of concrete, f_c', is to be based on 28-day tests performed in accordance with specifications ASTM C172, ASTM C31 and ASTM C39.

Early Loadings

For structures that are subjected to loadings before the end of the 28-day hardening period of concrete, the compressive strength of concrete is to be taken at the actual age of concrete at the time of loading.

Early Strength-Concrete

For early-strength concrete, the age for the tests for f_c' may be determined based on the cement manufacturer's certificate.

Modulus of Elasticity-Concrete

For the purposes of structural analyses and deflection checks, the modulus of elasticity, E_c, of normal weight concrete may be assumed as equal to 4733(f_c') 0.5 MPa (57(f_c') 0.5 ksi) or determined from stress–strain curves developed by tests (refer to Fig. 3). When the latter method is used, the modulus of elasticity is to be determined using the secant modulus for the stress equal to $0.5f_c'$.

Uniaxial Compression-Concrete

In lieu of tests, the stress–strain relation shown in Fig. 3 may be used for uniaxial compression of concrete.

Poisson Ratio

The Poisson ratio of concrete may be taken equal to 0.20.

Modulus of Elasticity—Reinforcement

The modulus of elasticity, E_s of non-prestressed steel reinforcement is to be taken as 200 × 103 MPa (29 × 103 ksi). The modulus of elasticity of prestressing tendons is to be determined by tests.

Uniaxial Tension—Reinforcement

The stress–strain relation of non-prestressed steel reinforcement in uniaxial tension is to be assumed as shown in Fig. 2. The stress–strain relation of prestressing tendons is to be determined by tests or taken from the manufacturer's certificate.

Yield Strength—Reinforcement

If the specified yield strength, f_y, of non-prestressed reinforcement exceeds 420 MPa (60 ksi), the value of f_y used in the analysis is to be taken as the stress corresponding to a strain of 0.35%.

4.5.3 Analysis of Plates, Shells, and Folded Plates

In all analyses of shell structures, the theory employed in analysis is not to be based solely on membrane or direct stress approaches. The buckling strength of plate and shell structures is to be checked by an analysis that takes account of the geometrical imperfections of the structure, the inelastic behaviour of concrete and the creep deformations of concrete under sustained loading. Special attention is to be given to structures subjected to external pressure and the possibility of their collapse (implosion) by failure of concrete in compression.

4.5.4 Deflection Analysis

Immediate deflections may be determined by the linear structural analysis. For the purposes of deflection analysis, the member stiffness is to be computed using the material properties specified in the design and is to account for the effect of cracks in tension zones of concrete. The effect of creep strain in concrete is to account for the computations of deflections under sustained loadings.

4.5.5 Analysis and Design for Shear and Torsion

The analysis and design of members subjected to shear or torsion or to combined shear and torsion is to adhere with the applicable requirements of ACI 318 or their equivalent.

4.5.6 Analysis and Design for Bending and Axial Loads

Assumed Conditions

The analysis and design of members subjected to bending and axial loads are to be based on the following assumptions:

(1) The strains in steel and concrete are proportional to the distance from the neutral axis.
(2) Tensile strength of the concrete is to be neglected, except in prestressed concrete members under unfactored loads, where the requirements in 14/4.5 apply.
(3) The stress in steel is to be taken as equal to E_S (refer to 14/4.6.2) times the steel strain, but not larger than f_y (refer to 14/4.6.2).
(4) The stresses in the compression zone of concrete are to be assumed to vary with strain according to the curve given in Fig. 1 or any other conservative rule. Rectangular distribution of compressive stresses in concrete specified by ACI 318 may be used.
(5) The maximum strain in concrete at the ultimate state is not to be larger than 0.30%.

Failure

The members in bending are to be designed in such a way that any section yielding of steel occurs prior to compressive failure of concrete.

4.6 Design Details

4.6.1 Concrete Cover

General

The following minimum concrete cover for reinforcing bars is required:

(1) Atmospheric zone not subjected to salt spray: 50 mm (2 in.).
(2) Splash and atmospheric zones subjected to salt spray and exposed to soil: 65 mm (2.5 in.).

(3) Submerged zone: 50 mm (2 in.).
(4) Areas not exposed to weather or soil: 40 mm (1.5 in.).
(5) Cover of stirrups may be 13 mm (0.5 in.) less than covers listed above.

Tendons and Ducts

The concrete cover of prestressing tendons and post-tensioning ducts is to be increased by 25 mm (1 in.) above the values listed in 12/4.5.

Sections Less Than 500 mm (20 in) Thick

In sections less than 500 mm (20 in.) thick, the concrete cover of reinforcing bars and stirrups may be reduced below the values listed in 12/4.5. However, the cover is not to be less than the following:

(1) 1.5 times the nominal aggregate size.
(2) 1.5 times the maximum diameter of reinforcement, or 19 mm (0.75 in.).
(3) Tendons and post-tensioning duct covers are to have 12.5 mm (0.5 in.) added to the above.

4.6.2 Minimum Reinforcement

The minimum requirements of ACI 318 are to be satisfied. In addition, for loadings during all phases of construction, transportation, and operation (including design environmental loading) where tensile stresses occur on a face of the structure, the following minimum reinforcement is to be provided.

$$A_s = \left(\frac{f_t}{f_y}\right) b d_e$$

where

$A_s =$ total cross-section area of reinforcement
f_t = mean tensile strength of concrete
$f_y =$ yield stress of the reinforcing steel
$b =$ width of structural element
d_e = effective tension zone, to be taken as $1.5c + 10d_b$ = cover of reinforcement
d_b = diameter of reinforcement bar

d_e is to be at least 0.2 times the depth of the section, but not greater than $0.5(h-x)$, where x is the depth of the compression zone prior to cracking and h is the section thickness. At intersections between structural elements, where transfer of shear forces is essential to the integrity of the structure, adequate transverse reinforcement is to be provided.

4.6.3 Reinforcement Details

Lapped joints and mechanical splices are to be avoided in structural members subjected to significant fatigue loading. Where lapped splices are used in members subjected to fatigue, the development length of reinforcing bars is to be twice that required by ACI 318, and lapped bars are to be tied with tie wire. Where mechanical splices are used in members subjected to fatigue, the coupled assembly of reinforcing bars and the mechanical coupler are to demonstrate adequate fatigue resistance by test. Where lapped bars are expected to be subjected to tension during operation, through-slab confinement reinforcement is to be considered at the splices. Where longitudinal bars are subjected to tension during operation, special consideration is to be given to number of reinforcements with splices at a single location. Reinforcing steel is to comply with the chemical composition specifications of ACI 359 if welded splices are used. For anchorage of shear and main reinforcement, mechanically-headed bars (T-headed bars) may be used if their effectiveness has been verified by static and dynamic testing. Shear reinforcement is to be full length without splices. Entire close-up stirrups are to be anchored by hooks or bends of at least 90° followed by a straight leg length of a minimum 12 bar diameters.

4.6.4 Post-tensioning Ducts

Ducting for post-tensioning ducts may be rigid steel or plastic (polyethylene or polystyrene). Steel tubing is to have a minimum wall thickness of 1 mm. Plastic tubing is to have a minimum wall thickness of 2 mm. Ducts may also be semi-rigid steel, spirally wrapped, of minimum thickness of 0.75 mm, and is to be grout-tight. All splices in steel tubes and semi-rigid duct are to be sleeved and the joints sealed with heat-shrink tape. Joints in plastic duct are to be fused or sleeved and sealed. The inside diameter of ducts is to be at least 6 mm (0.25 in.) larger than the diameter of the post-tensioning tendon to facilitate grout injection. Flexible ducts are to be used only in special areas where the rigid or semi-rigid duct is impracticable, such as at sharp bends. A mandrel is to be inserted into the ducts to prevent them from dislocating during concreting.

4.6.5 Post-tensioning Anchorages and Couplers

Anchorages for unbonded tendons and couplers are to develop the specified ultimate capacity of the tendons without exceeding the anticipated set. Anchorages for bonded tendons are to develop at least 90% of the specified ultimate capacity of the tendons, when tested in an unbonded condition without exceeding anticipated set. However, 100% of the specified ultimate capacity of the tendons is to be developed after the tendons are bonded in the member. Anchorage and end fittings are to be permanently protected against corrosion. Post-tensioning anchorages are to preferably be recessed in a pocket which is then filled with concrete. The fill is to be mechanically tied to the structure by reinforcements as well as bonded by epoxy or polymer. Anchor fittings for unbonded tendons are to be capable of transferring to the concrete a load equal to the capacity of the tendon under both static and cyclic loading conditions.

4.6.6 Embedded Metals in Concrete

Consideration is to be given to the prevention of corrosion on exposed faces of steel embedment. This embedment is to be separated from the reinforcing steel. Effects of dimensional changes due to factors such as prestressing and temperature changes which may result in fractures near embedment may require provisions to prevent deformation.

4.7 Construction and Detailing

4.7.1 General

Construction methods and workmanship are to follow accepted practices as described in ACI 301, ACI 318, ACI 357 or other relevant standards. Additional requirements relevant to concrete floating substructure are included below.

4.7.2 Mixing, Placing, and Curing of Concrete

Mixing

Mixing of concrete is to conform to the requirements of ACI 318 and ASTM C94.

Cold Weather

In cold weather, concreting in air temperatures below 2 °C (35 °F) is to be carried out only if special precautions are taken to protect the fresh concrete from damage by frost. The temperature of concrete at the time of placing is to be at least 4 °C (40 °F) and the concrete is to be maintained at this or a higher temperature until it has reached a strength of at least 5 MPa (700 psi). Protection and insulation are to be provided to concrete where necessary. The aggregates and water used in the mix are to be free from snow, ice, and frost. The temperature of fresh concrete may be raised by heating the mixing water or the aggregates or both. Cement is never to be heated, nor is it to be allowed to encounter water at a temperature greater than 60 °C (140 °F).

Hot Weather

During hot weather, proper attention is to be given to ingredients, production methods, handling, placing, protection and curing to prevent excessive concrete temperatures or water evaporation which will impair the required strength or serviceability of the member or structure. The temperature of concrete as placed is not to exceed 30 °C (90 °F) and the maximum temperature due to heat of hydration is not to exceed 65 °C (145 °F).

Curing

Special attention is to be paid to the curing of concrete to verify maximum durability and to minimise cracking. Concrete is to be cured with fresh water, whenever possible, to keep the concrete surface wet during hardening. Care is to be taken to avoid the rapid lowering of concrete temperatures (thermal shock) caused by applying cold water to hot concrete surfaces.

Sea Water

Sea water is not to be used for curing reinforced or prestressed concrete, although, if demanded by the construction programme, "young" concrete may be submerged in sea water provided it has gained sufficient strength to withstand physical damage. When there is doubt about the ability to keep concrete surfaces permanently wet for the whole curing period, a heavy-duty membrane curing compound is to be used.

Temperature Rise

The rise of temperature in concrete, caused by the heat of hydration of cement, is to be controlled to prevent steep temperature stress gradients which could cause cracking of concrete. Since the heat of hydration may cause significant expansion, members must be free to contract, so as not to induce excessive cracking. In general, when sections thicker than 610 mm (2 ft) are concreted, the temperature gradients between internal concrete and external ambient conditions are to be kept below 20 °C (68 °F).

Joints

Construction joints are to be made and located in such a way as not to impair the strength and crack resistance of the structure. Where a joint is to be made, the surface of concrete is to be thoroughly cleaned, and all laitance and standing water removed. Vertical joints are to be thoroughly wetted and coated with neat cement grout or equivalent enriched cement paste or epoxy coating immediately before placing of new concrete.

Watertight Joints

Whenever watertight construction joints are required, in addition to the above provisions, the heavy aggregate of existing concrete is to be exposed, and an epoxide-resin bonding compound is to be sprayed on just before concreting. In this case, the neat cement grout can be omitted.

4.7.3 Reinforcement

The reinforcement is to be free from loose rust, grease, oil, deposits of salt or any other material likely to affect the durability or bond of the reinforcement. The specified cover to the reinforcement is to be maintained accurately. Special care is to be taken to correctly position and rigidly hold the reinforcement to prevent displacement during concreting.

4.7.4 Prestressing Tendons, Ducts, and Grouting

General

Further guidance on prestressing steels, sheathing, grouts, and procedures to be used when storing, making up, positioning, tensioning and grouting tendons can be found in the relevant paragraphs of ACI 318, Prestressed Concrete Institute (PCI) publications, *Federation Internationale de la Precontrainte* (FIP) Recommended Practices, and the specialist literature.

Cleanliness

All steel for prestressing tendons is to be clean and free from grease, insoluble oil, deposits of salt or any other material likely to affect the durability or bond of the tendons.

Storage

During storage, prestressing tendons are to be kept clear of the ground and protected from weather, moisture from the ground, sea spray and mist. No welding, flame cutting, or similar operations are to be carried out on or adjacent to prestressing tendons under any circumstances where the temperature of the tendons could be raised, or weld splash could reach them.

Protective Coatings

Where protective wrappings or coatings are used on prestressing tendons, they are to be chemically neutral so as not to produce chemical or electrochemical corrosive attack on the tendons.

Entry of Water

All ducts are to be watertight, and all splices carefully taped to prevent the ingress of water, grout, or concrete. During construction, the ends of ducts are to be capped and sealed to prevent ingress of sea water. Ducts may be protected from excessive rust using chemically neutral protective agents such as vapor phase inhibitor powder.

Grouting

Where ducts are to be grouted, all oil or similar material used for internal protection of the sheathing is to be removed before grouting. However, water-soluble oil used internally in the ducts or on the tendons may be left on, to be removed by the initial portion of the grout.

Air Vents

Air vents are to be provided at all crests in the duct profile. Threaded grout entries, which permit the use of a screwed connector from the grout pump, may be used with advantage.

Procedures

For long vertical tendons, the grout mixes, admixtures, and grouting procedures are to be checked to verify that no water is trapped at the upper end of the tendon due to excessive bleeding or other causes. Suitable admixtures known to have no injurious effects on the metal or concrete may be used for grouting to increase workability and to reduce bleeding and shrinkage. The temperature of members is to be maintained above 10 °C (50 °F) for at least 48 h after grouting. General guidance on grouting can be found in the specialist literature. Holes left by unused ducts or by climbing rods of slipforms are to be grouted in the same manner as described above.

Chapter 15
Design of Stationkeeping Systems

1 General

Regardless of its type, the station keeping system is to be designed to:

(1) Maintain the position of the floating substructure within a specified limit from its reference position.
(2) Control the directional heading of the floating substructure if the orientation is important for safety or operational considerations.
(3) Assist in maintaining the acceleration and the tilting angle at tower top within a specified limit.

2 Designs of Station Keeping Systems

Typically, there are two types of position mooring systems: a conventional spread mooring system and a single point mooring system, as defined in Sects. 2.1 and 2.2. Station keeping for the TLP-type floating substructure relies on the tendon system as defined in Sect. 2.6. Dynamic positioning and thruster-assisted systems are defined in Sect. 2.7. The station keeping system may include mooring lines or tendons, connectors and hardware, winches, piles, anchors, and thrusters. For a single point mooring system, a turret, a turntable, buoys, and anchoring legs may also be part of the system. General design considerations for the station keeping system are described in Chap. 10, Sects. 2 and 8. The design conditions and the design criteria for the station keeping system of the Floating Offshore Wind Turbine are specified in this chapter. Where applicable, additional design requirements for a specific type of the station keeping system are to be in accordance with the following:

(1) Class guidance relating to the application of fibre rope for offshore for the station keeping system using synthetic fibre ropes.
(2) FPI Rules for the single point mooring system.

F. Aderouane, *Design, Construction, and Global Performance Analysis of Offshore Wind Turbines*, Green Energy and Technology, https://doi.org/10.1007/978-3-031-86010-2_15

(3) FPI Rules for the tendon system.

Use of innovative designs of the station keeping system (configuration, material, components, and equipment) that are not covered by this book or other existing industry standards will be specially considered by Class.

2.1 Spread Mooring System

A spread mooring system consists of multiple catenaries, semi-taut or taut mooring lines anchored to piles or drag anchors at the seabed. The other end of each line is individually attached to the hull or attached to winches or stoppers on the floating substructure through fairleads, as necessary. A mooring line may have one or more-line segments, in-line buoy(s) (spring buoy) or sinker(s) (clumped weight) along the line.

2.2 Single Point Mooring (SPM) System

A single point mooring system allows the floating substructure to weathervane. Three typical types of single point mooring systems are described below.

2.3 CALM (Catenary Anchor Leg Mooring)

A catenary anchor leg mooring system consists of a large buoy anchored by catenary mooring lines. The floating substructure is moored to the buoy by soft hawser(s) or a rigid yoke arm, which is a structure that only allows angular relative movement between the floating substructure and the buoy.

2.4 SALM (Single Anchor Leg Mooring)

A single anchor leg mooring system consists of an anchoring structure with built-in buoyancy at or near the water surface and is itself anchored to the sea floor by an articulated connection. The floating substructure is moored to the anchoring structure by soft hawser(s) or a rigid yoke arm.

2.5 *Turret Mooring*

A turret mooring system consists of several mooring legs attached to a turret that is designed to act as part of the floating substructure, allowing only angular movement of the floating substructure relative to the turret so that the floating substructure may weathervane. The turret may be mounted internally within the floating substructure or externally. Typically, a spread mooring arrangement connects the turret to the sea floor.

2.6 *Tendon System*

A tendon system provides a vertical mooring system to the floating substructure by linking the hull to the foundation system. Each tendon consists of a top section for attaching the tendon to the hull-mounted tendon porches, a tendon main body, and a bottom termination assembly for attaching the tendon to the foundation system. The tendon main body is commonly made up of steel tubulars. Any other form of tendons such as solid rods, bars or wire ropes and any other materials such as non-metallic materials and composites that meet the service requirements may also be specially considered by Class. The tendon main body may consist of several tendon elements connected by tendon connectors. Tendon connectors can be mechanical couplings, welded joints or other forms of structural connection that meets the service requirements. The tendons may also have ancillary components such as corrosion protection system components, tendon load and performance monitoring devices and VIV suppression devices. The term “tendon,” as used in this chapter, refers to the main body of the tendon system between the hull mounted porch and the foundation system.

2.7 *Redundant Mooring Systems*

A mooring system is considered as a redundant mooring system where the strength criteria in Sects. 3, 4, and 5, as applicable, for the intact and one-line damaged conditions are satisfied and the offset and motions of the Floating Offshore Wind Turbine are within the allowable design limits.

2.8 *Non-redundant Mooring Systems*

A non-redundant mooring system may be adopted in the design of the Floating Offshore Wind Turbine. The mooring system is non-redundant if, under the damaged condition with one broken line as defined in Sect. 3.2.2, the Floating Offshore Wind Turbine cannot maintain its position required by the Operating Manual or satisfy the strength criteria for the redundant mooring system as specified in this chapter.

3 Design Conditions

This chapter outlines loading conditions and design conditions to be applied in the design of the station keeping system of the Floating Offshore Wind Turbine.

3.1 Loading Conditions

3.1.1 Design Load Cases

The design load cases for the strength and fatigue assessment of the station keeping system are to be in accordance with Chap. 12, Sect. 3.2.

3.1.2 Survival Load Cases

The station keeping system is to be designed to withstand the survival load cases, as specified in Chap. 12, Sect. 3.3, without compromising its intended functions.

3.2 Design Conditions

3.2.1 Intact Condition

Intact Condition is the design condition of the station keeping system, where all components of the system are intact while the Floating Offshore Wind Turbine is subjected to the design load cases.

3.2.2 Damaged Condition with One Broken Line

Damaged Condition is the design condition of the redundant mooring system, where any one of mooring lines or tendons is assumed to have been broken or removed, while the Floating Offshore Wind Turbine is subjected to the design load cases as defined in Chap. 12, Sect. 3.2.4. The floating substructure is assumed to oscillate around a new equilibrium position determined after accounting for the effect of a broken line. Breakage of the mooring line or tendon that sustains the maximum load

Table 1 Safety factors for steel mooring lines or tendons

Loading condition	Redundancy of the station keeping system	Design condition of the station keeping system	Safety factor
Design load cases	Redundant	Intact	1.67
		Damaged condition with one broken line	1.05
		Transient condition with one broken line	1.05
	Non-redundant	Intact	2.0
Survival load cases	Redundant or Nonredundant	Intact	1.05

Notes (1) Safety factors are to be applied to the minimum breaking strength (MBS) of the mooring line or tendon. (2) Additional strength design criteria for tendons made up of steel tubulars are to comply with API RP 2T. (3) Requirements of tendon minimum tension check are to comply with API RP 2T

in the intact condition may not lead to the worst broken line case. The worst scenario is to be determined by analysing several cases of broken line, including lead line broken and adjacent line broken cases. In addition, the damaged condition with one broken line may cause out-of-plane bending (OPB) of the chain link of the remaining mooring lines or tendons. Where relevant, the effect of such bending on the mooring or tendon interface structure on the hull is to be assessed. Damaged Condition does not apply to a non-redundant station keeping system, for which an increased safety factor is required (refer to Table 1). For a system utilising the SALM concept, the damaged condition with one broken line is not relevant. A load case considering loss of buoyancy in a damaged compartment of the SALM structure is to be analysed to assess the integrity of station keeping capability.

3.2.3 Transient Condition with One Broken Line

Transient Condition is the design condition of the station keeping system, where breakage of a mooring line or tendon, usually the lead line, causes the moored floating substructure to exhibit transient motions (also known as overshooting) before it settles at a new equilibrium position. The transient condition could be an important design consideration when proper clearance between the moored floating substructure and nearby structures or facilities is required. Global performance analyses for this transient condition subjected to the design load cases defined in Chap. 12, Sect. 3.2.4 are to be performed. The effect of increased line tensions due to overshooting upon failure of one line are also to be considered. Transient Condition is to be applied to a redundant TLP tendon system made of chains, wire ropes, and fibre ropes, but does not apply to a TLP tendon system made of tubular members. Transient Condition does not apply to the non-redundant station keeping system.

3.2.4 Design Life

The design life of the station keeping system, which constitutes part of the Floating Offshore Wind Turbine, is to be in accordance with Chap. 10, Sect. 3.

4 Steel Mooring and Tendon Systems

4.1 Analysis Methods

The Floating Offshore Wind Turbine is a dynamic system that is subjected to steady forces of wind, current and mean wave drift, as well as wind, current and wave-induced dynamic forces. Calculations of the maximum station keeping system loading are to consider various relative directions of the wind, current and wave forces appropriate to the site conditions. Global performance analyses for the purpose of designing the station keeping system of the Floating Offshore Wind Turbine are to comply with Chap. 13. The dynamic analysis method in either the frequency domain or the time domain, or the combination of both is to be employed to determine characteristics of dynamic responses of the Floating Offshore Wind Turbine. The calculation of steady forces due to wind, current and wave are outlined in Chap. 12, Sect. 4. Fatigue analysis is in general to follow the procedure described in API RP 2T for the tendon system and API RP 2SK for other types of station keeping system. Fatigue life of each mooring line or tendon is to be evaluated.

4.2 Strength Criteria for Steel Mooring Lines and Tendons

The steel mooring line or tendon is to be designed with the safety factors specified in Table 1, which are to be applied to the minimum breaking strength of the mooring line or tendon. The maximum tension of a steel mooring line or tendon is not to exceed its minimum breaking strength divided by an applicable safety factor specified in Table 1. The safety factors in Table 1 apply when the 'partial safety factors' for load effects from all load categories are 1.0. These safety factors are defined for various combinations of

(1) The loading conditions described in Sect. 3.1
(2) The design conditions described in Sect. 3.2.
(3) Redundancy of the station keeping system.

Additional strength design criteria for tendons made up of steel tubulars are to be in accordance with API RP 2T. Requirements of tendon minimum tension check are to be in accordance with API RP 2T. For a TLP tendon system made of chain, tendon slack may be acceptable provided the snap loads are considered in the design. The

Table 2 Fatigue design factors (FDFs) for fatigue life of steel mooring lines or tendons

Redundancy of the station keeping system	Inspectable and repairable	Fatigue design factors
Redundant	Yes	2
	No	5
Non-redundant	Yes	3
	No	10

tendon slack is to be modelled in the global performance analysis. The snap loads are to be considered in the design of chains, connectors, connection to foundation, foundation, and hull interface structure. Allowances for corrosion and abrasion of a mooring line are to be taken into consideration following the recommendations in API RP 2SK.

4.3 *Fatigue Criteria for Steel Mooring Lines and Tendons*

The calculated fatigue life of the mooring line or tendon is not to be less than the design life of the station keeping system times the fatigue design factors (FDFs) listed in Table 2. The fatigue resistance of the mooring line or tendon is to be determined in accordance with the appropriate Class guidance on position mooring systems. Suitable T-N or S–N curves are to be selected, with special attention being given to the application method and limitations of those curves. The load combination for fatigue assessment is described in Sect. 3.1.1 and Chap. 12, Sect. 3.2.4. A minimum set of Design Load Cases (DLCs) for fatigue assessment is specified in Table 1 of Chap. 12, where "F" in the column entitled "Type of Analysis" designates the fatigue assessment. The station keeping system is to be considered intact in global performance analyses for the fatigue design load cases. For the tendon, fatigue damage due to a single extreme environmental event with a return period of 50 years, as described in Chap. 12, Sect. 3.3, is part of the robustness check. The unfactored damage accumulated during this event is to be equal to or less than 0.02. Such fatigue damage is not to be combined with the fatigue damage accumulation incurred by long-term environmental and operational loading.

4.4 *Design of Components and Equipment*

Typical station keeping system components and equipment for the Floating Offshore Wind Turbine may include winches, windlasses, chain, wire, in-line buoys, fairleads, and chain stoppers. The foundation components for the station keeping system may include drag anchors, pile anchors, vertically loaded anchors (VLAs) or suction piles. Gravity boxes, grouted piles, templates, etc., may also be used. Anchor design is addressed in Sect. 6. Other station keeping system components and equipment are to be in accordance with the applicable Class publications and recognised industry standards as listed below:

- Buoyancy Tanks: ASME Boiler and Pressure Vessel Code.
- Chain and Accessories: Class guidance relating to the certification of offshore mooring chain.
- Single Point Mooring Equipment: Class guidance on position mooring systems.
- Winches and Windlasses: Class guidance on position mooring systems.
- Wire Rope: API Spec 9A and API RP 9B.
- Tendon Foundation Components: FPI Rules and API RP 2T.
- Gravity boxes, grouted piles, templates: API RP 2A.
- Clump Weight: API RP 2SK.

The design load cases for the strength and fatigue analysis of the station keeping system components are to be in accordance with Sect. 3.1.1. When clump weights are used as mooring line components to provide desired mooring system restoring properties, the function of the clump weight is to be maintained throughout the design life of the station keeping system. Supporting documents and test reports are to be submitted to Class for review. The connectors, if used to attach the clump weight to the mooring line, are to be designed with the anticipated loads, and the requirements for chain and accessories apply. In general, the design load for the chain stopper, fairlead or tendon porch and its connection to the floating substructure is the breaking strength of the mooring line or tendon. Such design load may be reduced in the case that the mooring line or tendon breaking strength is over-designed due to other design considerations than the strength requirement. Hull structural interface with the station keeping system is to be designed in accordance with Chap. 14, Sect. 3.7.2. For the non-redundant station keeping system, a 20% increase is to be applied to those safety factors of strength design criteria defined for components of the redundant station keeping system under the intact design condition. For the robustness check of strength using the survival load cases as referenced in Sect. 3.1.2, the safety factor is to be at least 1.05. The fatigue life of the chain stopper, fairlead or tendon porch and its connection to the floating substructure, and tendon foundation components is not to be less than the design life of the station keeping system times the fatigue design factors (FDFs) specified in Table 2. The chain stoppers are to be function tested at the specified proof load to the satisfaction of the attending Surveyor.

4.5 Corrosion and Wear

Protection against chain corrosion and wear is normally provided by increasing chain diameters. Current industry practice is to increase the chain diameter by 0.2–0.4 mm per service year in the splash zone and in the dip or thrash zone on the hard seabed. A diameter increase of 0.1–0.2 mm per service year is typically applied to other areas of the chain. In the strength analysis, the chain diameter for determination of required Minimum Breaking Strength (MBS) is not to include corrosion and wear margins. In the fatigue analysis, the chain diameters associated with various service periods within the design life can be established if the corrosion rate is predictable. The chain diameter for a given service period is the nominal diameter minus the expected corrosion and wear at the end of that period. If the corrosion rate is uncertain, a conservative approach using the chain diameter excluding corrosion and wear margins is to be considered in the fatigue analysis. Corrosions of wire rope at connections to sockets could be excessive due to the galvanised wire acting as an anode for adjacent components. It is recommended that either the wire be electrically isolated from the socket, or the socket be isolated from the adjacent component. Additional corrosion protection may be achieved by adding sacrificial anodes to this area. For the steel tendon system, a corrosion protection and control system are to be used. Guidance on design of corrosion protection and control system for steel tendon system can be found in API RP 2T.

5 Synthetic Fibre Ropes

5.1 Analysis Methods

Global performance analyses for the purpose of designing the station keeping system using synthetic fibre ropes as the mooring lines are to comply with Sect. 7. The dynamic analysis method in either the frequency domain or the time domain, or the combination of both is to be employed to account for characteristics of dynamic responses of the Floating Offshore Wind Turbine. The calculation of steady forces due to wind, wave and current are outlined in Chap. 12, Sect. 4. Additional analysis guidance specific to synthetic fibre ropes is provided in the Class guidance relating to the application of fibre rope for offshore mooring.

5.2 Design Criteria

The design load cases for the strength and fatigue analysis of the station keeping system using synthetic fibre ropes as the mooring lines of the Floating Offshore Wind Turbine are to be in accordance with Sect. 3.1.1. The strength design is to satisfy the requirements in the Class guidance relating to the application of fibre rope for offshore mooring. For the non-redundant station keeping system, a 20% increase is to be applied to those safety factors of strength design criteria defined in the Class guidance relating to the application of fibre rope for offshore mooring for the redundant station keeping system under the intact design condition. The fatigue life of synthetic fibre ropes used as the mooring lines subjected to tension-tension cyclic loads is not to be less than the design life of the station keeping system times the fatigue design factors (FDFs) specified in Table 2. Additional design requirements regarding creep and compressive fatigue are to be in accordance with the Class guidance relating to the application of fibre rope for offshore mooring. For the robustness check of the strength of the mooring lines made of synthetic fibre ropes using the survival load cases as specified in Sect. 3.1.2, the safety factor is to be at least 1.05. For a synthetic fibre rope connected with a torque steel wire rope, the torque match is to satisfy the applicable requirement as specified in the Class guidance relating to the application of fibre rope for offshore mooring.

5.3 Design of Components and Equipment

Design requirements in Sect. 4.4 are to be satisfied, as appropriate, for the components and equipment of the station keeping system using synthetic fibre ropes as the mooring lines. Additional guidance specific to synthetic fibre ropes is provided in the Class guidance relating to the application of fibre rope for offshore mooring.

6 Anchor Holding Power

6.1 Drag Anchor

For a catenary mooring system with drag anchors, the mooring line length is to be sufficiently long such that there is in general no angle between the mooring line and the sea floor in any design condition, as described in Sect. 3. For an anchoring site with soft clay condition, a small angle for the damaged condition with one broken line may be considered by Class on a case-by-case basis. Drag anchor holding power depends on the anchor type, as well as the condition of an anchor deployed regarding penetration of the flukes, opening of the flukes, depth of burial, stability of the anchor during dragging, soil behaviour of the flukes, etc. The designer is to submit the performance data for a specific anchor type and site-specific soil conditions for the estimation of the ultimate holding capacity (UHC) of an anchor design. Because of uncertainties and wide variations of anchor characteristics, exact holding power is to be determined after the anchor is deployed and load tested. The anchor load, F_{anchor}, is to be calculated as follows. The maximum anchor load is to be determined by evaluating all design conditions described in Sect. 3. The safety factors for the holding capacity of a drag anchor are specified in Table 2.

$$F_{anchor} = P_{line} - W_{sub}D_{water} - F_{friction}$$

$$F_{friction} = fLW_{sub}$$

Table 3 Coefficient of friction of the mooring line on the sea floor

	Coefficient of friction (f)	
	Static	Sliding
Chain	1.00	0.70
Wire rope	0.60	0.25

where:

F_{anchor} = anchor load, in N (lb).
P_{line} = mooring line tension at any design condition, in N (lb).
W_{sub} = submerged unit weight of the mooring line, in N/m (lb/ft).
D_{water} = water depth, in m (ft).
$F_{friction}$ = holding power of the mooring line on the sea floor, in N (lb).
f = coefficient of friction of the mooring line on the sea floor, (dimensionless).
L = length of the mooring line on the sea floor, not to exceed 20% of the total length of a mooring line, in m (ft).

Note The above equation for F_{anchor} is strictly correct only for a single line having constant Wsub and without buoys or clump weights. Appropriate adjustments are required for other cases.

The coefficient of friction, f, depends on the soil condition and the type of mooring line. For soft mud, sand and clay, the values of f recommended by API RP 2SK for wire ropes and chains are listed in Table 3. The static (starting) friction coefficients are normally used to determine the holding power of the mooring line on the sea floor, while the sliding friction coefficients are normally used to compute the friction force on the mooring line during mooring deployment. Further guidance on design and insulation of drag anchors can be found in the Class guidance on the design and installation of drag anchors and plate anchors.

6.2 *Vertically Loaded Drag Anchor (VLA)*

VLAs can be used in a taut line mooring system with a 35-to-45-degree angle between the sea floor and the mooring line. These anchors are designed to withstand both the vertical and horizontal loads imposed by the mooring line. The structural and geotechnical holding capacity design of the VLA are to be submitted for review. The submitted design documentation is to include the ultimate holding capacity and the anchor's burial depth beneath the sea floor. Additionally, the fatigue analysis of the anchor and the connectors joining the VLA to the mooring line is to be submitted for review. Further guidance on design and insulation of VLA can be found in the Class guidance on the design and installation of drag anchors and plate anchors. The safety factors of VLA anchors' holding capacity are specified in Table 4.

Table 4 Factor of safety for anchor holding capacities

Loading condition	Redundancy of the station keeping system	Anchor type	Design condition of the station keeping system	Safety factor
Design load cases	Redundant	Drag anchors	Intact	1.5
			Damaged condition with one broken line	1.0
		Vertically loaded anchors (VLAs) Suction and dynamically embedded plate anchor	Intact	2.0
			Damaged condition with one broken line	1.5
		Pile anchors	Intact	To be in accordance with API RP 2T for the tendon foundation or API RP2 SK otherwise
			Damaged condition with one broken line	To be in accordance with API RP 2T For the tendon foundation or API RP2 SK otherwise
		Suction Piles	Intact	1.5–2.0* For the tendon foundation, refer to API RP 2 T
			Damaged condition with one broken line	1.2 to 1.5*
	Non-redundant	Any anchor type	Intact	20% increase in the safety factor required for the redundant system using the same type of anchors and under the intact design condition
Survival load cases	Redundant or Nonredundant	Any anchor type	Intact	1.05

Note * The safety factor to be used in the design is to be based on the extent of the geotechnical investigation, confidence in the prediction of soil-pile behaviour, experience in the design and behaviour of suction piles in interest, and the inclination of the mooring line load

6.3 Conventional Pile

Conventional pile anchors are capable of withstanding uplift and lateral forces at the same time. Structural analysis of the pile anchor is to be submitted for review. The analyses for different types of soil using representative soil resistance and deflection $(p - y)$ curves are to follow the Offshore Installations Rules, API RP 2A, API RP 2SK and API RP 2T, as applicable. The fatigue analysis of the pile is to be submitted for review. The safety factors for the holding capacity of a conventional pile anchor are specified in Table 4.

6.4 Suction Pile

Suction pile anchors are caisson foundations that are penetrated to the target depth by pumping out the water inside the pile to create under-pressure within the pile. Suction piles have larger diameters and are shorter in length than conventional piles. They typically consist of a stiffened cylindrical shell with a cover plate at the top and an open bottom. These piles can be designed to have a permanent top or a retrievable top depending on the required vertical holding capacity. The pad eye for the mooring line connection can be at the top or at an intermediate level depending on the application of a suction pile. Suction pile anchors are capable of withstanding uplift and lateral forces. Due to distinctive geometry of suction pile anchors, soil failure modes relevant to the suction pile anchors may be different than those applicable for slender conventional piles. The safety factors for the suction piles' holding capacity are specified in Table 4. Geotechnical holding capacity and structural analyses for the suction piles are to be submitted to demonstrate the adequacy of the suction piles to withstand in-service and installation loads. Fatigue analysis of the suction piles is also to be submitted for review. Additionally, installation analyses are to verify that the suction piles can achieve the design penetration and, if necessary, can be retrieved. It is recommended that the ratio of the force that would cause uplift of the soil-plug inside the pile to the effective pile installation force be at least 1.5 in the penetration analysis. Installation analyses results for the suction pile anchors are to be submitted for review.

6.5 Suction Embedded Plate Anchor

Suction embedded plate anchors and VLAs are both considered as plate anchors, which can be broadly categorised into drag-in and push-in plate anchors. Like VLAs, suction embedded plate anchors can be used in a taut line mooring system with an angle of 35 to 45 degrees between the sea floor and the mooring line. Suction embedded plate anchor's fluke is embedded in a vertical position. An adequate fluke rotation is achieved during the keying process by pulling on the mooring line. Suction embedded plate anchors are designed to withstand both vertical and horizontal loads imposed by the mooring line. The design documentation is to include the ultimate holding capacity and the anchor's burial depth beneath the sea floor. Additionally, strength and fatigue assessment of the anchor and the connectors joining the suction embedded plate anchor to the mooring line are to be performed. Further guidance on design and insulation of suction embedded plate anchors can be found in the Class guidance on the design and installation of drag anchors and plate anchors. The safety factors for suction embedded plate anchors' holding capacity are specified in Table 4.

6.6 Dynamically Embedded Plate Anchor

Like suction embedded plate anchors, dynamically embedded plate anchors are categorised as push-in plate anchors. Like VLAs and suction embedded plate anchors, dynamically embedded plate anchors can be used in a taut line mooring system with an angle of 35–45° between the sea floor and the mooring line. Dynamically embedded plate anchors are designed to withstand both vertical and horizontal loads imposed by the mooring line. The design documentation is to include the ultimate holding capacity and the anchor's burial depth beneath the sea floor. Additionally, strength and fatigue assessment of the anchor and the connectors joining the dynamically embedded plate anchor to the mooring line are to be performed. Further guidance on design and insulation of dynamically embedded plate anchors can be found in the Class guidance on the design and installation of drag anchors and plate anchors. The safety factors for dynamically embedded plate anchors' holding capacity are specified in Table 4.

6.7 Dynamically Installed Pile Anchor

Dynamically installed pile anchors are capable of withstanding uplift and lateral forces at the same time. Strength assessment of the pile anchor structure is to be performed. Fatigue assessment may also be required in particularly for the mooring or tendon line attachment pad eye or lug. The holding capacity of a dynamically installed pile anchor is to be determined using a suitable method, such as finite element method, with consideration of three-dimensional pile-soil interactions. Model test results may be used to augment engineering analyses. When the pile includes fins and/or appendages to increase its holding capacity, an equivalent pile diameter appropriate for the loading direction may be derived for the holding capacity analysis. When the pile reaches its ultimate capacity, the pile axial deformation is in general not to exceed 10% of the pile diameter or, if applicable, the equivalent pile diameter. In addition, the lateral deformation is in general not to exceed 10% of the pile's main body width/diameter. Further guidance on design and installation of dynamically installed pile anchors can be found in the Class guidance on the design and installation of dynamically installed piles. The safety factors for the holding capacity of dynamically installed pile anchors are the same as those for conventional piles as specified in Table 4.

6.8 Factor of Safety

The factors of safety for holding capacity of drag anchors, VLAs, suction and dynamically embedded plate anchors, conventional pile anchors, and suction piles are specified in Table 4, where the factor of safety is defined as the anchor holding capacity divided by the maximum anchor load. The required ultimate holding capacity is to be determined based on mooring line or tendon loads derived from dynamic analyses as described in Sects. 4.1 or 5.1.

7 Field Test

After the mooring system is deployed, each mooring line is in general required to be test loaded in accordance with the FPI Rules. For all types of anchors, the attainment of design-required minimum soil penetration depth is to be verified at the site. The field test requirements for the following anchor are to comply with the FPI Rules:

- Drag anchor including high-efficiency drag anchors.
- Conventional piles.
- Suction piles.
- Plate anchor including Vertically Loaded Anchors (VLAs), and suction and dynamically embedded plate anchor.
- Dynamically installed anchors.

Class will determine the necessity of a maximum intact design tension pull test depending on the extent of the geotechnical investigation, the magnitude of loading, analytical methods used for the geotechnical design and the experience with the soils in interest. If the maximum intact design tension pull tests are waived, preloading of each anchor is to reach such a level that the applied load is sufficient to develop the ultimate holding capacity of the anchor and is not to be less than the mean intact design tension. The integrity and alignment of the mooring line are also to be verified.

Chapter 16
Stability and Watertight/Weathertight Integrity

1 General

The Floating Offshore Wind Turbine is to comply with the requirements of stability and watertight and weathertight integrity specified in this chapter. It is suggested that the Coastal State or other governmental authorities having jurisdiction over the Floating Offshore Wind Turbine to be installed be contacted to obtain any further applicable criteria.

2 Stabilities

The stability (righting stability) requirements specified in this chapter are applicable to the floating substructures with the hull types. Any unconventional stability issues that may be specific to an innovative configuration of the Floating Offshore Wind Turbine are subject to special considerations by Class. The concept of the requirement of the area ratio of the righting and overturning moment curves may be considered for application to the floating substructure having an unconventional hull configuration. The dynamic-response-based intact stability analysis may also be acceptable if model tests are performed to validate the intact analysis results. The floating substructure, which may carry a partial or an entire RNA depending on a specific afloat condition, is to have positive metacentric height (GM) in calm water equilibrium position for all afloat conditions during its pre-service and in-service phases. For determining compliance with the stability requirements contained herein, the floating substructure is to be assumed free of mooring restraints except for TLP-type floating substructures. However, detrimental effects of catenary mooring systems or of the thrusters for dynamically positioned installations are to be considered as appropriate. For TLP-type of floating substructure in the in-service condition, the stability is assessed based on global performance analysis for all DLCs defined in Table 1 of Chap. 12. The

F. Aderouane, *Design, Construction, and Global Performance Analysis of Offshore Wind Turbines*, Green Energy and Technology, https://doi.org/10.1007/978-3-031-86010-2_16

metacentric height is to be specified for each mode of operation and guidance is to be included in the Operating Manual on the procedure to determine and satisfy the expected metacentric height. This may be accomplished by including the minimum metacentric height in the calculation of the allowable distance from the keel to the centre of gravity (KG).

3 Stabilities in Pre-service Phases

3.1 Wet Tow

An intact and damage stability analysis for the wet tow to location is to be performed. The stability criteria are to be submitted for approval by Class. The criteria are to consider the specific aspects of the transit operation, including environmental conditions and Coastal State requirements. Dismissal of damage stability criteria may be considered by Class, provided the operation does not constitute a hazard to life, the environment, and operations of other vessels.

3.2 Installation and Commissioning

During the installation and commissioning operations, the Floating Offshore Wind Turbine is to have a positive metacentric height. All relevant afloat conditions during installation or commissioning operations are to be considered. If the Floating Offshore Wind Turbine is to accommodate personnel during these operations, the stability is to comply with in-service stability specified on Sect. 4 for the case in which the RNA is in the parked condition while the RNA, the tower and the floating substructure are subjected to the extreme storm condition with a return period of 50 years. The installation analysis is to be submitted for review.

4 Stabilities in In-service Phases

For the stability analysis, consideration is to be given to relevant unfavourable effects, including, but not limiting to, those resulting from the following:

(1) Environmental conditions, such as wind, wave (including green water effects, if applicable), current, and snow and ice accumulation.
(2) Applicable damage scenarios (including owner-specified requirements).
(3) Motions of the floating substructure in six degrees of freedom.
(4) Effects of various turbine RNA operation conditions.
(5) Effects of the turbine control system.

(6) Effects of the station keeping system.
(7) Free-surface effects in ballast tanks.

Stability calculations are to reflect the actual configuration of the Floating Offshore Wind Turbine while afloat. Free flooding compartments are not to be accounted for in the stability assessment. In the case where the permanent ballast is installed in a free flooding compartment, the net weight of the permanent ballast (using the in-water density) is to be included in the loading calculation.

4.1 *Intact Stability*

The intact stability of the floating substructure carrying the RNA, and the tower is to satisfy either the intact stability criteria specified in Sect. 4.1, or the dynamic-response-based intact stability criteria specified in Sect. 4.1.1. Other standards may be acceptable for uncommon operations, provided consideration is given to hazards associated with those operations. Such operations and standards are to be submitted to Class for approval.

4.1.1 Intact Stability Criteria

The floating substructure carrying the RNA, and the tower is to have sufficient stability to withstand the overturning moment equivalent to the one produced by any operational and extreme design load conditions, which are represented by the Design Load Cases (DLCs) for the in-service phase and with the 'Type of Analysis' designated as 'S' in Table 3 of Chap. 12. As a minimum, the following design conditions are to be considered in the calculation of the overturning moment.

(1) The RNA is in power production and subjected to the wind speed V_r and V_{out}, , respectively, and the extreme turbulence model (ETM). refer to DLC1.3 in Table 3 of Chap. 12. An example overturning moment curve is depicted in Fig. 1 for the case where the turbine RNA is initially in the power production mode but shuts down after the inclination angle exceeds its operating limit. Where relevant, the effect of shutdown (DLC 4.1, 4.3, and 12/1 in Table 1 of Chap. 12) on the overturning moment curve during the transition from the power production mode to the parked condition is to be included.
(2) The RNA is in the parked condition. The RNA, the tower and the floating substructure are subjected to the extreme storm condition with a return period of 50 years (refer to DLC6.2 in Table 3 of Chap. 12). An example overturning moment curve shown in Fig. 2 represents the case where the RNA is parked and subjected to the storm wind condition.
(3) The RNA is in the parked condition with fault. The RNA, the tower and the floating substructure are subjected to the extreme storm condition with a return

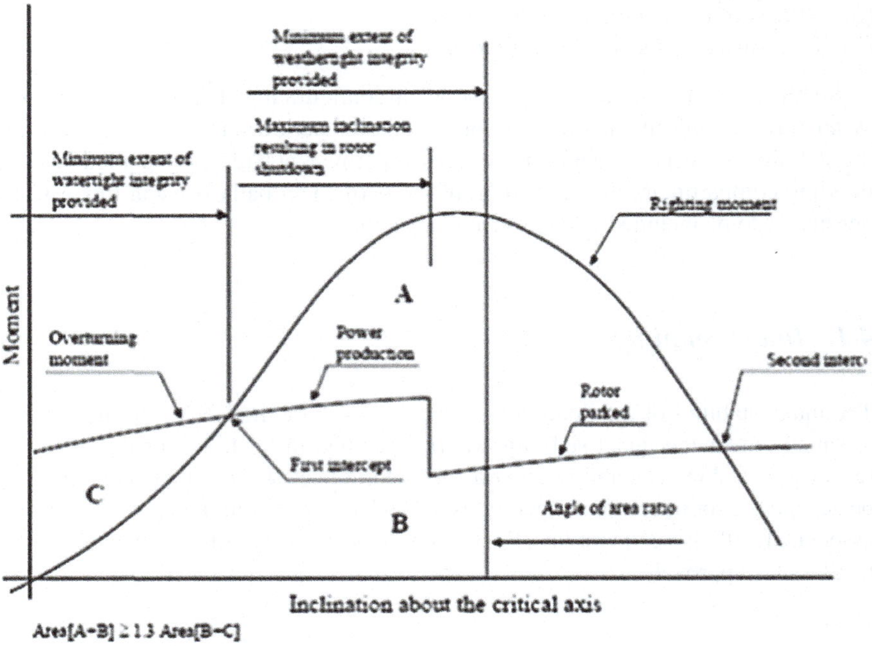

Fig. 1 Intact stability curve for the RNA in power production

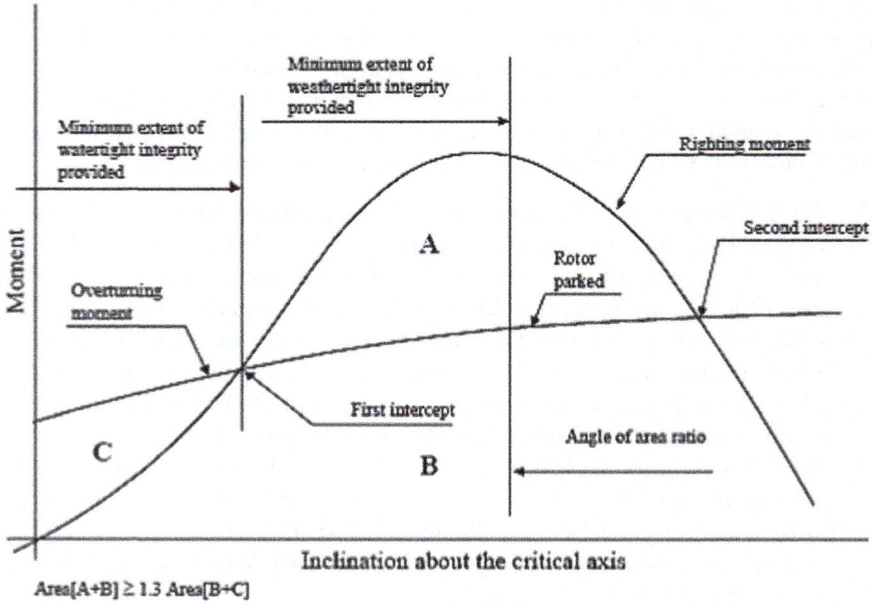

Fig. 2 Intact stability curve for the parked RNA

period of 1 year (refer to DLC7.1 in 5–2/3.5 Table 1). An example overturning moment curve is shown in Fig. 2.

The overturning moment is to be calculated in accordance with Sect. 4.3. For the Spar-type floating substructure, the righting energy (area under the righting moment curve) at the inclination angle of 30° is to reach a value of not less than 30% more than the area under the overturning moment curve to the same limiting angle. In all cases, the righting moment curve is to be positive over the entire range of angles from upright and all down flooding angles are to be greater than 30°. For other types of the floating substructure under free-floating conditions, the righting energy (area under the righting moment curve) at or before the angle of the second intercept of the righting and the overturning moment curves or the down flooding angle, whichever is less, is to reach a value of not less than 30% more than the area under the overturning moment curve to the same limiting angle. In all cases, the righting moment curve is to be positive over the entire range of angles from upright to the second intercept angle. For the TLP-type floating substructure in in-service conditions, the stability is typically provided by the pretension and stiffness of the tendons, rather than by the righting moments. The positive tendon tension is to be maintained. Where applicable, the effect of tendon slacking is to be considered. The intact condition is to include the full range of centre of gravity variations permitted by acceptable operating procedures. Additional considerations, when applicable, are to be given to the inclination limits imposed by the design or operational requirement of the RNA and tower, the global performance requirement, or the design requirement of the station keeping system.

4.1.2 Dynamic-Response-Based Intact Stability Criteria

As an alternative to the intact stability criteria specified in Sect. 4.1, the dynamic-response-based intact stability criteria provides a rational safety margin against capsize and down flooding by incorporating the dynamic motion response characteristics into the stability criteria. The dynamic-response- based criteria described in his Subparagraph are not applicable for the TLP-type floating substructure in in-service conditions, where the stability is typically provided by the pretension and stiffness of the tendons. refer to Sect. 4.1. Dynamic motion analyses are to be conducted for the operational and extreme design load conditions, which are represented by the Design Load Cases (DLCs) for the in-service phase and with the 'Type of Analysis' designated as 'S' in Table 3 of Chap. 12. For each design load case associated with extreme storm conditions, both winds dominated (peak wind) and wave dominated (peak wave) cases, along with their associated other environmental conditions, are to be considered. Dynamic motion analyses for assessing capsize and down flooding in free-floating conditions are in general to follow the requirements specified in Chap. 13, except that the restoring effects of the station keeping systems are to be excluded for the DLCs related to the RNA in normal shutdown, emergency stop or

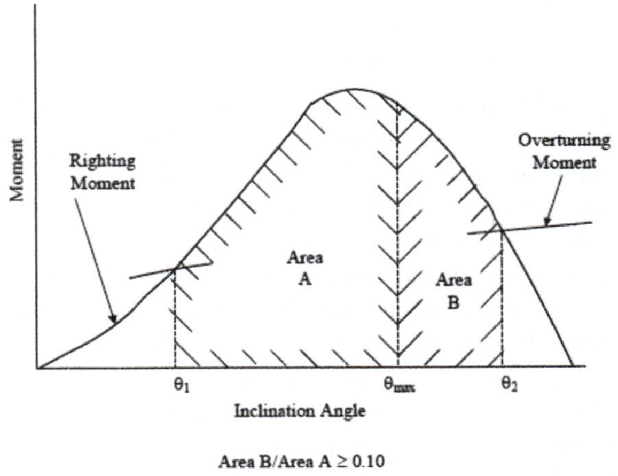

Fig. 3 Capsize criterion

parked conditions. The dynamic responses required as the input for the intact stability assessment include:

- For assessing the capsize criterion, maximum heel angle (θ_{max}) of the floating substructure.
- For assessing the down flooding criterion, maximum reduction in down flooding distance (RDFD) due to tilting of the floating substructure and motions of the floating substructure relative to wave.

Capsize

For all orientation angles, the area under the righting moment curve measured between θ_{max} and θ_2 (Area B as shown in Fig. 3) is not to be less than 10% of the area under the same curve measured between θ_1 and θ_{max} (Area A as shown in Fig. 3), where θ_1 and θ_2 are the inclination angles at the first and the second intercepts of the righting and the overturning moment curves. The overturning moment is to be determined according to Sect. 4.1.

Down Flooding

For all down flooding openings, the maximum reduction in down flooding distance, *RDFD*, is to be not greater than the initial down flooding distance, DFD_0, as depicted in Fig. 4. Down flooding openings are those openings which may be required to remain open, or which are not fitted with, as a minimum, an automatic weathertight closure.

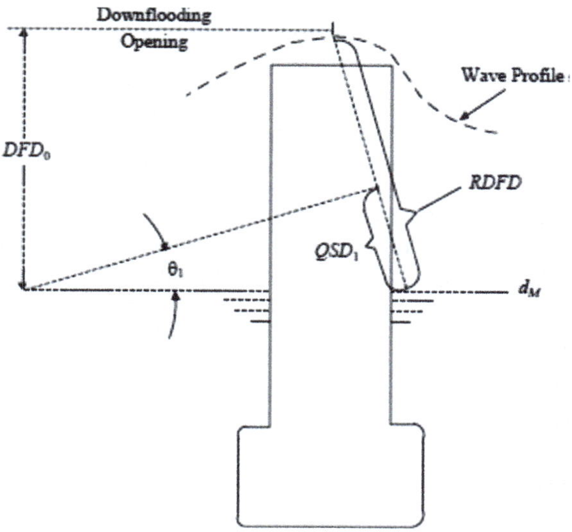

Fig. 4 Down flooding criterion. DFD_0: Down flooding distance to initial waterline. QSD_1: Reduction in down flooding distance due to static heeling. $RDFD$: Maximum reduction in down flooding distance

Model Tests

When comparably model test results are not available, model tests for a specific type of Floating Offshore Wind Turbine carrying a specific design of the RNA are to be performed to calibrate analytical tools and verify simulation results for the purpose of assessing the intact stability based on the dynamic-response-based criteria in Sects. 4.1a, b. The following information of model tests is to be submitted to Class for review:

- Description of model configurations, scaling methods, model RNA control scheme and instrumentation plan.
- Description of input and measured wind and wave spectra.
- Description of turbine operating conditions considered in model tests.
- Response spectra and Response Amplitude Operators (RAOs), presented in tabular form for an appropriate range of periods (or frequencies).
- Model test motion results, including mean, maximum, root mean square and significant values, for six degrees-of-freedom motion and relative motion for at least four reference points on the hull, along with the mooring line tensions corresponding to those motions.

4.2 Damage Stability

Any single watertight compartment, as listed in the following, located wholly or partially below the draft associated with any mode of operation afloat is to be assumed independently flooded, regardless of exposure and source of the assumed flooding:

(1) A compartment containing pumps used for the handling of water ballast.
(2) A compartment containing machinery with a sea water cooling system.
(3) A compartment adjacent to the sea.

For the TLP-type floating substructure, flooding in any one tendon compartment independently is also to be considered. Where access openings are fitted on watertight divisions between compartments to provide access to one or more compartments, the compartments are to be assumed flooded simultaneously, unless the following measures are taken to avoid operations that can result in the accidental flooding of such compartments when the covers are removed:

(1) Warning (or Notice) plates (e.g., "Watertight Door (or Hatch) – Keep Closed") are placed on the access opening covers.
(2) Instructions and warnings are provided in the Operating Manual.
(3) System locks are available to prevent unintentional ballasting operations, where either compartment is designed for use as a ballast tank.
(4) Documented procedures are in place addressing how to avoid progressive flooding and outlining actions necessary to maintain the stability of the floating substructure during repair operations offshore or during transit to a repair facility.

For assessing damage stability, the overturning moments equivalent to that exerted on the floating substructure, the tower, and the RNA in DLC 9.3 and DLC 10.3 are to be considered. The overturning moment is to be calculated according to Sect. 4.3. The final waterline of the damaged floating substructure, which is assumed to sustain the one-compartment damage and subjected to the overturning moment described above for assessing the damage stability, is to:

(1) Not exceed the level to which watertight integrity is to be maintained as shown on the diagrams submitted in accordance with 8/6.
(2) Not exceed the lowest point of the hull main deck or the top of the buoyant hull, whichever is lower.
(3) Be at least 1.5 m (5 ft) below any unprotected opening that could lead to further flooding of the hull.

For the TLP-type floating substructure, positive tendon tensions are to be maintained when the one compartment damage and the overturning moment described above in this chapter are applied. The ability to compensate for damage incurred by pumping out or ballasting other compartments is not to be considered when determining whether positive tendon tensions can be maintained. A description of the survival capabilities after damage is to be included in the Operating Manual.

4.3 Overturning Moment

Environmental conditions and RNA operating modes to be considered in calculating the overturning moment are specified in Sect. 4.1 for assessing the intact and damage stability. For stability calculations, the overturning moment due to wind loads calculated in accordance with Chap. 12, Sect. 4.2 is to be determined at several angles of inclination for each mode of operation. The rotor thrust force is to be calculated with consideration of characteristics of a specific design of turbine rotor, effects of RNA operating conditions, and actions of turbine's control system. The calculation is to be performed in a manner to reflect the range of stability about the critical axis. The lever arm for the overturning force is to be taken vertically from the centre of lateral resistance or, if available, the centre of hydrodynamic pressure of the underwater body to the centre of pressure of the areas subjected to wind loading. Overturning force and centre of pressure derived from wind tunnel tests using a representative model of the Floating Support Structure and the RNA may be considered alternatively. The wind profile adopted for wind tunnel tests is to comply with Chap. 11, Sect 3.2 or use a more conservative profile. In the case that the current force increases the overturning moment, the adverse effect of the current is to be considered in the overturning moment calculation. The current force exerted on the hull is to be calculated according to Chap. 12, Sect. 4.4. For the dynamically-positioned Floating Offshore Wind Turbine, the overturning moment is to be calculated using the sum of a wind force up to the aggregate thrust of the thruster system in each direction analysed with a lever arm equal to the distance from the centre of wind pressure to the centre of the thruster propeller disc. The overturning moment due to the remaining wind force (if any) is to be calculated using a lever arm equal to the distance from the centre of wind pressure to the centre of lateral resistance. For this purpose, the aggregate thrust needs not to be taken greater than the wind force. The overturning moment due to other unfavourable environmental effects, such as snow and ice accumulation and green water, and the unfavourable effect of the station keeping system, as described in Sect. 4 are to be considered as appropriate.

5 Weight Control

An inclining test or an equivalent weighing procedure established by a combination of a hull lightweight survey, measurement of permanent ballast and weighing of the major components is to be conducted to accurately determine the weight and the position of the centre of gravity of the Floating Offshore Wind Turbine. The lightweight and centre of gravity are in general to be determined by an inclining test. In some cases, the configuration of the Floating Offshore Wind Turbine can make inclining the structure unfeasible or impractical. In such case, the lightweight and its centre of gravity are to be determined by a combination of a thorough lightweight survey and calculations. If the inclining test is conducted, it is required for one

Floating Offshore Wind Turbine per each specific design to determine the lightweight and position of centre of gravity, when as near to completion as practical. An inclining test procedure is to be submitted for review prior to the test. The inclining test is to be witnessed by the Surveyor. For the same type of the Floating Offshore Wind Turbine, which are considered by Class to be identical regarding hull form and arrangement, except for minor changes in machinery, outfit, etc., detailed weight calculations showing only the differences of weight and centres of gravity will be satisfactory, provided the accuracy of the calculations is confirmed by a lightweight survey. The results of the inclining test, or lightweight survey and inclining test adjusted for weight differences, are to be submitted for review. Changes of onboard load conditions after the inclining test and during service are to be carefully accounted for.

The results of the inclining experiment and lightweight survey are to be broken into the independent components of the Floating Offshore Wind Turbine (columns, deck, tower, etc.) and are to clearly indicate the position of these components. Weighing is to be supported by calculated weights, and to include a combination of afloat lightweight surveys and use of certified load cells to the satisfaction of the attending Surveyor. When the vertical position of the centre of gravity of each component cannot be measured, it is to be placed on an indisputably conservative location with consideration of the calculated position. Where the permanent ballast is installed, at least two (2) separate measures of the permanent ballast are to be performed. A global weighing plan and procedures for each individual weighing, together with the weight calculations, are to be submitted for approval. Reports of the weighing are to be submitted for review. The Operating Manual is to provide guidance for the maintenance of a weight change log. For the TLP-type floating substructure, the Operating Manual is to provide guidance for the periodical correlation between calculated and measured tendon tension.

6 Watertight/Weathertight Integrity

6.1 General

A plan, identifying the disposition (open or closed) of all non-automatic closing devices and locations of all watertight and weathertight closures, and unprotected openings is to be submitted for review. Upon satisfactory review, the plan is to be incorporated into the Operating Manual.

6.2 Weathertight Integrity

All external openings whose lower edges are below the levels to which weathertight integrity is required to be maintained as shown on the diagrams submitted in accordance with 8/6 are to have weathertight closing appliances. Openings fitted with closing appliances to maintain weathertight integrity are to effectively resist the ingress of water due to intermittent immersion of the closure.

6.3 Watertight Integrity

Suitable closing appliances are to be fitted to achieve watertight integrity for all internal and external openings whose lower edges are below the levels to which watertight integrity is to be maintained for both intact and damaged conditions as shown on the diagrams submitted in accordance with 8/6. For additional watertight integrity requirements, refer to the MOU Rules.

Chapter 17
Machinery, Equipment and Systems

1 Design Requirements

This chapter provides requirements for onboard machinery, equipment, and systems.

2 Marine Piping Systems

Marine piping systems are those systems that may be required for maintaining the normal operations of the Floating Offshore Wind Turbine (such as power generation, bilge, ballast, tank venting and sounding, etc.). The MOU Rules provides the definitions of these systems. Marine piping systems are to be in accordance with the requirements of the MOU Rules, except as modified in this chapter.

2.1 Bilge System

The design of bilge systems is to meet the applicable requirements in the MOU Rules. If portable power-driven pumps are used, in lieu of a permanent bilge system, at least two such pumps are to be provided and are to be stored onboard or carried by the attending service vessel. All pumps and arrangements for pumping are to be readily accessible.

F. Aderouane, *Design, Construction, and Global Performance Analysis of Offshore Wind Turbines*, Green Energy and Technology, https://doi.org/10.1007/978-3-031-86010-2_17

2.2 Ballast System

The system fitted is to provide the capability to ballast and deballast all ballast tanks that are not used as permanent ballast tanks. All pumps and valves are to be fitted with a remote means of operation. The normal or emergency operation of the ballast system is not to introduce a greater risk of progressive flooding due to the opening of hatches, manholes, etc. in watertight boundaries.

2.2.1 Pumping Systems

Where ballast systems powered by pumping systems are installed, at least two ballast pumps are to be provided, one of which is to be permanently connected to the ballast system. The second pump may be a spare held in reserve or an eductor type arrangement permanently connected to the system. If submersible ballast pumps are installed in each ballast tank, one spare pump must be always stored onboard.

2.2.2 Compressed Air Systems

Where a ballast system powered by compressed air is installed, a satisfactory quantity of compressed air is to be always available to the system. If two compressors are installed, one compressor is to be powered by either the emergency switchboard or a dedicated engine. Each compressor is to be capable of providing 100% of the required quantity of compressed air, as specified below. If only one compressor is provided, this compressor is to be powered by either the emergency switchboard or a dedicated engine, and a quantity of stored compressed air equivalent to the required capacity specified below is to be provided. The required quantity of compressed air is the amount of air capable of bringing the installation from its worst-case flooding condition, as defined by 16/4.2, to the normal operating draft and inclination.

2.3 Vents and Sounds

Except for those compartments that are not fitted with a fixed means of drainage, vent pipes are to be fitted on all tanks, cofferdams, voids, tunnels, and compartments which are not fitted with other ventilation arrangements. The requirements for sounding are to comply with the MOU Rules. However, to prevent duplication of pipe runs, it will be acceptable to sound the void spaces through the vent lines. In the case of a sealed vent, a sounding plug is to be provided to permit void space sounding.

3 Electrical Systems

The design criteria of electrical systems associated with marine systems are to be in accordance with applicable requirements described in the MOU Rules. Floating Offshore Wind Turbines are not typically provided with an emergency generator. There is typically a small battery installation that supports the required emergency loads. Some floating substructures are column stabilised and are provided with ballast systems that trim the floating offshore wind turbine to optimise the wind turbine performance. The designer is to verify that the ballast system, or failure of the ballast system, does not affect safety or stability. Where verified, the eighteen hours for ballast pumps and control system emergency power is not applicable.

4 Firefighting Systems and Equipment

Firefighting systems and equipment are to be designed with consideration of the size, type and intended service of the Floating Offshore Wind Turbine. As a minimum, portable extinguishers are to be provided in the quantities and locations in accordance with the applicable requirements in the MOU Rules. In all cases, the selection of the fire extinguishing medium is to be based on the fire hazard for the space protected. The fire extinguishers are to be visible and readily accessible. The plans and specifications of firefighting systems and equipment are to be submitted according to 8/8. Attention is drawn to the relevant requirements of the Coastal State or other governmental authorities having jurisdiction over the Floating Offshore Wind Turbine to be installed.

5 Safety Outfitting

Safety outfitting and measures are to be designed with suitable consideration of the nature of unmanned installations. It is suggested that the Coastal State or other governmental authorities having jurisdiction over the Floating Offshore Wind Turbine to be installed be contacted to obtain applicable requirements. All accessible spaces that require inspection are to be provided with ladders for safe access.

6 Identification Marks

A unique name or number is to be assigned to each Floating Offshore Wind Turbine, which is to conform to requirements of the Coastal State or other governmental authorities having jurisdiction over the Floating Offshore Wind Turbine. This name or number is to be permanently displayed on the structure and will be entered in the Class Record. Draft marks are to be permanently marked in at least two (2) places on the outside of the floating substructure indicating maximum permissible draft.

Chapter 18
Surveys

1 Surveys During Construction, Installation and Commissioning

This chapter pertains to surveys and inspections during construction, installation and commissioning of the Floating Offshore Wind Turbine including all the items within the scope of classification (refer to Chap. 8, Sect. 2.1). The documentation requirements for review are given in Chap. 8 of this book. A general quality plan highlighting required surveys together with Class hold points is to be determined by the Fabricator or Owner and agreed upon by the attending Surveyor. For the RNA and tower, the survey scope includes the following two situations:

- The optional RNA notation is requested by the Owner. The RNA and tower including its end connections are type approved by Class and are within the scope of classification and survey.
- The optional RNA notation is not requested by the Owner. However, the tower and its connection to the RNA and/or to the floating substructure are not included in the wind turbine type certificate. In this scenario, the tower and its connection are within the scope of classification and survey. The RNA with a wind turbine type certificate is not within the scope of classification.

1.1 Construction Surveys

1.1.1 Scope

This chapter pertains to surveys and testing to be carried out during construction of the Floating Offshore Wind Turbine at the builder's yard or facility.

F. Aderouane, *Design, Construction, and Global Performance Analysis of Offshore Wind Turbines*, Green Energy and Technology, https://doi.org/10.1007/978-3-031-86010-2_18

1.1.2 Survey at Vendor's Shop

During construction of equipment components for a Floating Offshore Wind Turbine, the attending Surveyor is to have access to vendors' facilities to witness construction and/or testing, as required by this book. The vendor is to contact the attending Surveyor to make necessary arrangements. If the attending Surveyor finds reason to recommend repairs or additional surveys, notice will be immediately given to the Owner or Owner's Representative so that appropriate action may be taken. Survey requirements for equipment components and packaged units at the vendor's shop are summarised in relevant paragraphs of the applicable Class Rules and guides. Each vendor is to have an effective quality system which is to be verified by the attending Surveyor.

1.1.3 Structure Fabrication/Erection

A quality control programme compatible with the type and size of the planned Floating Offshore Wind Turbine is to be developed and submitted to Class for review. Class will review, approve and, as necessary, request modification of this programme. The Fabricator is to work with the attending Surveyor to establish the required hold points on the quality control programme to form the basis for future surveys at the fabrication yard. As a minimum, the items enumerated in the various applicable paragraphs below are to be covered by the quality control programme. The Surveyor will be assigned to monitor the fabrication of items within the scope of classification, and to verify that competent personnel are carrying out the tests and inspections specified in the quality control programme. It is to be noted that the monitoring provided by Class is a supplement to and not a replacement for inspections to be carried out by the Fabricator or Owner. During construction, the Surveyor is to have access to structures at all reasonable times. The attending Surveyor is to be notified as to when and where parts of the structure may be examined. If, at any visit, the Surveyor finds occasion to recommend repairs or further inspection, notice is to be made to the Fabricator or its representatives. The Fabricator is to maintain a system of material traceability to the satisfaction of the attending Surveyor for all Special and Primary Application Structures. Data as to place of origin and results of relevant material tests for structural materials are to be retained and made readily available during all stages of construction. Such data are to be available to the Surveyor upon request. Where equipment and components are assembled in blocks or modules, the Surveyor is to inspect the fit-up, piping, and electrical connections, and to witness the required tests on the completed assembly in guidance with the quality control programme and in accordance with the approved plans and Rule/guide requirements. The progress and suitability of structural fit-up and joining of constructed/fabricated blocks/modules are to be to the satisfaction of the attending Surveyor. All principal erection joints are subject to visual examination, proven tight, and the extent of Non-Destructive Examination (NDE) carried out is to be to the satisfaction of the attending Surveyor.

1.2 Surveys on Steel Structures

1.2.1 Quality Control Programme

The quality control programme for the construction of the steel Floating Offshore Wind Turbine is to include the following items, as appropriate.

(1) Material quality and traceability.
(2) Steel forming.
(3) Welder qualification and records.
(4) Welding procedure specifications and qualifications.
(5) Weld inspection.
(6) Tolerances alignments and compartment testing.
(7) Corrosion control systems.
(8) Tightness and hydrostatic testing procedures.
(9) Non-destructive testing.
(10) Verification of SCIPs.

The items which are to be considered for each of the topics mentioned above are outlined in Sect. 2.1.

1.2.2 Material Quality and Traceability

The properties of the material are to comply with Chap. 9 of this book. Manufacturer's certificates are to be supplied with the material. Verification of the material's quality is to be done by the Surveyor at the plant of manufacture, in accordance with relevant Class Rules and guides. Alternatively, material manufactured to recognised standards may be accepted in lieu of the above steel requirements provided the substitution of such materials is approved by Class. Materials used are to be in accordance with those specified in the approved design and all materials required for classification purposes are to be tested in the presence of the Surveyor. The Fabricator is to maintain a material traceability system for all the Primary and Special Application structures.

1.2.3 Steel Forming

When forming changes base plate properties beyond acceptable limits, appropriate heat treatments are to be carried out to re-establish required properties. Unless approved otherwise, the acceptable limits of the reestablished properties are to meet the minimums specified for the original material before forming. As required, Class will survey formed members for their compliance with the forming dimensional tolerances required by the design.

1.2.4 Welder Qualification and Records

Welders who are to work on the structure are to be qualified in accordance with the welder qualification tests specified in a recognised code or, as applicable, Chap. 9 of this book to the satisfaction of the attending Surveyor. Certificates of qualification are to be prepared to record evidence of the qualification of each welder qualified by an approved standard/code, and such certificates are to be available for the use of the Surveyor. If welders have been previously tested in accordance with the requirements of a recognised code and provided that the period of effectiveness of the previous testing has not lapsed, these previous welder qualification tests may be accepted.

1.2.5 Welding Procedure Specifications and Qualifications

Welding procedures are to be approved in accordance with Chap. 9 of this book. Welding procedures conforming to the provisions of a recognised code may, at the Surveyor's discretion, be accepted. A written description of all procedures previously qualified may be employed in the structure's construction provided it is included in the quality control programme and made available to the Surveyor. When it is necessary to qualify a welding procedure, this is to be accomplished by employing the methods specified in the recognised code, and in the presence of the Surveyor.

1.2.6 Weld Inspection

As part of the overall quality control programme, a detailed plan for the inspection and testing of welds is to be prepared. This plan is to include the applicable provisions of this book.

1.2.7 Tolerances and Alignments

The overall structural tolerances, forming tolerances, and local alignment tolerances are to be commensurate with those considered in developing the structural design. Inspections are to be carried out to verify that the dimensional tolerance criteria are being met. Particular attention is to be paid to the out-of-roundness of members for which buckling is an anticipated mode of failure. Structural alignment and fit-up prior to welding are to be monitored to promote consistent production of quality welds.

1.2.8 Corrosion Control Systems

The details of any corrosion control systems employed for the structure are to be submitted for review. Installation and testing of the corrosion control systems are to be carried out to the satisfaction of the attending Surveyor in accordance with the approved plans.

1.2.9 Tightness and Hydrostatic Testing Procedures

Compartments which are designed to be permanently watertight or to be maintained watertight during installation including all openings and penetrations are to be tested by a procedure approved by the attending Surveyor. The MOU Rules may be used where applicable. The testing is also to be witnessed by the attending Surveyor.

1.2.10 Non-destructive Testing

A system of non-destructive testing is to be included in the fabrication specification of the structures. The minimum extent of non-destructive testing is to comply with the Class guidance for non-destructive inspection or recognised design Code. All non-destructive testing records are to be reviewed and approved by the attending Surveyor. Additional non-destructive testing may be requested by the attending Surveyor if the quality of fabrication is not conforming to industry standards.

1.3 Surveys on Concrete Structures

1.3.1 Quality Control Programme

The quality control programme for a concrete structure is to cover the following items, as appropriate.

(1) Inspections prior to concreting.
(2) Inspection of batching, mixing, and placing concrete.
(3) Inspections of form removal and concrete curing.
(4) Inspection of prestressing and grouting.
(5) Inspection of joints.
(6) Inspection of finished concrete.
(7) Tightness and hydrostatic testing as applicable refer to 11-1/3.7.9).

The items which are to be considered for each of the topics mentioned above, except for (7), are indicated in Sect. 1.3.1.

1.3.2 Inspections Prior to Concreting

Prior to their use in construction, the manufacturers of cement, reinforcing rods, prestressing tendons and appliances are to provide documentation of the pertinent physical properties. These data are to be made available to the attending Surveyor for verification of conformity with the properties specified in the approved design. As applicable, at the construction site, the Surveyor is to be satisfied that proper consideration is being given to the support of the structure during construction, the storage of cement and prestressing tendons in weathertight areas, the storage of admixtures and epoxies to manufacturer's specifications, and the storage of aggregates to limit segregation, contamination by deleterious substances and moisture variations within the stockpile. Forms and shores supporting the forms are to be inspected to verify that they are adequate in number and type, and that they are in accordance with the approved plans. The dimensions and alignment of the forms are to be verified by the attending Surveyor. The measurements are to be within the allowable finished dimensional tolerances specified in the approved design. Reinforcing steel, prestressing tendons, post-tensioning ducts, anchorages and any included steel are to be checked, as appropriate to the planned structure, for size, bending, spacing, location, firmness of installation, surface condition, vent locations, proper duct coupling, and duct capping.

1.3.3 Inspection of Batching, Mixing, and Placing Concrete

The production and placing of the concrete are to employ procedures which will provide a well-mixed and well-compacted concrete. Such procedures are also to limit segregation, loss of material, contamination, and premature initial set during all operations. Mix components of each batch of concrete are to be measured by a method specified in the quality control programme. The designer is to specify the allowable variation of mix component proportions, and the Fabricator is to record the actual proportions of each batch. Testing during the production of concrete is to be carried out following the procedures specified in the quality control programme. As a minimum, the following concrete qualities are to be measured by the Fabricator.

(1) Consistency.
(2) Air content.
(3) Density or Specific Gravity.
(4) Strength.

Field testing of aggregate gradation, cleanliness, moisture content, and unit weight is to be performed by the Fabricator following standards and schedules specified in the quality control programme. The frequency of testing is to be determined with the consideration of the uniformity of the supply source, volume of concreting, and variations of atmospheric conditions. Mix water is to be tested for purity following methods and schedules specified in the quality control programme.

1.3.4 Inspections of Form Removal and Concrete Curing

The structure is to have sufficient strength to bear its own weight, construction loads and the anticipated environmental loads without undue deformations before forms and form supports are removed. The schedule of form removal is to be specified in the quality control programme, giving due account to the loads and the anticipated strength. Curing procedures for use on the structure are to be specified in the quality control programme. When conditions at the construction site cause a deviation from these procedures, justification for these deviations is to be fully documented and included in the construction records. Where the construction procedures require the submergence of recently placed concrete, special methods for protecting the concrete from the effects of salt water are to be specified in the quality control programme. Concrete is not to be submerged until 28 days after placing.

1.3.5 Inspection of Prestressing and Grouting

A schedule indicating the sequence and anticipated elongation and stress accompanying the tensioning of tendons is to be prepared. Any failures to achieve proper tensioning are to be immediately reported to the designer to obtain guidance on needed remedial actions. Pre- or post-tensioning loads are to be determined by measuring both tendon elongation and tendon stress. These measurements are to be compared. In the case that the variation of measurements exceeds the specified amount, the cause of the variation is to be determined, and any necessary corrective actions are to be accomplished. The grout mix is to conform to that specified in the design. The Fabricator is to keep records of the mix proportions and ambient conditions during grout mixing. Tests for grout viscosity, expansion and bleeding, compressive strength, and setting time are to be made by the Fabricator using methods and schedules specified in the quality control programme. Employed procedures are to verify that ducts are filled. Anchorages are to be inspected to verify that they are located and sized as specified in the design. Anchorages are also to be inspected to verify that they are provided with adequate cover to mitigate the effects of corrosion.

1.3.6 Inspection of Joints

Where required, leak testing of construction joints is to be carried out using procedures specified in the quality control programme. When deciding which joints are to be inspected, consideration is to be given to the hydrostatic head on the subject joint during normal operation, the consequence of a leak at the subject joint, and the ease of repair once the Floating Offshore Wind Turbine is in service.

1.3.7 Inspection of Finished Concrete

The surface of the hardened concrete is to be inspected for cracks, honeycombing, pop-outs, spalling and other surface imperfections. When such defects are found, their extent is to be reported to the Surveyor and to the designer for guidance on any necessary repairs. The structure is to be examined using a calibrated rebound hammer or a similar non-destructive testing device. Where the results of surface inspection, cylinder strength tests or non-destructive testing do not meet the design criteria, the designer is to be consulted regarding remedial actions to be taken. The completed sections of the structure are to be checked for compliance with specified design tolerances for thickness, alignment, etc., and to the extent practicable, the location of reinforcing and prestressing steel and post-tensioning ducts. Variations from the tolerance limits are to be reported to the designer for evaluation and guidance on any necessary remedial actions.

1.4 Piping Systems

All piping installation and testing is to be in accordance with Class-approved drawings and procedures. Welds are to be visually inspected and non-destructively tested, as required and to the satisfaction of the attending Surveyor. Upon completion of satisfactory installation, the piping system is to be proven tight by pressure testing in accordance with the applicable requirements in the MOU Rules. Where sections of pipes are hydrostatically tested at the fabrication shops, an onboard test is to be conducted to confirm proper installation and tightness of the flanged and/or welded connections.

1.5 Electrical Systems and Installations

All electrical wiring, equipment, and systems within the scope of classification are to be installed and tested in accordance with Class-approved drawings and procedures. Proper support for all cables and suitable sealing of cable entries to equipment are to be verified. Upon completion of wire connections, the affected sections of the equipment and cabling are to be insulation-tested and proven in order. All grounding is also to be verified in order.

1.6 Instrumentation and Control Systems

All instrumentation and control systems within the scope of classification are to be installed and tested in accordance with Class-approved drawings and procedures. All supports are to be verified. Upon completion, instrumentation and control systems are to be functionally tested and demonstrated to be in order.

1.7 Mechanical Equipment and Systems

All mechanical equipment and systems within the scope of classification is to be installed and tested in accordance with Class-approved drawings and procedures, including the grounding of the equipment. Upon completion, mechanical equipment and systems are to be functionally tested and proven in order.

1.8 Fire and Safety Features

All fire and safety features are to be installed and tested in accordance with Class-approved drawings and procedures and to the satisfaction of the attending Surveyor.

1.9 Station Keeping System

Fabrication tests for the station keeping system components, such as anchors, chains, wires, shackles, etc. are to satisfy the requirements of the Class Rules for materials and welding for the respective sizes of equipment. Compliance with the Class guidance relating to the certification of offshore mooring chain is also required. Requirement specific to synthetic fibre ropes is provided in the Class guidance on the application of fibre rope for offshore mooring. In addition, it is to be confirmed that the station keeping system equipment complies with the Owner's specification. Physical testing, including break, pull, dimensional and non-destructive testing, is required to be performed in accordance with the submitted specifications and to the satisfaction of the attending Surveyor. Back-up structure/foundations on the floating substructure are required to be surveyed in compliance with Class approved drawings, in the presence of and to the satisfaction of the attending Surveyor. Proper fit-up, alignment, and final weldments of hull foundations for mooring winches, fairleads and other station keeping system components are to be visually examined. Completed welds are to be examined using surface NDT (preferably Magnetic Particle Inspection) to the extent deemed necessary by the attending Surveyor.

1.10 Installation, Hook-Up, and Commissioning Surveys

1.10.1 Surveys During Installation and Hook-Up

For the floating substructure and the station keeping system, detailed requirements for surveys during installation and hook-up are to be in accordance with applicable requirements in the FPI Rules. In general, it is to include the following items:

- General description of the mooring system and the floating substructure.
- Pre-installation verification procedures for the sea-bed condition and contingency procedures.
- Pile/anchor and mooring line installation procedures.
- Tensioning and proof load testing procedures of the anchor piles or anchor-chain system.
- Hook-up of the anchor chain system to the floating substructure.
- Final field erection and levelling.
- Pre-tensioning of mooring system.
- Hook-up of piping, electrical, instrumentation and mechanical equipment.

Where the RNA and tower are not within the scope of classification, the valid type certificate of the RNA and the tower are to be examined by the attending Surveyors. The installation of the RNA and tower and the hook-up of the power cable system are not to damage the interface structure to the floating substructure. Where the RNA and the tower are within the scope of classification, the Surveyor is to witness the installation of at least one (1) RNA and one (1) tower per each wind turbine type. Where there are more than fifty (50) RNAs and towers of the same type, at least one (1) additional RNA and tower installation per every fifty (50) turbines of the same type is to be witnessed by the Surveyor. The selection of the RNA and the tower to be witnessed is to reflect having Surveys at the start and end of installation periods and the rate of installation within an installation period of the RNA and tower. The attending Surveyor is to verify that the RNA and the tower to be installed comply with relevant design documents for the Floating Offshore Wind Turbine. Deviations from approved design documents and plans or any incidents such as damage or overstress to the floating substructure or the station keeping system during the installation may require resubmission of supporting documentation to provide an assessment of the significance of deviation and any necessary remedial actions to be taken. Any anomalies noted during the installation of the tower (when classed), floating substructure, and mooring are to be recorded in the ISIP for reference during later surveys.

1.10.2 Commissioning Surveys

The commissioning date will be the date on which the Surveyor issues the Interim Classification Certificate for the Floating Offshore Wind Turbine. Commissioning of all Class Rules and guide-required systems is to be verified by the attending Class Surveyor. The commissioning is to be in accordance with the approved step-by-step commissioning procedures. The Surveyor is to be permitted access to critical/hold points to verify that the procedures are satisfactorily accomplished. The Surveyor is to observe the Floating Offshore Wind Turbine operating under various capacities and conditions. Where the RNAs are within the scope of classification, the Surveyor is to witness the commissioning of at least one (1) RNA per each wind turbine type. Where there are more than fifty (50) RNAs of the same type, the commissioning of at least one (1) additional RNA per every fifty (50) turbines of the same type is to be witnessed by the Surveyor. The selection of RNA commissioning to be witnessed is to reflect having Surveys at the start and end of RNA commissioning periods. Approved turbine RNA operations including emergency procedures are to be verified to the extent deemed necessary by the attending Surveyor. Overall performances of the turbine RNA are to be verified for compliance with the design parameters used in the design of the Floating Offshore Wind Turbine. Records of all these performances are to be maintained and made available to class.

1.10.3 Personnel Safety

Verification of personnel safety features which include checks of operational readiness of all lifesaving, fire detection and firefighting equipment, emergency shutdown systems, unobstructed escape routes, lighting and establishment of communication procedures are to be taken during commissioning and are required to be verified by the attending Surveyor.

2 Surveys After Installation and Commissioning

2.1 General

This chapter pertains to periodical surveys after installation and commissioning for the maintenance of classification of the Floating Offshore Wind Turbine. For the RNA and the tower, the survey scope includes the following two situations:

- The optional RNA notation is requested by the Owner. The RNA and tower including its end connections are type approved by Class and are within the scope of classification and survey.
- The optional RNA notation is not requested by the Owner. However, the tower and its connection to the RNA and/or to the substructure are not included in the wind turbine type certificate. In this scenario, the tower and its connection are within the scope of classification and survey. The RNA with a wind turbine type certificate is not within the scope of classification.

2.1.1 Notification and Availability for Survey

The Surveyor is to be provided access to the classed Floating Offshore Wind Turbine at all reasonable times. The Owners or their representatives are to notify the Surveyor on all occasions when the Floating Offshore Wind Turbine can be examined on site. The Surveyor is to undertake all surveys on the classed Floating Offshore Wind Turbine upon request, with adequate notification, of the Owners or their representatives and are to report thereon to Class. Should the Surveyor find occasion during any survey to recommend repairs or further examination, notification is to be given immediately to the Owners or their representatives in order that appropriate action may be taken. The Surveyor is to avail themselves of every convenient opportunity for performing periodical surveys in conjunction with surveys of damages and repairs to avoid duplication of work. Also refer to the applicable Class Rules relating to the conditions for classification of offshore units and structures.

2.2 In-Service Inspection Programme (ISIP)

The ISIP as required by 8/14.2 is to cover all surveys for the design life of the Floating Offshore Wind Turbine. The In-Service Inspection Programme (ISIP) is a comprehensive programme that outlines the procedures to be followed and the inspection frequency of a Floating Offshore Wind Turbine. All Floating Offshore Wind Turbines are to be surveyed in accordance with a Class approved ISIP plan. For further details, refer to the FPI Rules. Structure Critical Inspection Points (SCIPs) are to be included in the ISIP. The ISIP plan is also to include:

(1) A copy of the Class review letter agreeing to the defined Structural Critical Inspection Points.
(2) Structural Wastage allowances.
(3) Details of access to structures and/or remote inspection techniques employed.

Depending on the type, configuration and structural design of a floating substructure, the reference can be made to the FPI Rules for typical structural areas and joints that are critical inspection areas. In addition, intersection structures of the floating substructure supporting the tower are Special Application Structures. Where the tower and its interface structures to the RNA are within the scope of classification, they are Special Application Structures. For Floating Offshore Wind Turbines surveyed under Continuous Survey, the due dates shown in the Class Survey Status are to be per the "ISIP Master Inspection Plan." An outline of the installation inspection schedule and frequency of examination for all the hull structure and mooring system components are to be included in the ISIP plan. In general, this outline may be called the "ISIP Master Inspection Plan" and cover the entire life-cycle of the field. For further requirements of the ISIP, refer to the FPI Rules, where applicable. Inspection requirements for the UWILD are to be included in the approved In-Service Inspection Programme (ISIP). The ISIP is also to discuss the means of access for each compartment, detailing how the periodic examinations will be completed, including general examination and access for close examinations of Structural Critical Inspection Points. Proposed alternative techniques of inspection should be detailed in the ISIP. The Attending Surveyor is to review the ISIP for progression of the Special Periodical Survey and for Floating Offshore Wind Turbines under Continuous Survey at each Annual Survey. This review is to verify that the ISIP is being maintained per the approved schedule and any adjustments are to take into consideration before the due date of the Special Survey. An ISIP is mandatory for Class surveys on all Floating Offshore Wind Turbines. As an alternative to a time-based inspection programme, a Risk Based Inspection (RBI) plan approved in accordance with Sect. 1.1 of this book may be considered so long as it includes all components of the ISIP. The RBI plan is to be submitted and agreed by Class.

2.3 Surveys

2.3.1 Annual Survey

An Annual Survey of the Floating Offshore Wind Turbine is to be carried out within three (3) months before or after each annual anniversary date of the crediting of the previous Special Periodical Survey or original construction date. Where the Surveyor is engaged in the survey of a group of structures of similar design and location, and where requested by the Owner, special consideration will be given to the timing of Annual Surveys and Special Periodical Surveys such that all periodical survey due dates can be harmonised. For the Floating Offshore Wind Turbine on Continuous Survey, all Continuous Survey requirements for those parts (items) due are to be completed each year. The Annual Survey will not be credited, and the Certificate of Classification will not be endorsed unless Continuous Survey items which are due or overdue at the time of the Annual Survey are either completed or granted an extension. For further requirements of the Annual Survey, refer to the MOU Rules, where applicable, as follows:

(1) For the floating substructure and, where applicable, the tower; all units over 5 years of age are to have the continued effectiveness of corrosion control methods in uncoated ballast tanks verified.
(2) For machinery, equipment, and systems and, where applicable, the RNA.

The Annual Survey of the station keeping systems is to follow the applicable requirements in the FPI Rules. For column tanks when the ballast is permanent and not coated, confirmation of corrosion condition of test coupon is to be made during Annual Survey. For the Annual Survey of a wind farm with a group of wind turbines having a similar design, the group of wind turbines can be considered as a set and examined in accordance with the approved ISIP.

2.3.2 Special Periodical Survey

A Special Periodical Survey of the Floating Offshore Wind Turbine is to be carried out within five (5) years of the initial Classification Survey, and at five-year intervals thereafter. The Special Periodical Survey may be commenced at the fourth Annual Survey and be continued with completion by the fifth anniversary date. Where the Special Periodical Survey is commenced prior to the fourth Annual Survey, the entire survey is to be completed within 15 months if such work is to be credited to the Special Periodical Survey. A Special Periodical Survey will be credited as of the completion date of the survey but not later than five years from date of build or from the date recorded for the previous Special Periodical Survey. If the Special Periodical Survey is completed within three (3) months prior to the due date, the Special Periodical Survey will be credited to agree with the effective due date. Special consideration may be given to Special Periodical Survey requirements in unusual cases. Consideration may be given for extensions of Rule-required Special Periodical Surveys under extreme circumstances. The Special Periodical Survey requirements are in addition to the Annual Survey requirements stated in Sect. 2.3.1 of this book. For further requirements of the Special Periodical Survey, refer to the MOU Rules, where applicable, as follows:

(1) For the floating substructure and, where applicable, the tower.
(2) General Visual Inspection (GVI), Close Visual Inspections (CVI) and Non-destructive Testing (NDT) requirements detailed in the In-Service Inspection Programme (ISIP) are to be followed.
(3) Requirements of the MOU Rules (Lightship Displacement) are not applicable. However, units are to be verified floating at the expected draft.
(4) For machinery, equipment, and systems and, where applicable, the RNA The Special Periodical Survey of the station keeping systems is to follow the applicable requirements in the FPI Rules.

For permanent ballast tanks, internal examination requirements will be specially considered for tanks used exclusively for permanent ballast which are fitted with an effective means of corrosion control in accordance with the Class Rules for the conduct of surveys after construction. For Special Periodical Surveys of an offshore wind farm with a group of wind turbines having a similar design, the group of wind turbines can be surveyed as a set, in accordance with the ISIP. Continuous Surveys can be applied (refer to Sect. 2.3.4).

2.3.3 Underwater Inspection in Lieu of Drydocking Survey (UWILD)

UWILD Surveys are to be carried out by a company approved by Class as an External Specialist. The Survey is to be completed once in any five-year period and in association with Special Periodical Survey. Consideration may be given for extensions of the UWILD Survey due dates under special circumstances. An examination of the underwater parts of each Floating Offshore Wind Turbine, including floating substructure, corrosion protection system, power cable system (if classed), and mooring system are to be made at UWILD Survey. Prior to examination, all power cables, mooring, and anchoring attachments are to be cleaned, including all openings to the sea, if any. Cables and mooring lines, including connecting hardware, are to be examined over the full length from the lowest exposed point at the seabed to the connection point at the Floating Offshore Wind Turbine. Inspection requirements for the UWILD are to be included in the approved In-Service Inspection Programme (ISIP). Any underwater GVI/CVI or NDT requirements included in the In-Service Inspection Programme (ISIP) are to be followed. For further requirements of the UWILD Surveys, refer to the FPI Rules, where applicable. For the UWILD Surveys for an offshore wind farm with a group of wind turbines of similar design, the group of wind turbines can be surveyed as a set, in accordance with the ISIP. A Continuous Survey can be applied (refer to Sect. 2.3.4).

2.3.4 Continuous Survey Programme

At request of the Owner, and upon approval of the proposed arrangements, a system of Continuous Surveys may be undertaken, whereby the Special Periodical Survey requirements are carried out in regular rotation to complete all the requirements of the Special Periodical Survey within a five-year period. The proposed arrangements are to provide for survey of approximately 20% of the total number of survey items during each year of the five-year period. Reasonable alternative arrangements may be considered. Each part (item) surveyed becomes due again for survey approximately five (5) years from the date of the survey and the due parts (items) are to be completed each year. For Continuous Surveys, a suitable notation will be entered in the Record and the date of the completion of the cycle published. Class may withdraw its approval for Continuous Survey if the Surveyor's recommendations are not complied with.

2.3.5 In-Line Surveys and Timing of Surveys

All items required to undergo the Special Periodical Survey are to be carried out at the same time and interval in order that they are recorded with the same crediting date. In cases where damage has involved extensive repairs and examination, the survey may, where approved by Class, be accepted as equivalent to the Special Periodical Survey. Surveys are to be completed within three (3) months of their due dates, unless extended by agreement with Class. Surveys carried out within this three-month window period will be credited and due at the same anniversary date in subsequent cycle. When so desired by the Owner, any part of the Floating Offshore Wind Turbine may be offered for survey prior to the three-month window and the survey will be credited as of the date it has been surveyed.

2.4 Survey Report File

All survey reports and records of all abnormalities found are to be compiled into the Survey Report File that is to be always kept by the Owner for reference during any survey. The records to be kept include, but are not limited to, the following:

(1) Approved In-Service Inspection Programme (ISIP), as required by Sect. 2.2.
(2) The updated status records of all class surveys.
(3) The records of all abnormalities found that are to include all videos and photographic records.
(4) The records of all repairs performed on any abnormalities found and any further repetitive abnormalities found after the repairs.
(5) Records of all corrosion protection system maintenance, including records of all cathodic potential readings taken, records of depletion of all sacrificial anodes, anode replacement, impressed current maintenance records, such as voltage and current demands of the system, coating breaks and the monitoring records of the steel material wastage in way of the coating break areas.
(6) All classification reports pertaining to the Floating Offshore Wind Turbine.
(7) Reports of thickness measurements of the floating substructure.
(8) Reports of all NDT and thickness measurements performed.
(9) Underwater examination reports including examination of hull and mooring.
(10) Mooring integrity documentation.

2.5 Incomplete Surveys

When a survey is only partially completed, the Surveyor is to report immediately upon the work completed so that the Owners and Class may be advised of the parts still to be surveyed.

2.6 Alterations

No alterations that affect or may affect classification are to be made to the classed Floating Offshore Wind Turbine, unless plans of the proposed alterations are submitted and approved by Class before the work of alterations is commenced. Such work, when approved, is to be performed to the satisfaction of the Surveyor. Nothing contained in this chapter or in a rule or regulation of any government or other administration or the issuance of any report or certificate pursuant to this chapter or such a rule or regulation is to be deemed to enlarge upon the representations expressed in the Class Rules relating to the conditions for classification of offshore units and structures. The issuance and use of any such reports or certificates are to be governed in all respects by the Class Rules relating to the conditions for classification of offshore units and structures.

2.7 Damage and Repairs

If the Floating Offshore Wind Turbine that has been classed suffers any damage to structure, machinery, piping, equipment, mooring, etc. which may affect classification, Class is to be notified, and the damage examined by the Surveyor. Details of intended repairs are to be submitted for approval, and the work is to be carried out to the satisfaction of the attending Surveyor. When a piece of machinery, piping, equipment, mooring, etc. suffers a premature or unexpected failure and is subsequently repaired or replaced without Surveyor attendance, details of the failure, including damaged parts, where practicable, are to be retained onboard for examination by the Surveyor during the next scheduled visit. Alternatively, the part or parts may be landed ashore for further examination and testing, as required. If failures noted in the paragraph above are deemed to be a result of inadequate or inappropriate maintenance, the maintenance plan and the survey and inspection plan are to be amended and resubmitted for approval.

2.8 Welding and Replacement of Materials

2.8.1 Ordinary and Higher Strength Structural Steels

Welding or other fabrication performed on the structural steels is to follow the requirements of Chap. 9.

2.8.2 Special Materials

Welding or other fabrication performed on other steels of special characteristics or repairs or renewals of such steel or adjacent to such steel is to be accomplished with procedures approved for the special materials involved. The procedures are to consider the information provided in Chap. 9.

2.8.3 Substitutions and Alterations

Substitution of steel differing from that originally installed, alteration of original structural configuration, or change from bolted to welded joint is not to be made without approval by Class.

2.9 *Inspection for Concrete Structures*

When there are no requirements provided in the book, inspection for concrete structures is to be based on ISO 19903, Chap. 14.

2.10 *Preparations for Safe Execution of Surveys*

The Owner is to provide the necessary facilities for a safe execution of the surveys to be carried out by the attending Surveyor. Refer to the MOU Rules for the requirements for preparations for safe execution of surveys.

2.11 *Maintenance of Rotor-Nacelle Assembly (RNA)*

The Owner is to submit an annual report for review by the attending Surveyor attesting to the following:

(1) Maintenance has been carried out by authorised and qualified personnel in accordance with the maintenance manual.
(2) The control settings have been checked regarding conformance with the limiting values specified in design documentation of the classed Floating Offshore Wind Turbine.
(3) All repair, modification and replacement have not altered values of RNA parameters specified in design documentation of the classed Floating Offshore Wind Turbine.

2.12 *Survey for Extension of Use*

When an existing classed Floating Offshore Wind Turbine is to remain in service at the same location for an extended period beyond its original design life, it is to be subject to inspections and testing, as deemed necessary, to determine its condition. The extent of the inspections and testing will depend on the completeness of the existing survey documents and inspection records. Class will review and verify maintenance manual, logs, and records. Any alterations, repairs, or installation of equipment since installation are to be included in the records. Those survey requirements in Sect. 2.3.2 for the Special Periodical Survey are to be included in the survey for extension of use. The surveys cover examination of splash zone, inspection of above water and underwater structural members and welds for damages and deteriorations, inspection of the station keeping system, examination and measurements of corrosion protection systems and marine growth, sea floor condition survey, examination of secondary structural attachments such export electrical cable support structures, service decks, etc. Special attention is to be given to the following critical areas:

(1) Areas of high stress.
(2) Areas of low fatigue life.
(3) Damage incurred during installation or while in service.
(4) Repairs or modifications made while in service.
(5) Abnormalities found during previous surveys.

An inspection report of the findings is to be submitted to Class for review and evaluation of the condition of the Floating Offshore Wind Turbine. The need for more frequent future periodical surveys is to be determined based on the calculated remaining fatigue life and the expected extension of service life.

2.13 Relocation of Existing Installations

An existing Floating Offshore Wind Turbine that is classed for a specified location requires special consideration when relocation to a new site is proposed. The Owner is to advise Class of the proposal to change locations, addressing removal, transportation, and re-installation aspects of the change. Survey requirements described in Sects. 1 and 2.12, wherever applicable, are to be complied with in addition to engineering analyses required to justify the integrity of the installation for the design life at the new location.

2.14 Certification on Behalf of Coastal States

When Class is authorised to perform surveys on behalf of a Coastal State or a governmental authority, and when requested by the Owner, items as specified by the Coastal State, the governmental authority or Owner will be surveyed. Reports indicating the results of such surveys will be issued accordingly. Where the periodicity and types of surveys on behalf of a Coastal State or a governmental authority differ from those required by the applicable portions of this chapter, Coastal State or governmental authority's requirements take precedence.

3 Risk-Based Surveys for Maintenance of Class

3.1 General

Risk-Based Inspection (RBI) provides an alternative means to Classification rule-based or calendar-based inspection regimes. A properly developed risk-based inspection plan or reliability centred maintenance plan may be credited as satisfying requirements of Surveys for Maintenance of Class for the Floating Offshore Wind Turbine. The plan is to be developed in accordance with the Class guidance for the risk-based inspection of floating offshore installations or the Class guidance for surveys based on machinery reliability and maintenance techniques.

3.1.1 Applicability

While this chapter provides risk-based survey requirements as an alternative for maintenance of Class, the chapters on the classification process contained in this book remain applicable. Where no specific references or guidance are given in this chapter, the relevant requirements of conventional Rules/guides remain valid.

3.1.2 Survey Periods

Because of the diverse nature and purposes of Floating Offshore Wind Turbines and the varied contents of inspection plans likely to be developed as part of an Owner's risk-based approach to Classification, the survey schedule of individual items will be established by the RBI plan. The plan will include Annual Surveys and Special Surveys.

3.2 Certification on Behalf of Coastal States

The application of the publications referenced in Sect. 3 does not cover any statutory survey requirements that may apply to the Floating Offshore Wind Turbine being considered. Only when the Coastal States and/or other governmental authorities accept and authorise Class for certification based on risk-based inspection techniques, Class will carry out such surveys, as authorised. The Owner is to confirm that in developing the inspection plan, consideration is given to applicable requirements external to Class. If the Coastal States and/or other governmental authorities do not accept a risk-based approach, surveys will be carried out in a conventional, prescriptive manner in compliance with Sect. 2.

Appendix 1 Wind Spectra and Coherence Functions

Kaimal Wind Spectrum and Exponential Coherence Model

A modified version of the Kaimal wind spectrum is provided in IEC 61,400–1 (2019). The power spectral densities for the wind fluctuations in three dimensions are given as follows:

$$\frac{f \cdot S_k(f)}{\sigma_k^2} = \frac{\frac{4fL_k}{V_{hub}}}{(1 + 6fL_k/V_{hub})^{\frac{5}{3}}}$$

where:

$S_k(f)$ = spectral energy density at frequency f, in m^{2}s^{-2}/Hz (ft^{2}s^{-2}/Hz).
f = frequency, in Hz.
k = index referring to the direction of wind speed component (i.e., 1 = longitudinal, 2 = lateral, and 3 = upward.
σ_k = standard deviation of turbulent wind speed component (refer to Table 1)

$$\left[\int_0^\infty S_k(f)df\right]^{0.5}$$

L_k = integral parameter of turbulent wind speed component (refer to Table 1).

Along with the Kaimal spectrum, an exponential coherence model is provided in IEC 61400-1 (2019) to account for the spatial correlation of the longitudinal wind speed:

$$Coh(r,f) = \exp\left[\sqrt{\left(f \cdot \frac{r}{V_{hub}}\right)^2 + \left(0.12 \cdot \frac{r}{L_c}\right)^2}\right]$$

where:

$Coh(r,f)$ = coherence function at frequency f.
f = frequency, in Hz.

Table 1 Spectral parameters for the Kaimal model

	Wind speed direction		
	k = 1 (longitudinal)	k = 2 (lateral)	k = 3 (upward)
Standard deviation (σ_k)	σ_1	$0.8\sigma_1$	$0.5\sigma_k$
Integral Scale (L_k)	$8.1\Lambda_1$	$2.7\Lambda_1$	$0.66\Lambda_1$

Note: σ_1 = standard deviation of longitudinal turbulent wind speed; Λ_1 = scale parameter of the turbulence

r = magnitude of the projection of the separation vector between the two points on to a plane normal to the average wind direction, in m (ft).
V_{hub} = 10-min mean wind speed at hub height, in m/s (ft/s).
L_c = coherence scale parameter, in m (ft) $= 8.1\Lambda_1$.

NPD (Frøya) Wind Spectrum and Coherence Model

For the extratropical storm wind, the NPD (Frøya) wind spectrum in conjunction with the two-point coherence function recommended by API RP 2MET (2019) may be applied.

Wind Spectrum

The following wind spectrum is defined for the energy density of the longitudinal wind speed fluctuations:

$$S(f) = \frac{320\phi^2\left(\frac{U_0}{10\phi}\right)^2\left(\frac{z}{10\phi}\right)^{0.45}}{\left(1+\tilde{f}^n\right)^{\left(\frac{5}{3n}\right)}}$$

$$\tilde{f} = 172f\left(\frac{z}{10\phi}\right)^{\frac{2}{3}}\left(\frac{U_0}{10\phi}\right)^{-0.75}$$

where:

$S(f)$ = spectral energy density at frequency f, in m^2s^{-2}/Hz (ft^2s^{-2}/Hz).
f = frequency, in Hz.
U_0 = 1-h mean wind speed at 10 m (32.8 ft) above the SWL, in m/s (ft/s).
$n = 0.468$.
z = height above the SWL, in m (ft).
ϕ = unit conversion factor.
= 1 when using SI units (m, m/s).
= 3.28 when using US Customary units (ft, ft/s).

Spatial Coherence

The squared correlation between the spectral energy densities $S(f)$ of the longitudinal wind speed fluctuations between two points $(xj, yj, zj), j = 1, 2$ in space is described by the two-point coherence function as follows:

Table 2 Coefficients and distances for the three-dimensional coherence function

i	Δ_i	q_i	p_i	r_i	α_i
1	$\lvert x_2 - x_1 \rvert$	1.00	0.4	0.92	2.9
2	$\lvert y_2 - y_1 \rvert$	1.00	0.4	0.92	45.0
3	$\lvert z_2 - z_1 \rvert$	1.25	0.5	0.85	13.0

$$Coh(f) = \exp\left\{ -\frac{1}{U_0/\phi} \cdot \left[\sum_{i=1}^{3} A_i^2 \right]^{\frac{1}{2}} \right\}$$

$$A_i = \alpha_i \ f^{ri} \left(\frac{\Delta_i}{\phi} \right)^{qi} \left(\frac{\sqrt{z_1 z_2}}{10\phi} \right)^{-pi}$$

where the coefficients α_i, p_i, q_i, r_i and the distances Δ_i are specified in Table 2, and

$Coh(f)$ = coherence function at frequency f.
f = frequency, in Hz.
U_0 = 1-h mean wind speed at 10 m (32.8ft) above the SWL, in m/s (ft/s).
(x_j, y_j, z_j) = spatial coordinates of two points ($j = 1, 2$) where x_j is in the longitudinal (along wind) direction, y_j is in the lateral (across wind) direction and z_j is the height above the SWL in the upward direction, in m (ft).
ϕ = unit conversion factor.
= 1 when using SI units (m, m/s).
= 3.28 when using US Customary units (ft, ft/s).

Tropical Cyclone Wind Spectrum

For tropical storm wind, the following wind speed spectrum for the energy density of the longitudinal wind speed fluctuations is recommended by API RP 2MET (2019):

$$S(f, z) = \frac{4 I_u^2 U_{w,1h}(z) L_{u,x}(z)}{\left[1 + 70.8 \left(\frac{f L_{u,x}(z)}{U_{w,1h}(z)} \right)^2 \right]^{\frac{5}{6}}}$$

where:

$S(f, z)$ = spectral energy density at frequency, f, and height z, in m^2s^{-2}/Hz (ft^2s^{-2}/Hz).
f = frequency, in Hz.
z = height above the SWL, in m (ft).
$U_{w,1h}(z)$ = 1-h average wind speed at a height z above the MSL, in m/s (ft/s).
$I_{u,x}(z)$ = turbulence intensity at the height z above the MSL, (dimensionless).

$L_{u,x}(z)$ = integral length scale at the height z above the MSL, in m (ft)

$$L_{u,x}(z) = \frac{50\phi\left(\frac{z}{\phi}\right)^{0.35}}{\left(\frac{z_0}{\phi}\right)^{0.063}}$$

Z_0 = surface roughness length for tropical cyclone winds, in m (ft).
ϕ = unit conversion factor, (dimensionless).
= 1 when using SI units (m, m/s).
= 3.28 when using US Customary units (ft, ft/s).

Appendix 2 Tropical Cyclone Wind Speed Profile, Standard Deviation, Turbulence Intensity and Gust Factor

Introduction

For tropical storm wind, the following wind speed spectrum for the energy density of the longitudinal wind speed fluctuations is recommended by API RP 2MET (2019):

$$V_{1hr}(z) = \frac{u*}{k}\ln\left(\frac{z}{z_0}\right)$$

where:

$V_{1hr}(z)$ = one hour mean longitudinal wind speed at the height z above the sea surface, in m/s (ft/s).
$u*$ = friction velocity, in m/s (ft/s).
z = height above the sea surface, in m (ft).
z_0 = surface roughness length, in m (ft).
k = von Karman constant, (dimensionless) = 0.4.

For mean wind speed $V_{1hr}(z_{ref})$ lower than approximately 27 m/s (52.5 knots), the surface roughness length z_0 over the open ocean may be calculated using the Charnock equation:

$$z_0 = \frac{\alpha u^2*}{g} = \frac{\alpha}{g}\left[\frac{kV_{1hr}(z)}{\left(ln\frac{z}{z_0}\right)}\right]^2$$

where:

α = 0.011, as suggested in IEC 61,400–3, for the open ocean.

z_{ref} = 10 m (33 ft) reference height above the sea surface, in m (ft).
$V_{1hr}(z_{ref})$ = 1-h mean wind speed at the reference height z_{ref} above the sea surface, in m/s (ft/s).
g = acceleration of gravity, in m/s^2 (ft/s^2).

Alternatively, the surface roughness length z_0 over the open ocean can be determined from the boundary layer wind profile and expressed in terms of the surface drag coefficient $C_d(z_{ref})$ at the reference height $z_{ref} = 10$ m (33 ft) through the surface shear stress $\tau_0 = pu_*^2 = \rho C_d(z_{ref}) V_{1hr}^2(z_{ref})$, where ρ is the air density and $C_d(z_{ref})$ is the surface drag coefficient at the reference height z_{ref} above the sea surface, such that:

$$u* = \sqrt{C_d(z_{ref})} V_{1hr}(z_{ref})$$

$$z_0 = 10 \times \exp\left(-\frac{k}{\sqrt{C_d(z_{ref})}}\right)$$

where the surface drag coefficient $C_d(z_{ref})$ may be calculated by:

$$C_d(z_{ref}) = [0.49 + 0.065 V_{1hr}(z_{ref})] \times 10^{-3}$$

The values of $C_d(z_{ref})$ and z_0 determined by the equations given above will increase along with the mean wind speed. However, it is found that the $C_d(z_{ref})$ levels off for the wind speed $V_{1hr}(z_{ref})$ exceeding approximately 22–30 m/s (43.8–58.3 knots) and may even decrease slightly at very high wind speeds, probably due to wave breaking and increasingly frothy sea surface. The maximum $C_d(z_{ref})$ over the open ocean is estimated in the range of 0.0019–0.0025 depending on the size of the tropical cyclone and the maximum wind speed. As a results, the surface roughness length z_0 over the open ocean is in the range of 0.001–0.0034 m (0.0033–0.011 ft). For typical tropical cyclones over the open ocean, the following limiting value should be applied:

$$C_d(z_{ref}) \leq 0.0023$$

Standard Deviation of Tropical Cyclone Wind Speed and Turbulence Intensity

The standard deviation of instantaneous tropical cyclone wind speed within one hour time duration is determined using the formulation suggested in ESDU 83045 Strong Winds in the Atmospheric Boundary Layer, Part 2: Discrete Gust Speeds (2002), in conjunction with the friction velocity $u*$ and the surface roughness length z_0 determined in accordance with this appendix:

$$\sigma_u(z) = \frac{7.5\eta\left[0.538 + 0.09ln\left(\frac{z}{z_0}\right)\right]^{\eta 16}}{1 + 0.156ln\left(\frac{u*}{fz_0}\right)} u*$$

where:

$\sigma_u(z)$ = standard deviation of the instantaneous longitudinal wind speed at the height z above the sea surface, in m/s (ft/s).
η = scaling parameter, (dimensionless) $= 1 - \frac{6fz}{u*}$.
f = Coriolis parameter, in rad/s $= 1.458 \times 10^{-8} sin\phi$.
ϕ = local angle of latitude, in degrees.

The turbulence intensity, defined as the standard deviation of the wind speed divided by the mean wind speed at the same height, is given by:

$$I_u(z) = \frac{\sigma_u(z)}{V_{1hr}(z)} = \frac{\sigma_u(z)}{u*}\frac{u*}{V_{1hr}(z)}$$
$$= \frac{7.5\eta\left[0.538 + 0.09ln\left(\frac{z}{z_0}\right)\right]^{\eta 16}}{1 + 0.156ln\left(\frac{u*}{fz_0}\right)} \frac{k}{ln\left(\frac{z}{z_0}\right)}$$

where the frication velocity $u*$ can be determined either by the surface wind profile, or the relation derived from surface shear stress. The turbulence intensity $I_u(z)$ is weakly dependent on the mean wind speed and, for strong winds, can be considered independent of the mean wind speed.

Tropical Cyclone Wind Gust Factor

The longitudinal gust wind speed, $V_\tau(z)$, which represents the maximum value of longitudinal wind speed occurring in any given observation period averaged over a shorter period, is related to the mean longitudinal wind speed in the same observation period through the gust factor as defined below:

$$V_\tau(z) = G(z, \tau, T_0)V_{\tau o}(z)$$
$$= \left[1 + g(z, \tau, T_0)I_u(z)\right]V_{To}(z)$$

where:

$V_\tau(z)$ = longitudinal gust wind speed at the height z above the sea surface with averaging time duration of τ,, in m/s (ft/s).

$V_{\tau o}(z)$ = mean longitudinal wind speed at the height z above the sea surface in any given reference observation period T_o, in m/s (ft/s).
$G(z, \tau, T_0)$ = gust factor, (dimensionless).
$g(z, \tau, T_0)$ = peak factor, (dimensionless).
z = height above the sea surface, in m (ft).
τ = gust averaging time duration, in seconds.
T_o = reference observation time duration ($T_o > \tau$), in seconds.
$I_u(z)$ = turbulence intensity at the height z above the sea surface, (dimensionless).

For the one-hour observation period, the peak factor can be calculated by:

$$g(z, \tau, T_0 = 3600s) = \frac{\sigma_u(z, \tau, T_o = 3600s)}{\sigma_u(z)}\left[\sqrt{2\ln(3600v)} + \frac{0.577}{\sqrt{2\ln(3600v)}}\right]$$

$$\frac{\sigma_{u(z,\tau,T_0=3600s)}}{\sigma_u(z)} = 1 - 0.193\left(\frac{T_u}{\tau} + 0.1\right)^{-0.68}$$

where:

$\sigma_{u(z,\tau,T_0)}$ = standard deviation of the wind speed at the height z above the sea surface with averaging time duration of τ observed during a period of T_o,, in m/s (ft/s).
$\sigma_u(z)$ = standard deviation of the instantaneous wind speed at the height z above the sea surface, in m/s (ft/s).
υ = zero up-crossing frequency of wind speed, in Hz.
$= \frac{\left[0.007+0.213\left(\frac{T_u}{\tau}\right)^{0.654}\right]}{T_u}$.
T_u = longitudinal integral length time scale, in seconds = $3.12z^{0.2}$.

The gust factor associated with one hour reference period is relate to the peak factor through:

$$G(z, \tau, T_0 = 3600s) = 1 - g(z, \tau, T_0 = 3600s)I_u(z)$$

As an example, the maximum 10-min mean wind speed observed in a one-hour time duration can be obtained by:

$$V_{10min}(z) = G(z, \tau = 600s, T_0 = 3600s)V_{1hr}(z)$$

For $z = z_{ref} = 10$ m (33 ft) above the sea surface,

$$V_{10min}\left(z_{ref}\right) = 1.03V_{1hr}\left(z_{ref}\right)$$

For a reference period $T_o(> \tau)$ smaller than 3600s, the gust factor can be calculated by:

$$G(z, \tau, T_0) = \left[0.2193ln(log_{10}T_0) - 0.7242\right] \times G(z.\tau, T_0 = 3600s)$$

Appendix 3 Fatigue Analysis for Floating Support Structures

Introduction

Floating Offshore Wind Turbines are subjected to complex loading conditions that involve environmental loads, particularly those from wind, wave and current, and actions of the control, safety, and protection systems. The loading conditions for the fatigue analysis of the Floating Support Structure are specified in 12/3. The fatigue analysis method suggested in this appendix assumes that the structural response of the Floating Support Structure can be represented by three components in distinctive frequency ranges relative to the predominant wave frequencies:

- Low-frequency component that comprises the responses to wave and wind induced low-frequency motions.
- Wave-frequency component caused by direct wave loads or wave induced motions in the range of frequencies associated with most of the wave energy.
- High-frequency component due to second and higher-order wave loads and aerodynamic responses of the wind turbine RNA.

The following paragraphs in this appendix provide suggested methods for the fatigue analysis of the Floating Support Structures, with consideration of the frequency range separation and the nonlinearity of the loading conditions. Alternative fatigue analysis methods may also be used.

Fatigue Analysis of the Floating Substructure in the Low Frequency

Range

The wind and wave induced low-frequency motions depend on natural frequencies of rigid-body motions of the floating substructure. They are also affected by the type of the station keeping system and the mass of the structure and should preferably be analysed using time-domain global motion analysis. The frequency range of the low-frequency motions should have a distinctive separation from the lowest predominant wave-frequency, which should be used as the upper limit frequency of the low-pass filter. If the separation of frequency range is not obvious, the upper limit frequency of the low-pass filter may be assumed to be 0.05 Hz. The time increment used in global motion analysis should be carefully selected such that the peak of motion response can be captured. For the Column-Stabilised and Spar-type floating substructure, natural frequencies of pitch and roll motions are usually smaller than predominant wave frequencies. The low-frequency pitch and roll motions could introduce significant structural loads in the following hull structural components:

- Bracing members and connections in a Truss Spar-type floating substructure.
- Bracing members connecting the columns and pontoons in a Column-Stabilised floating substructure.

The low-frequency pitch and roll motions could also cause significant structural loads in the Tower. The pitch and roll motions derived from time-domain global motion analysis can be represented by Power Spectral Density (PSD) functions, which is further applied to the stress transfer functions for Column-stabilised or Spar-type floating substructure to obtain the stress PSD in the low frequency range. To derive the stress transfer functions, linear static global finite element analyses may be performed using the floating substructure with a quasi-static unit pitch or roll motion. The inclined position may be achieved by imposing an assumed wind on the tower and current on the hull structure. Therefore, in addition to the gravity and buoyancy loads at the initial un-displaced position, the low-frequency loads include the assumed current and wind load, gravity load due to inclinations, varying buoyancy, and mooring/cable tension on the floating substructure at the rotated position. Stresses calculated at selected fatigue critical locations are the stress transfer functions. Appropriate stress concentration factors (SCFs) in accordance with the Class guidance relating to fatigue assessment of offshore structures should be incorporated into the stress transfer functions.

Fatigue Analysis of the Floating Substructure in the Wave Frequency

Range

Wave-frequency responses of the floating substructure subjected to wave forces are normally the main cause of fatigue damage in the hull structures. Wave forces on large floating bodies can be obtained using the diffraction/radiation analysis. Additional drag and inertia load on slender bracing members may be computed by Morison Equations and should be appropriately considered in the motion analysis. Six degree-of-freedom rigid-body motion Response Amplitude Operators (RAOs) can be obtained from either frequency-domain or time-domain global motion analysis. RAOs of sectional loads and moments that are critical for structural designs can also be derived. The spectral-based fatigue analysis method (refer to the Class guidance relating to fatigue assessment of offshore structures) can be used for fatigue analysis of the hull structure subjected to wave forces in the wave frequency range.

The stress transfer functions should be determined through finite element analyses for the floating substructure subjected to wave forces and motions associated with a unit wave amplitude and various wave periods and directions.

Fatigue Analysis of the Floating Substructure in the High Frequency

Range

A time-domain global motion analysis should be performed to identify the frequency range of high-frequency responses that should in general have a distinctive separation from the highest predominant wave-frequency to be used as the lower limit frequency of the high-pass filter. If the separation of frequency range is not obvious, the lower limit frequency of the high-pass filter may be assumed to be 0.25 Hz. High-frequency loading may come from various sources such as the high-frequency wave loads, high-frequency aerodynamic loads transmitted from the tower, and vortex-induced-vibration (VIV) loads transmitted from the mooring lines or tendons. Those high-frequency loads may cause hull structural vibrations, in the TLP-type floating substructure, where the load frequency is close to the structural natural frequency. Such vibrations and structural dynamic effects should be assessed and their effects on fatigue damage accumulation should be appropriately considered. The rigid-body motions of the floating substructure due to high-frequency loads may be neglected for the Column-Stabilised or the Spar-type floating substructure. For the TLP-type floating substructure, high-frequency rigid-body motions of the floating substructure may be significant and, if necessary, the inertia forces in structural components induced by high-frequency rigid-body motions should be considered. The stress PSDs in the high frequency range can be determined through the acceleration PSDs derived from time-domain global motion analysis and the transfer functions of stress components. To derive the stress transfer functions, linear static finite element analyses can be carried out for the floating substructure under a quasi-static inertial force associated with a unit acceleration and balancing loads at the tower-hull connection. Other loads from wind, wave, current, gravity load, buoyancy, and mooring/cable tension may be neglected. For the TLP-type floating substructure, balancing loads may also be applied at the tendon porches. Appropriate stress SCFs in accordance with the Class guidance relating to fatigue assessment of offshore structures should be incorporated into the stress transfer functions.

Calculation of Fatigue Damage of the Floating Substructure

Subjected to Broadband Spectral Loading

The Floating Offshore Wind Turbine typically has broadband responses that do not follow the Rayleigh distribution. Fatigue damage of the floating substructure can be calculated in various ways after the stress response spectra in the three distinctive frequency ranges are obtained through load/motion spectra and stress transfer functions. It is recommended that Dirlik's Method be used in the calculation of fatigue damage of the floating substructure. Either the summation of the stress spectra associated with the three-frequency range or the summation of spectral moments of stress spectra associated with three frequency range may be used in the implementation of Dirlik's Method.

Time-Domain Fatigue Analysis for the Heave Plate and the Tower

Due to nonlinear aerodynamic loads and coupling effects of wind and wave loads, the time-domain fatigue analysis is preferred for the wind turbine Tower. By using this method, the motions and loads at the Tower top and base are first calculated through the time-domain global motion analysis of the Floating Offshore Wind Turbine. The structural dynamic analysis of the Tower can then be performed in the time domain with the motions and loads applied at the Tower top and base as boundary conditions. Heave plates may be used in the Column-Stabilised or Spar-type floating substructure to improve the motion performance of the hull. The main load on the heave plate is hydrodynamic pressures induced by waves and motions of the hull. Substantial nonlinear hydrodynamic loads are exerted on the heave plate due to the nonlinear viscous drag and relative hull motions. Where the heave plate is connected to other hull structures through bracing members, the global deformation of those bracing members due to bending and shear may also have a significant impact on the heave plate design. To determine hydrodynamic loads on the heave plates, a global motion analysis should be performed in the time domain. The pressure distribution on the heave plate is typically not uniform, and the centre of pressure could be away from the centre of the plate. The pressure distribution is affected by the heave plate size and configuration, hull structure global pitch and roll motions, and combined wave and current loads. To calculate the pressure distributions that are required for the structural design, the heave plate is normally divided into a collection of smaller sub-panels. In the time domain analysis, these subpanels can be modelled using Morison elements with equivalent drag and inertia coefficients. Once the hydrodynamic forces on these equivalent Morison elements are obtained, they can be mapped to the finite element model of the heave plate for structural analyses, along with appropriate boundary conditions mapped from the global deformation of the structures connecting the heave plate to the main hull.

Part III
Global Performance and Integrated Load Analysis for Offshore Wind Turbines

Chapter 19
Introduction

1 General

Global performance analysis for an offshore wind turbine is primarily conducted to calculate the global responses including displacements, motions, and loads that can be used for the design and design evaluation of the offshore wind turbine and its subsystems and components. The subsystems of an offshore wind turbine typically include the rotor-nacelle assembly (RNA), support structure, and if deemed relevant, power cables. For a bottom-founded offshore wind turbine, the support structure comprises the tower, substructure, and foundation. For a floating offshore wind turbine, the support structure comprises the tower, floating substructure, and station keeping system. The complexity and fidelity of these subsystems in the global performance analysis model may vary at distinctive design stages. In the design process of an offshore wind turbine, wind turbine manufacturers design the wind turbine generator (WTG) including the RNA and the tower. Since the wind turbine generator is often type certified to the IEC standards, a site-specific design evaluation should be performed to verify the wind turbine generator's suitability for a wind farm project. For the substructure and foundation or station keeping system of an offshore wind turbine, there are typically two design stages:

(1) ***Stage 1: Preliminary Design Phase***. Wind turbine manufacturers are often not directly involved. For the purposes of the design analysis, a generic wind turbine model or a simplified wind turbine model is normally used.
(2) ***Stage 2: Detailed Design Phase***. Wind turbine manufacturers are involved. The design analysis is typically based on an integrated design approach using an integrated analysis model.

Each design stage may involve several design cycles in which analysis methodologies and modelling strategies of various fidelity levels are used for global performance analysis. These methodologies include the coupled and uncoupled time-domain analysis approaches as well as the frequency domain analysis approaches. Integrated Load

F. Aderouane, *Design, Construction, and Global Performance Analysis of Offshore Wind Turbines*, Green Energy and Technology, https://doi.org/10.1007/978-3-031-86010-2_19

Analysis (ILA) as an integrated time domain analysis approach may be required by the regulators, Class, or certification bodies. The terms "integrated load analysis" and "coupled analysis" are often used interchangeably in the literature. Although the integrated load analysis usually uses the coupled analysis approach, the integrated load analysis also refers to a specific design process for offshore wind turbines. In the context of offshore wind turbine design standards such as IEC 61400-3-1 and IEC 61400-3-2, an integrated load analysis refers to the design analysis method and procedure for load and load effect calculations for the entire offshore wind turbine comprising the RNA and the support structure. The design evaluation of the RNA components and the support structure is performed using the results from the integrated load analysis.

2 Applications

This book provides guidance on global performance analysis and integrated load analysis (ILA) methodologies, modelling strategies, and suggestions for numerical simulations for Bottom-founded Offshore Wind Turbines and Floating Offshore Wind Turbines. Since integrated load analysis (ILA) is required for certification by the IEC 61400 series standards and recommended for class, this book also provides guidance on the integrated load analysis procedures for bottom-founded and floating offshore wind turbines to comply with the following standards where applicable:

- IEC 61400 series standards for offshore wind turbines.
- Other referenced industry standards applicable to offshore wind turbines.

It should be noted the intention here is not to set additional design requirements and criteria other than those specified in the Part 1 and Part 2, and IEC 61400 series standards.

3 Analysis Methodologies

3.1 General

Global performance analysis for offshore wind turbines in this book refers to:

(1) Global load analysis for bottom-founded offshore wind turbines.
(2) Global performance analysis for floating offshore wind turbines.

Different analysis methodologies may be used in a global performance analysis as summarised below in this chapter.

3.2 *Frequency-Domain and Time-Domain Analysis for Offshore Wind Turbine*

3.2.1 Frequency-Domain Analysis

Frequency-domain analysis refers to calculation of loads and responses in the frequency domain by solving the equations of motion using methods of harmonic analysis or methods of Laplace and Fourier transformations.

Frequency-Domain Analysis for Bottom-Founded Offshore Wind Turbines

For Bottom-founded Offshore Wind Turbines, the frequency-domain analysis approach may be used for the fatigue analysis of the substructure subject to wave loads. The equations of motion are linearised in the frequency domain analysis. The wind and wave load effects are decoupled. The dynamic interaction between the RNA and the support structure caused by combined wind and wave excitations is not accurately modelled. The critical dynamic interaction is aerodynamic damping. The frequency domain approach is thus less accurate than the time-domain approach. In the frequency-domain analysis, the support structure is modelled with a finite element (FE) model, and structural dynamics equations are solved in the frequency domain. The RNA is typically modelled as an equivalent concentrated mass on top of the tower. Aerodynamic damping as a function of wind speed and wind-wave misalignment should be separately evaluated and applied in the analysis. The frequency-domain approach can reduce simulation time significantly compared to the time-domain approach. Due to accuracy limitations, the frequency-domain method is usually used in the preliminary design phase.

Frequency-Domain Analysis for Floating Offshore Wind Turbines

For Floating Offshore Wind Turbines, the frequency-domain analysis includes the following types of analysis in the frequency domain for the floating substructure and station keeping system:

- Wave load calculation.
- Motion analysis.
- Air gap analysis.
- Mooring analysis.

In the calculation of Floating Offshore Wind Turbine wave frequency responses, linear wave theory is usually employed, while more sophisticated methods may be employed to model finite amplitude waves. The low frequency motion analysis should be conducted to evaluate the responses to wind dynamics and wave drift forces. The damping levels used in such analyses should be properly determined and documented. For the Tension Leg Platform (TLP)-type floating support structure, where second order sum-frequency effects are deemed significant, the high frequency springing responses of the floating substructure and tendons should be evaluated. Frequency-domain analyses for evaluating aerodynamic responses of the RNA and effects of turbine control systems should be properly formulated. Preferably, combined aerodynamic, hydrodynamic, and control system actions in the frequency domain are used in the calculation of Floating Offshore Wind Turbine dynamic responses. Frequency-domain analyses cannot capture nonlinear dynamic interactions among the components and subsystems of the Floating Offshore Wind Turbine. Methods of approximating nonlinearity in the frequency domain, and their limitations, should be investigated to show they can provide acceptable solutions for the intended application. Frequency-domain analyses are also unable to account for transient responses as well as nonlinear aerodynamic and hydrodynamic load effects. Because of these limitations, most currently available simulation software for Floating Offshore Wind Turbines is based on the time-domain analysis approach. Frequency-domain analyses are normally performed to calculate the hydrodynamic coefficients which are used as input to time-domain analyses.

3.2.2 Time-Domain Analysis

Time-domain analysis refers to calculation of the loads and responses in the time domain.

Time-Domain Analysis for Bottom-Founded Offshore Wind Turbines

Two time-domain approaches may be utilised for the global load analysis for Bottom-founded Offshore Wind Turbines. One is the integrated load analysis, which usually uses a coupled time-domain analysis approach. Another is the sequential analysis approach, which is a semi-coupled time-domain analysis approach.

Time-Domain Analysis for Floating Offshore Wind Turbines

Time-domain analysis for Floating Offshore Wind Turbines includes the following types of analysis in the time domain:

- Integrated load analysis.
- Motion analysis.
- Air gap analysis.
- Mooring analysis.

Time-domain analyses consist of numerically solving the equations of motion in the time domain for the Floating Offshore Wind Turbine subjected to environmental conditions and the RNA operational loads. As the input to time-domain analyses, time series of wind and wave conditions are generated to simulate turbulent wind conditions and stochastic wave elevations and kinematics. Time-domain analyses are the preferable approach for Floating Offshore Wind Turbine global performance analyses, primarily because they can provide a rational means of modelling the nonlinear and transient effects in Floating Offshore Wind Turbine global responses. These nonlinear effects include, but are not limited to, hydrodynamic drag forces, finite wave amplitude effects, nonlinear restoring forces from moorings, and effects of motion suppression devices or components (e.g., heave plates). Time-domain analyses also allow modelling of the coupling effects among responses of the turbine RNA, the tower, the floating substructure, the station keeping system, and if needed, the power cable. Time-domain analyses should be conducted for a sufficient period to achieve stationary statistics, particularly for low frequency responses. Multiple realisations of an individual set of stochastic site conditions may be necessary to generate adequate data for statistical analysis and to verify the consistency of the simulation. The most probable maximum responses should be predicted using appropriate distribution curve fitting or other recognised statistical techniques. For the TLP-type floating substructure, the ringing (the high frequency vertical vibration excited by impulsive loading) and the springing (the high frequency vertical vibration excited by cyclic loading at or near the resonant periods) responses of the TLP hull and the tendon should be considered as appropriate. Further guidance on high frequency ringing and spring analyses can be found in API RP 2T. The effect of Vortex-Induced Motions (VIM) for the floating substructure in the form of Spar, single column TLP, or other types of deep-draft hull structure should be considered as appropriate.

3.2.3 Combined Time-Domain and Frequency-Domain Analysis

Combined time-domain and frequency-domain analysis refers to the approach where first-order wave frequency responses are computed in the frequency domain while other dynamic responses are computed in the time domain. Total dynamic responses are the combination of relevant responses obtained from frequency-domain and time-domain analyses.

Combined Time-Domain and Frequency-Domain Analysis for Bottom-Founded Offshore Wind Turbines

Combined time-domain and frequency-domain analysis may be used in fatigue analysis for the substructure in the preliminary design stage. In this approach, the wave and wind load effects are decoupled. The fatigue analysis is performed through a combination of the time domain analysis of aerodynamic responses and the frequency domain analysis of hydrodynamic responses of the substructure. The contribution of wind to the combined responses can be simulated for a wind turbine in a calm sea, considering the added mass as the only hydrodynamic effect. The wave responses can be determined in the frequency-domain. Aerodynamic damping should be separately evaluated and applied in the analysis. The equivalent stress range of the combined fatigue for each load case can be obtained by combining the equivalent stress range of fatigue induced by wind and wave responses using direct quadratic superposition or the weighted quadratic superposition.

Combined Time-Domain and Frequency-Domain Analysis for Floating Offshore Wind Turbines

This approach can be used in the following analysis for the floating substructure and station keeping system of Floating Offshore Wind Turbines:

- Motion analysis.
- Air gap analysis.
- Mooring analysis.

To reduce the complexity and computational effort associated with full time-domain simulations, combined time-domain and frequency-domain analyses are often employed. Typically, the mean and low frequency responses (hull displacements, mooring line tensions, anchor loads, etc.) are computed in the time domain while the wave frequency responses are solved separately in the frequency domain. The frequency-domain solution of wave frequency responses is normally processed to obtain either statistical peak values or time series, which are then superimposed on the mean and low frequency responses.

3.3 Quasi-static and Dynamic Analysis for Foundation and Station Keeping System

3.3.1 Quasi-static Analysis

The quasi-static analysis approach refers to the analysis method where the foundation of Bottom-founded Offshore Wind Turbines or the station keeping system of Floating Offshore Wind Turbines is modelled quasi-statically.

Quasi-static Analysis for the Foundation of Bottom-Founded Offshore Wind Turbines

The soil-structure interaction (SSI) effects are modelled as linear or nonlinear elastic springs. Soil damping effects may be added separately.

Quasi-static Analysis for the Station Keeping System of Floating Offshore Wind Turbines

In quasi-static analyses, the mooring system is modelled quasi-statically and wave actions are accounted for by statically offsetting the hull using wave-induced hull motions. Dynamic actions on the mooring lines associated with mass, damping and fluid accelerations are neglected. Experience has shown that the reliability of mooring system designs based on quasi-static analyses can vary widely depending on the hull type, water depth and mooring line configuration.

3.3.2 Dynamic Analysis

Dynamic analysis refers to the analysis method where the foundation of Bottom-founded Offshore Wind Turbines or the station keeping system of Floating Offshore Wind Turbines is modelled dynamically.

Dynamic Analysis for the Foundation of Bottom-Founded Offshore Wind Turbines

The dynamic analysis approach considers the dynamics of the soil-structure interaction (SSI) through the modelling of the effects of damping, nonlinear stiffness, and effects of added-mass from the soil.

Dynamic Analysis for the Station Keeping System of Floating Offshore Wind Turbines

Dynamic analysis of the mooring system accounts for the time-varying effects due to mass, damping, and fluid accelerations. Time-varying fairlead motions are calculated from the hull structure's surge, sway, heave, roll, pitch, and yaw motions. Dynamic models are used to predict the mooring line responses to fairlead motions. Either frequency-domain or time-domain analyses can be used to predict dynamic responses of the mooring system.

3.4 Coupled, Semi-coupled, and Uncoupled Analysis

3.4.1 Coupled Analysis

A coupled analysis considers the interactions or coupling effects between subsystems of an offshore wind turbine.

Coupled Analysis for Bottom-Founded Offshore Wind Turbines

There are several types of interactions, also known as coupling, among subsystems of a Bottom-founded Offshore Wind Turbine. A coupled analysis implements the time-domain approach. More specifically, the complete system of equations accounting for the structural model of the substructure, elastic models of the tower and the RNA, the foundation model, as well as the control system, are solved simultaneously using a nonlinear time-domain dynamic analysis. The foundation can be modelled using either the quasistatic or dynamic analysis method. When a coupled model is used for global performance analyses, coupling effects among responses of the turbine RNA, the tower, the substructure, and the foundation can be considered at each incremental analysis time step. A more accurate simulation of the effects of the turbine control system and the turbine's operating conditions can also be achieved using this approach.

Coupled Analysis for Floating Offshore Wind Turbines

There are several types of interactions, also known as coupling, among subsystems of the Floating Offshore Wind Turbine. A coupled analysis implements the time-domain analysis approach. More specifically, the complete system of equations accounting for the rigid body model of the hull structure, elastic models of tower and turbine RNA, the slender body model for the cables and mooring lines, as well as the control system are solved simultaneously using a nonlinear time domain approach for dynamic analysis. In a coupled analysis, the mooring system is modelled using the dynamic analysis method. When a coupled model is used for global performance analyses, coupling effects among responses of the turbine RNA, the tower, the floating substructure, the station keeping system, and the subsea power cable can be considered at each incremental analysis time step. A more accurate simulation of the effects of the turbine control system and turbine's operating conditions can also be achieved using this approach.

3.4.2 Semi-coupled Analysis

A semi-coupled analysis considers part of the interactions or coupling effects between subsystems of an offshore wind turbine.

Semi-coupled Analysis for Bottom-Founded Offshore Wind Turbines

The sequential analysis approach for Bottom-founded Offshore Wind Turbines is a semi-coupled analysis in the time domain. The finite element model of the substructure is reduced to a super element connected to the tower supporting the RNA model. Through the coupled aero-servo-elastic simulation, the interface loads at the tower base can be obtained. The obtained interface loads together with the wave loads on the substructure are used as the input for further dynamic structural analysis. Aerodynamic damping is not simulated consistently and simultaneously along with the wind and wave loads. The aerodynamic damping should be evaluated separately and applied in the analysis.

Semi-coupled Analysis for Floating Offshore Wind Turbines

In a semi-coupled analysis for a Floating Offshore Wind Turbine, the coupling effects between aero-elastic and aero control (servo) are neglected. The tower and the RNA are modelled as part of the hull as a rigid body. The aerodynamic loads are modelled by the wind forces applied at the top of the tower. Effects of the control system should be suitably considered in deriving the aerodynamic loads on the RNA. The aero-control (servo) coupling effects may be represented by the rotor thrust force whose magnitude is expressed as a function of wind speed and the RNA's operational condition. In the parked condition, nacelle yaw misalignment may be included by using appropriate wind force drag coefficients and windage areas. The hull, mooring system and, if needed, power cables are dynamically coupled in this approach.

3.4.3 Uncoupled Analysis

An uncoupled analysis approximates the tower and the RNA as a concentrated mass attached to the substructure. For Floating Offshore Wind Turbines, the station keeping system and cables are modelled quasi-statically in uncoupled analysis.

Uncoupled Analysis for Bottom-Founded Offshore Wind Turbines

The uncoupled analysis is normally not used for Bottom-founded Offshore Wind Turbines. The flexibility of the tower and turbine blades may affect the dynamic responses of the overall system significantly.

Uncoupled Analysis for Floating Offshore Wind Turbines

In an uncoupled analysis for a Floating Offshore Wind Turbine, the system of equations accounting for the rigid body motions is solved in the time domain. The tower and the RNA are modelled as part of the hull as a rigid body. Effects of the control system should be suitably considered in deriving the aerodynamic loads on the RNA. The aero-control (servo) coupling effects may be represented by the rotor thrust force whose magnitude is expressed as a function of wind speed and the RNA's operational condition. The effects of the mooring and subsea cable systems are modelled quasi-statically using nonlinear springs based on the quasi-static restoring force characteristics. All other coupling effects between the hull and the mooring system (e.g., contributions from damping and current loading on the mooring lines) need to be pre-calculated and provided as direct input to the analysis. The same load model may be applied for the hull in coupled, semi-coupled or uncoupled analyses.

3.5 Integrated Load Analysis

An integrated load analysis performs load and load effect calculations for the entire offshore wind turbine comprising the RNA and the support structure. The design evaluation of the RNA components and the support structure is performed using the results from the integrated load analysis. The integrated load analysis is normally conducted at the detailed design stage with involvement of the support structure designer and the wind turbine manufacturer. It also involves an independent third party to perform independent verification of the integrated load analysis. The integrated load analysis is usually performed using the coupled analysis approach in the time domain.

4 Software Tools for Integrated Load Analysis

Software tools commonly used by the industry to perform the ILA for offshore wind turbines can perform the coupled aero-hydro-servo-elastic analysis in the time-domain. In these ILA software tools, the aerodynamics, elasticity of the blade and tower, hydrodynamics on the substructures, and the control system are solved simultaneously in the time domain. The analysis methods for different functional modules commonly used in these software tools are summarised in Table 1.

These software tools can be categorised into two categories:

- ***Category 1***: The main function includes the integrated aero-hydro-servo-elastic load analysis for wind turbine generator (WTG) and substructure and foundation or station keeping system. These software tools also have the capability of modelling the floating substructures and mooring/cable systems and may provide optional interfaces to other commercial software.
- ***Category 2***: The main function includes the aero-servo-elastic load analysis for wind turbine generator (WTG) including RNA, tower, and turbine controller. These software tools may have limited capability of modelling the substructures and foundation or station keeping system. They normally can be linked to other commercial software with the capability of modelling the substructure hydro-elastic interaction, floating substructure, mooring/cable system, and foundation. For Category 2 software tools, the aero servo-elastic programme is normally

Table 1 Analysis methods in software tools for offshore wind turbines

Functional modules	Functions	Analysis methods
Aerodynamics (Aero)	Aerodynamic loads for RNA	BEM + DW + DS/FVW
Hydrodynamics (Hydro)	Hydrodynamic loads for substructure	Panel[(1)] + Morison element
Control system (Servo)	Control system responses and actuation loads	DLL
Structural dynamics (Elastic)	Modelling flexibility of blade, tower, substructure, and drive train responses	Modal, Multibody, FEM

Note (1) Results for the Panel model is input from a standalone diffraction-radiation software tool. In general, these software tools were initially developed for Bottom-founded Offshore Wind Turbines. The current start-of-the-art software tools have the capability of performing integrated load analysis for both Bottom-founded Offshore Wind Turbines and Floating Offshore Wind Turbines. Software tools may also have an interface to other commercial software for modelling floating substructure, substructure hydro-elastic coupling, mooring/cable system, and complex foundations including Soil–Structure Interaction (SSI)

connected to commercial software though an interface dynamic link library (DLL) as shown in Fig. 1. In these software tools, there are several different approaches for developing the aerodynamic model, hydrodynamic model, control system interfaces, and structural dynamic model as summarised below.

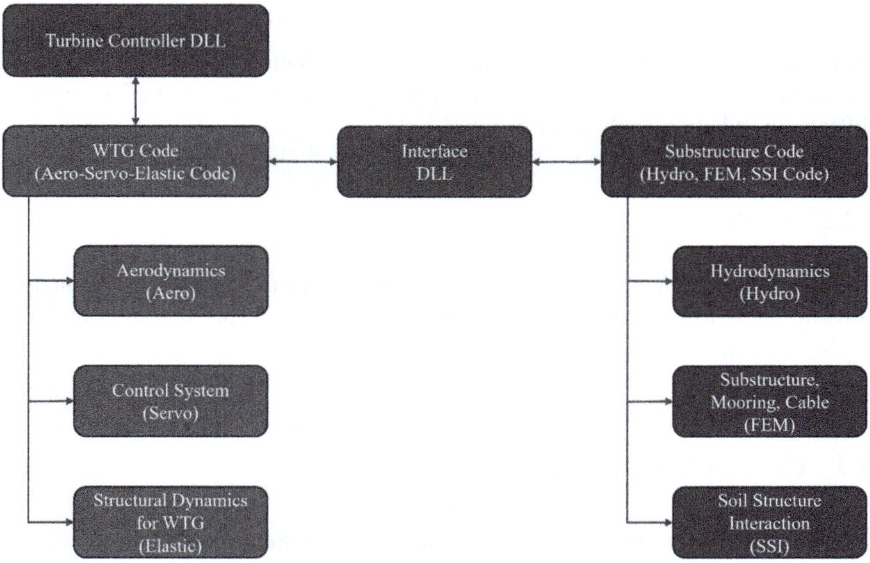

Fig. 1 Typical interface between WTG (RNA + tower) software and substructure software

4.1 Aerodynamic Models

4.1.1 Low Fidelity Engineering Model

The blade element momentum method (BEM), together with the dynamic wake (DW) (also called dynamic inflow) model and the dynamic stall (DS) model, is typically used for aerodynamic load calculation.

4.1.2 Higher Fidelity Model

ILA software tools may also apply a higher fidelity model based on the free vortex wake (FVW) method for aerodynamic load calculation.

4.2 Hydrodynamic Models

Hydrodynamic models usually include the panel model along with the Morison element model.

4.2.1 Panel Model

ILA software tools normally import the diffraction-radiation results of a panel model for hydrodynamic loads on large volume substructure solved through standalone diffraction-radiation analysis software.

4.2.2 Morison Element Model

The Morison element model is normally available within the integrated load analysis software tools to calculate the hydrodynamic loads for slender members and drag loads of the substructure.

4.3 Control System Interfaces

Control systems for the wind turbine can be modelled through an external dynamic link library (DLL) for the turbine controller. Some software tools may allow users to define a subroutine or an interface with specific control design software.

4.3.1 Dynamic Link Library

The control system algorithm is normally compiled into a dynamic link library (DLL) and integrated into an ILA software tool. Such a DLL may be integrated through a predefined subroutine within the ILA software tool together with a predefined number of control parameters and a predefined data transfer format. DLL can also be linked to the ILA software tools through an application programming interface as an external function. The application programming interface can provide flexibility for accommodating the control parameters and data transfer format in integrating the user developed DLL.

4.3.2 User Defined Subroutine

For some ILA software tools, control systems can be programmed as a user-defined subroutine. In such a case, the source code and the user defined subroutine may need to be compiled together. This approach is available in open-source programmes.

4.3.3 Interface to Control Design Software

Some ILA software tools have an interface to specific control design software for control systems testing and design.

4.4 Structural Dynamic Models

Different methods including multibody-dynamics formulation, finite element method (FEM) and modal superposition method are used to model the structural dynamics of the entire system for rigid bodies motions and flexibility member deflections and loads.

4.4.1 Multibody-Dynamics Formulation

Multibody dynamics formulations are used to account for the degrees of freedom (DOFs) of the overall system. These include rigid body motions, flexible structural model with modal superposition, or FEM for blade and tower, torsional motion of wind turbine drive train.

4.4.2 Finite Element Method (FEM)

FEM is usually used for slender members such as blade, tower, shaft, mooring lines, and cables. Blades may be modelled using both the modal superposition method and the finite element method.

4.4.3 Modal Superposition

The modal superposition method is usually used to model the blade and tower mode shapes for the first several vibration modes. A modal analysis is necessary to obtain the modal shapes.

Chapter 20
Integrated Load Analysis for Bottom-Founded Offshore Wind Turbines

1 Introduction

1.1 General

The integrated load analysis (ILA) is performed for the following purposes:

- Classification based on Part 1 when ILA is recommended.
- Certification and verification in compliance with local regulatory requirements, IEC certification scheme based on IECRE OD-502, and IEC 61400 series standards or other industry standards for offshore wind turbines.

The ILA involves aero-hydro-servo-elastic analyses that can account for the coupled dynamics of a Bottom-founded Offshore Wind Turbine including the RNA, control system, tower, substructure, and foundation modelled and solved simultaneously in the time domain. In contrast to the sequential analysis methods, the ILA accounts for the coupling effects and damping effects including aerodynamic and hydrodynamic damping in a consistent manner. It can provide a more rational calculation of global motions and loads for the detailed design evaluation of the RNA components, tower, substructure, and foundation. The software tools for ILA should be compatible with the modelling approaches for individual components of a Bottom-founded Offshore Wind Turbine and the load transfer strategy for detailed design evaluations for each component. The numerical modelling and simulation methods should comply with the applicable classification or certification requirements. Where there are no specific requirements, the guidance in this book may be applied.

1.2 Design Data, Reports and Drawings

The following design data, reports, and drawings are necessary for the ILA:

F. Aderouane, *Design, Construction, and Global Performance Analysis of Offshore Wind Turbines*, Green Energy and Technology, https://doi.org/10.1007/978-3-031-86010-2_20

- Site conditions and environmental data (such as water depth, seabed bathymetry, soil data, wind, waves, current, water level, wind/wave misalignment, and fatigue sea states).
- RNA modelling data (such as geometry, dimensions, air foil aerodynamic data, drive train properties, blade material properties, stiffness, damping, mass distribution, natural frequencies, turbine power, and thrust curves).
- Tower modelling data (such as geometry, dimensions, material properties, stiffness, damping, mass distribution, and natural frequencies).
- Wind turbine control system modelling data (such as description of the wind turbine control software, the controller dynamic link library (DLL) to integrate to the load analysis software, and operation procedures).
- Substructure modelling data (such as geometry, dimensions, material properties, stiffness, damping, and mass distribution).
- Foundation modelling data (such as geometry, dimensions, material properties, stiffness, damping, and mass distribution).
- Other design considerations pertinent to the Bottom-founded Offshore Wind Turbines, if applicable.

2 Integrated Load Analysis Procedures

2.1 General

The ILA consists of the following key steps:

(1) Build a global performance analysis model.
(2) Define environmental load models.
(3) Set up data files for each Design Load Case (DLC).
(4) Perform ILA simulations.
(5) Obtain time histories of loads and motions.
(6) Post-process and report results.

2.2 ILA Procedure for Bottom Founded Offshore Wind Turbines

The ILA procedure for Bottom-founded Offshore Wind Turbines is illustrated in Fig. 1. The ILA model for a Bottom-founded Offshore Wind Turbine includes the WTG (RNA and tower) model, substructure model and foundation model. The substructure may be simplified as a beam element model. The nonlinearity of the foundation stiffness should be included in the ILA model. Time histories of aerodynamic loads and the associated wave and current loads from the ILA can be used as inputs for the subsequent detailed structural analysis and design code check. The

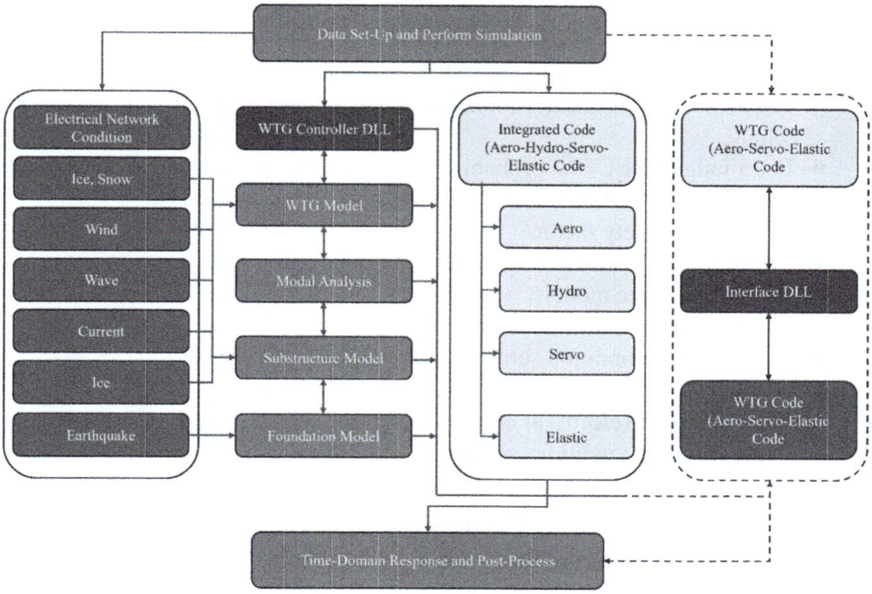

Fig. 1 ILA procedure for bottom-founded offshore wind turbines

WTG loads and foundation loads calculated through ILA are used for the WTG and foundation design evaluations. Depending on the ILA software capability, a detailed substructure model together with the WTG model and foundation model can also be developed for the ILA. The calculated loads are used directly within the software tools for the time-domain dynamic structural analysis and design code check for the substructure. The ILA for a Bottom-founded Offshore Wind Turbine typically includes the following steps:

(1) Build a global performance analysis model:

- WTG model including the RNA and tower models.
- Wind turbine controller model.
- Foundation model.
- Substructure model.
- Mode shapes calculated through the modal analysis.

(2) Define environmental load models:

- Wind load models.
- Wave load models.
- Current load models.
- Modelling of ice and snow accumulation induced loads, if applicable.
- Modelling of sea/lake ice loads, if applicable.
- Modelling of seismic loads, if applicable.

(3) Set up data files for each Design Load Case (DLC):

- Simulation time duration, time step, and number of random seeds.
- Random seed for each simulation requiring applying random waves and winds.
- For a fatigue DLC, the probability of occurrence or number of occurrences of fatigue design conditions.
- Partial load safety factor.
- Wind, wave, current and water level models, and parameters.
- Wind farm wake models.
- Wind turbine controller setup.
- Wind turbine condition setup.
- Nacelle yaw misalignment setup.
- Gird condition (electrical network condition) setup.
- Other external conditions setup.

(4) Perform ILA simulations.
(5) Obtain time histories of loads of the RNA, tower, substructure, and foundation.
(6) Post-process and report results.

3 Modelling of Bottom-Founded Offshore Wind Turbines

3.1 General

In general, dynamic analyses should be conducted to evaluate dynamic responses of an offshore wind turbine subjected to site-specific external conditions and operating conditions. While various simplified analysis methods may be used in the preliminary design, a coupled analysis approach is recommended for the final detailed design. Prototype tests and model tests may be used to supplement the load calculation. A global performance analysis model should be developed for the ILA and include the following aspects:

- Modelling of foundation.
- Modelling of the substructure.
- Modelling of flexibility of the tower.
- Modelling of rotor blades and control and safety systems.
- Modelling of dynamics of drive trains.
- Modelling of wind farm wake effects.
- Modelling of mode shapes through the modal analysis.

The external conditions to be considered in the ILA include:

- Environmental conditions.
- Electrical network conditions.

In addition, various fault conditions relevant to the electrical system, control and safety systems, and mechanical components should be properly modelled in the analysis.

3.2 Modelling of Foundation

3.2.1 General

Foundation modelling can affect the global dynamics of Bottom-founded Offshore Wind Turbines. The soil-structure interaction (SSI) should be developed in the foundation model. Typical types of foundations for Bottom-founded Offshore Wind Turbines include:

- Gravity-based foundation (GBF).
- Monopile foundation (large diameter pile).
- Pile foundation (tripod, jacket).
- Other foundations.

The gravity-based foundation may be used for soils with a high load-bearing capacity. For soils with low load-bearing capacity, deep foundations such as piles are commonly used. Pile foundations can be either monopiles or pile foundations of jackets or tripods. For TLP-type Floating Offshore Wind Turbines with tendon foundation, the simplified soil-structure interaction (SSI) model is usually employed in the global model. The foundation model should include the following two aspects:

- Stiffness model.
- Damping model.

Foundation stiffness affects the natural frequencies of a Bottom-founded Offshore Wind Turbine. Foundation damping can affect the magnitude of the global dynamic responses. In power production conditions, aerodynamic damping is normally more important than soil damping. While in parked conditions, aerodynamic damping is low, especially in the side-to-side directions, and soil damping may be important for parked and wave-wind misalignment conditions. The foundation stiffness model should consider the nonlinear stiffness of the soil that can represent the soil reactions ranging from small displacements required for the calculation of natural frequencies of a wind turbine structure and fatigue load cases to the large displacement response required for the calculation of the extreme load cases. The foundation model may only include the stiffness model, while the damping effects may be separately evaluated and applied to the model. Where no data is available for soil damping, the soil damping may be taken as a percentage of 0.25–1.5% of the critical damping. To be conservative, the foundation damping may also be neglected. In the time-domain analysis using the finite element method, foundation damping can be added as Rayleigh structural damping. For the modal superposition method, an overall foundation damping may be added to each of the modes. When the added mass of soil is of importance, the effects can be added separately through an added mass matrix at the top of the foundation.

3.2.2 Gravity-Based Foundation

The shallow gravity-based foundation may be assumed rigid with respect to the underlying soil. For large foundations, the added soil mass is often large and should be included in the model. In general, the finite element method should be used for the flexible foundation structure. In the global performance analysis model, the following two modelling techniques can be applied:

- Macro-element model.
- Distributed springs model.

A macro-element model is used to represent a 6-DOF nonlinear force–displacement relations at a point of the foundation in the vertical, horizontal, and rotational directions. A macro-element model can be introduced at the base of the foundation to account for soil-structure interaction. The macro-element model can simulate the 3D behaviour of a rigid shallow foundation under static and cyclic loading and plasticity and uplift mechanisms. More detailed guidance for a macro-element model is given in "Advanced Foundation Models". In the distributed springs model, several uncoupled nonlinear horizontal and vertical springs are used to model the soil-structure interaction. The vertical spring is distributed under the foundation to capture foundation rotations, which are often the dominant contributor to the SSI effects. Horizontal springs are attached to the sides of the foundation. The vertical springs model the vertical load–displacement behaviour, while horizontal springs model the horizontal load displacement * behaviour against the side of a foundation and horizontal sliding behaviour at the base of the foundation. In addition, dampers can be included to capture foundation damping. The vertical springs can be modelled to represent compressive yield and detachment at the point of uplift. Foundation stiffnesses and bearing capacity can be calculated from semi-empirical equations based on foundation dimensions and properties, soil properties, soil, and foundation testing, or from finite element analysis (FEA).

For soil properties and soil and foundation testing, refer to API RP 2GEO and ISO 19901-4 for guidance. The nonlinear spring stiffness relations may be obtained based on direct analysis using finite element analysis as discussed in "Advanced Foundation Models". Horizontal force and overturning moment are applied to the foundation top. The loads are increased at a constant ratio while translation and rotation of the foundation are recorded. Force–displacement and moment-rotation relations can then be obtained. An elastic–plastic soil model may be used to evaluate the soil damping ratio under cyclic loading. The analysis is performed with a loading–unloading-reloading cycle statically or dynamically. The energy dissipation in a loading cycle can be used to estimate the soil damping as a function of loading amplitude.

3.2.3 Monopile Foundation

Foundation models for the global load analysis of monopile-based Bottom-founded Offshore Wind Turbines may use one of the following types of techniques as shown in Fig. 2:

- Apparent fixity model.
- Coupled springs model (i.e., stiffness matrix).
- Distributed springs model.

Simplified Linear Models

(1) ***Rigid Foundation***. In the rigid foundation model, the pile is clamped rigid at the mudline. This simplified method provides higher stiffness than a real foundation. It is used in some software tools as a modelling option for preliminary analysis of support structures.

(2) ***Apparent Fixity Model***. In the apparent fixity model, the monopile is extended to a certain depth below the seabed and is rigidly fixed there. The apparent fixity model assumes that the substructure is fixed (cantilevered, without surrounding soil) at a depth below the original seabed, with a fixity depth and beam properties determined such that it matches the same lateral displacement and rotation at the seabed as the one resulting from the pile embedded in the actual soil profile. The geometry and material properties of the pile above seabed are used for the extended length of the pile. The extended length is determined to provide the same fundamental eigenfrequency as the pile foundation. The soil damping, if considered, is included separately. Due to the nonlinear nature of the actual foundation, the response from a linear model is accurate under a particular load level. This method can be used in preliminary design, and it is in general not suitable for detailed design.

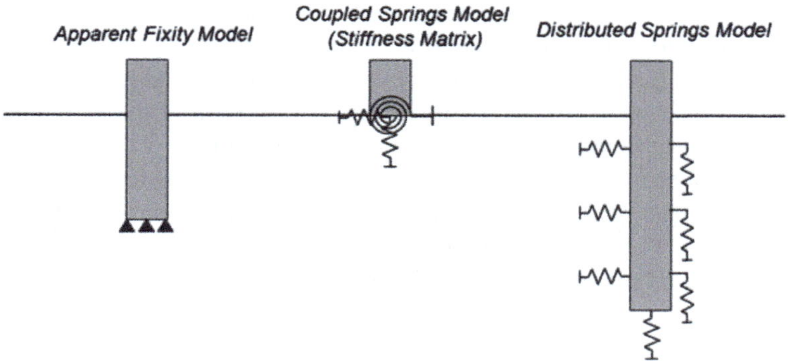

Fig. 2 Monopile foundation models for ILA

(3) ***Coupled Springs Model (Stiffness Matrix)***. The coupled springs model or stiffness matrix model is used to model the foundation as a 6×6 stiffness matrix. The 6×6 stiffness matrix represents a set of linear coupled or uncoupled springs for three translational and three rotational DOFs applied at a point on the pile at the mudline. The stiffness matrix is obtained through linearisation of the pile soil stiffness around a given load level. FEA may be used to calibrate the stiffness matrix. The first and second natural frequencies can be more accurately estimated using a stiffness matrix rather than the apparent fixity model. The coupled springs model replaces the soil with one linear stiffness matrix located at the seabed that accounts for the six rigid-body DOFs (or fewer) of the base of the monopile. This approach can be extended to include a viscous damping matrix to account for the energy dissipated by the foundation. For a nonlinear foundation, this method only accounts for soil stiffness at one load level. This method can be used in preliminary design, and it is in general not suitable for detailed design.

Distributed Springs Model Based on $p - y$ Curves Method

(1) **American Petroleum Institute (API)** $p-y$ **Curves Method**. In current industry design practice, the monopile foundation is modelled based on the $p - y$ curves method. The pile foundation is modelled by a series of discrete lateral and axial springs distributed along the length of the pile. Most ILA software tools can model linearly distributed springs. Some allow for uncoupled nonlinear elastic springs. In these software tools, the spring models usually represent the API $p - y$, $t - z$, and $Q - z$ curves. The $p - y$ curves model is constructed based on soil and pile test data. The $p - y$ curves can also be calibrated to FEA results of the soil and the foundation. Details about this method can be found in Sect. 3.2.4. The API $p-y$ curves method is developed for the slender piles used for offshore oil and gas platforms. This model was developed based on the measurements from test piles with an outer diameter (*D*) no greater than 2 m (6.5 ft) and an aspect ratio (length to diameter ratio *L/D*) larger than 10, which is much greater than those for monopiles. Due to the larger diameter (*D*) and smaller length to diameter (*L/D*) ratio of the monopiles, the geometry and load pattern differ from slender piles. In addition to the lateral deflection of the pile, monopiles can also have rigid body motions. Therefore, the API $p - y$ curves derived for flexible piles may not be accurate for modelling soil-structure interaction for large diameter short monopiles with the diameter (*D*) typically ranging from 3.5 to 8 m (11.5 to 26 ft) and the aspect ratio (*L/D*) between 5 and 12 due to the associated pile rigid motions. Studies using finite element analysis of large diameter piles showed that the API $p - y$ curves method tends to overestimate soil stiffness for large diameter monopiles, particularly at greater soil depths. Using the modified $p - y$ curves based on FEA can improve the results. It is suggested that the $p - y$ curves be calibrated with FEA results or large-scale test data. The Pile Soil Analysis (PISA) method includes three other components

of soil reaction through distributed rotational springs and lateral and rotational springs at the base in addition to the nonlinear lateral ($p - y$) springs. It can achieve improved predictions of soil monopile interaction. The details of the PISA method are provided in "Distributed Springs Model Based on $p - y$ Curves Method". The foundation damping may be modelled by hysteresis loop created by loading and unloading $p - y$ curves. In software tools using the $p - y$ curves model, the soil is typically modelled with elastic springs with no foundation damping. Foundation damping values from 0.25 to 1.5% of critical damping may be used in the integrated load analysis for the monopile foundation. Foundation damping can also be calibrated using finite element analysis.

(2) **PISA p-y Curves Method**. Pile Soil Analysis (PISA) method is a new methodology for the design of offshore wind turbine monopile foundations based on field testing and advanced finite element analysis. The PISA design method retains the advantages of the traditional $p - y$ method, while incorporating enhancements to improve the performance of the modelling approach for low length to diameter ratio monopiles. In the PISA design method, soil reaction curves are defined as four-parameter functions in which the parameters are depth-dependent. In addition to a distributed lateral load, which is modelled by $p - y$ curves, three other components of soil reaction are assumed to act on the monopile (i.e., distributed moment, lateral force, and moment at the base). Nonlinear soil reaction curves for lateral loading, rotation, base shear, and base rotation used in the PISA design method are calibrated by 3D FEA with soil properties from site investigation and validated with field tests. A procedure is developed for the calibration of soil reaction curves using 3D FEA results and can be used for the monopile design. Numerical soil reaction curves are determined from the 3D FEA results using a process to extract the nodal forces acting at the soil-pile interface, and stresses in the interface elements between the pile and the soil. In addition to the design procedure based on soil reaction curves calibrated from the 3D FEA results, there are predefined soil reaction curves within the PISA design method. These curves may be adopted for design where site investigation data are limited at an early stage of the design process. The PISA design method is developed for monopiles under monotonic lateral loading. The PSI effects of soil-stiffness reduction and soil damping under cyclic loading can be considered in the design separately.

Advanced Foundation Models

(1) **Macro-element Models**. In the macro-element model, the response of a pile and the surrounding soil is condensed to a force–displacement relation at the seabed. The models have input parameters that are predefined based on numerical analyses of the pile and soil or from model testing. The macro-element adds six (6) DOFs (3 rotational and 3 translational) to the full structural model, fewer than the degrees of freedom required for distributed springs method. Complex macro-element models can account for several fundamental effects such as the

effect of cyclic loading in the soil, change in stiffness, gapping, and damping. The computational burden will depend on the complexity of the macro-element. The main disadvantage of the macro-element approach is that the pile itself is excluded from the analyses and needs to be modelled in a separate analysis.

(2) **Finite Element (FE) Model**. An FE model of the soil and the foundation in an aero-servo-hydro-elastic simulation tool for ILA in the time domain provides the most accurate modelling approach if used together with appropriate constitutive soil models. The main disadvantage is the large computational cost of time-domain 3D FEA with continuum elements, which makes the approach impractical for ILA in the design process. The FE-based advanced foundation methods are normally used to calibrate the foundation model used for ILA. The results of the finite element analysis can also be used to construct a set of $p-y$ curves used for the pile foundation model. Guidance on the development of $p-y$ curves using finite element analysis is provided in ISO 19901-4 Annex A.

The FE-based model can be used to evaluate the soil stiffness reduction and damping effects under cyclic loading. The typical nonlinear spring approach used in the ILA normally considers these effects separately. The effects of soil stiffness reduction and damping can be considered in the nonlinear spring approach used in the ILA through 3D FEA based on the elastoplastic model.

3.2.4 Pile Foundation

The pile foundations for long and slender tripod and jacket piles are usually modelled using the $p-y$ curves method based on API RP 2A-WSD. In the $p-y$ curves method, a series of discretely distributed, uncoupled, nonlinear elastic springs at nodal points along the pile are applied to represent the pile-soil lateral, axial, and tip resistances through the lateral soil resistance deflection $p-y$, axial shear transfer $t-z$, and mobilised end bearing capacity $Q-z$ curves as given below:

- $p-y$ **Curve**: Relationship between the lateral soil resistance per unit length of pile (p) and pile deflections due to the resistance (y), as a function of depth below the original seafloor (z).
- $t-z$ **curve:** Relationship between mobilised soil-pile shear transfer (t) and local pile deflection (z) at any depth.
- $Q-z$ **curve:** Relationship between mobilised end bearing resistance (Q) and axial pile tip displacement (s).

where

P = lateral soil resistance per unit length of pile, in N/m (lb/ft).
t = mobilised soil-pile shear transfer, in Pa (lb/ft^2).
Q = mobilised end bearing resistance, in N (lb).
y = deflections of pile due to the lateral soil resistance, in m (ft).
z = depth below the original seafloor, in m (ft).

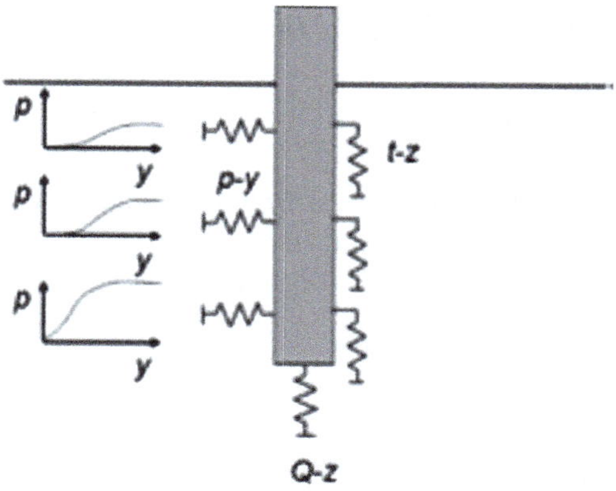

Fig. 3 Pile-soil interaction model

= axial local pile deflection, for axial shear transfer t-z curves, in m (ft)
= axial pile tip displacement, for $Q - z$ curves, in m (ft).

The nonlinear curves ($p - y$ curve, $t - z$ curve, and $Q - z$ curve) can be constructed using stress–strain data from laboratory soil samples and dated from pile load tests based on the procedures provided in API RP 2A-WSD, API RP 2GEO and ISO 19901-4. The $p - y$ curves model can also be calibrated using the finite element model of the soil and foundation. Foundation design software for oil and gas platforms typically has the capability to model the pile soil interaction based on the $p - y$ curves method. Pile-soil interaction can account for pile finite deflection ("P-delta" effect) and nonlinear soil behaviour in both transverse and axial directions. The pile is modelled by beam-column elements subject to axial and bending loads. The soil around the pile is modelled by a series of discrete springs. The pile-soil interaction model based on the $p - y$ curves method is shown in Fig. 3.

3.2.5 Other Foundations

Modelling methods as given in Sect. 3.2 may be applied to other types of foundations such as suction buckets. General guidelines on foundation modelling can be found in IEC 61400-6, API RP 2A, ISO 19902, ISO 19903, and API RP 2T. Geotechnical and foundation engineering applicable to a broad range of offshore structures can be found in API RP 2GEO and ISO 19901-4. The design of pile foundations for fixed offshore steel structures is detailed in API RP 2A and ISO 19902. Requirements for the design of shallow gravity foundations for fixed offshore concrete structures are detailed in ISO 19903. Foundation design for TLP-type Floating Offshore Wind Turbines can be found in API RP 2T.

3.3 Modelling of the Substructures

3.3.1 General

The substructures of a Bottom-founded Offshore Wind Turbine can be modelled using finite element method or a rigid body with 6 DOFs of motions. Modelling of the substructures should consider the following loads and load effects:

- Hydrostatic loads.
- Gravitational and inertial loads.
- Wind, wave, and current loads.
- Vortex-induced vibration (VIV).
- Other loads (such as ice loads, ice and snow accumulation induced loads, earthquake loads, impact loads, etc.).

Global structural models may comprise equivalent beams, space frame models, or combined shell element/beam element models, as appropriate. Models should accurately represent the global stiffness of the substructure and the relative stiffness of the major structural components. In the case where local structures significantly affect the global stiffness and/or response, such effects should be adequately considered when developing the global structural model. A combined global or local structural model may be warranted. The extent of the global structural model should be defined such that boundary conditions and loads can be imposed at well-defined or well-understood interfaces. Space frame structures consisting of slender members can be analysed using the 3D frame analysis to calculate internal member forces and moments. The effects of joint eccentricity and flexibility, where significant, should be accounted for. The structural members in the substructure can be modelled based on linear elastic theory. However, nonlinear relationships between loads and load effects should be properly accounted for, when deemed important. This chapter provides guidance on structural modelling techniques for the following types of substructures:

- Gravity based substructure.
- Monopile substructure.
- Tripod substructure.
- Jacket substructure.
- Other substructures.

Guidance on modelling of environmental loads on substructures is provided in Sect. 4.

3.3.2 Gravity-Based Substructure

Gravity-based substructures can be modelled as a 6-DOF rigid body. To model the natural frequencies of the tower, the structural flexibility of the connection of the substructure to the tower should be included in the analysis.

3.3.3 Monopile Substructure

The monopile is normally modelled using a beam finite element method. The model should account for the distributed stiffness and mass properties along the axial direction of the monopile. The moment of inertia about the axial direction should be accounted for. For grouted piles, the stiffness and mass of the grouted material should be considered in the global model where applicable. A detailed local structural analysis using a refined finite element model with beam, shell, and solid elements can be performed with the global loads from the global load analysis for both strength and fatigue analyses of the monopile substructure.

3.3.4 Tripod Substructure

The tripod substructure is normally modelled using a simplified beam finite element model in the global model considering the geometry of the members, stiffness, and mass distribution of each member. Joint flexibility may also be included in the model. A detailed FE model is needed for the subsequent dynamic structural analysis using the global loads determined through the global load analysis. Alternatively, a detailed FE model can be directly incorporated into the global model. The detailed FE model can comprise beam, shell, and solid elements. Joint flexibility should be considered to model the overall stiffness of the substructure. The stiffness of the connection of the tower to the tripod substructure should be appropriately modelled. In general, joints should be modelled using shell elements.

3.3.5 Jacket Substructure

The jacket substructure can be modelled as a 3D space frame using a simplified beam finite element model. The beam FE model should model the geometry of the members, stiffness, and mass distribution of each member. Joint flexibility may also be included in the model. A detailed FE model is needed for the subsequent time-domain dynamic structural analysis using the global loads determined through the global load analysis. Alternatively, a detailed FE model can be directly incorporated into the global load model. The detailed FE model can comprise beam, shell, and solid elements. Joint flexibility should be considered to model the overall stiffness of the substructure. The stiffness of the connection of the tower to the jacket substructure should be appropriately modelled. In general, joints should be in general modelled using shell elements.

3.3.6 Other Substructures

The substructure can be modelled as one or multiple rigid bodies considering multibody dynamics, finite element models, or a combination of both. For general modelling methods, guidance can be found in API RP 2A, ISO 19902 and ISO 19903.

3.4 Modelling of Flexibility of Tower

The tower can be modelled based on linear elastic theory. However, nonlinear relationships between loads and load effects should be properly accounted for, when deemed important. Structural damping of the tower should be properly determined. The wind loads on the tower should be included. The influence of the tower shadow or tower upwind effects on the wind field perturbation should be properly modelled. The tower structures can be modelled using the finite element method. The model should account for the distributed stiffness and mass properties along the axial direction of the tower. The moment of inertia about the axial direction should be accounted for. The dynamic structural response of the tower may also be calculated by the modal superposition method. Sufficient modes should be included in the analysis. As a minimum, the first few orders of bending modes should be included. If the tower does not have sufficient torsional stiffness, torsional modes should also be included in the modelling. Modal analyses of the tower should consider the influences of substructure and foundation for a Bottom-founded Offshore Wind Turbine. Guidance on the tower modal analysis is provided in Sect. 3.4.

3.5 Modelling of Dynamics of Drive Trains

Drive train dynamics should be carefully considered in the ILA. As a minimum, the torsional mode of the drive train should be included in the analysis. The drive train includes all torque-transmitting components from the rotor to the generator, as well as the elastic mounting of the drive train. The parameterisation assumptions of the drive train model used in the global load calculation should be verified by the calculation using more detailed drive train models. In most cases, the controlling parameters are "resulting drive train stiffness" and "moment of inertia of generator rotor." The verification can also be conducted through the comparison of the first eigenfrequency obtained from the detailed drive train model to the corresponding value derived from the global load simulation model. Other verification techniques can also be used, if appropriate.

3.6 Modelling of Rotor Blade and Control and Safety Systems

3.6.1 General

The rotor blades can in general be modelled based on linear elastic theory. However, nonlinear relationships between loads and load effects should be properly accounted for, when deemed important. As a minimum, the edgewise and flap wise bending degrees of freedom should be considered in the evaluation of aero-elastic responses. Structural damping of the rotor blades should also be carefully selected. The blade structural model can be modelled using the finite element method. The model should account for the distributed stiffness and mass properties along the axial direction of the blade. The dynamic structural response of the blade may also be modelled using the modal superposition method. Modelling of rotor blades and control and safety systems should include the following loads and load effects:

- Gravitational and inertial loads.
- Aerodynamic loads.
- Actuation loads (or operational loads).
- Other loads (such as wake loads, impact loads, ice load, etc.), if applicable.

The influence of the control system on the loads, especially aerodynamic loads, should be properly modelled. Resonance and dynamic amplification of motions due to control system actions should be avoided. For a three-blade horizontal axis wind turbine, the combined effects of inertial, gravitational, and aerodynamic loads of the RNA include the following frequency components:

- Rotor rotation frequencies (1P).
- Blade passing frequencies (3P).
- Harmonics of 1P and 3P (2P, 6P, 9P, etc.).
- Natural frequencies of rotor blades.
- Natural frequencies of other RNA components.

Guidance on the blade modal analysis is provided in Sect. 3.9.

3.6.2 Gravitational and Inertia Loads

Gravitational and inertial loads are static and dynamic loads that can be induced by gravity, vibrations, rotations, and seismic activities. In dynamic analyses, structural dynamics properties and the coupling of vibratory modes should be properly modelled. The following items should be considered:

- Elasticity of the blades.
- Elasticity of the drive train and generator (drive train dynamics).
- Elasticity of the tower.
- Global motions of the floating support structures.
- Mass eccentricity of the rotor.

- Helideck dynamics (if relevant).
- Stiffness of the floating support structure and mooring system (if relevant).
- Elastic mounting of the machinery, vibration dampers (if relevant).

3.6.3 Aerodynamic Loads

The aerodynamic loads on rotor blades are dependent upon:

- Rotational speed of the rotor.
- Average wind speed across the rotor plane.
- Intensity of turbulence.
- Density of the air.
- Aerodynamic shapes of the offshore wind turbine components and their interactive effects, including aeroelastic effects.

The calculation method for aerodynamic loads on rotor blades is normally based on the blade element momentum theory (BEM). Other methods such as the potential flow method, computational fluid dynamics (CFD), etc., may also be used. In addition, aerodynamic loads on the nacelle should be considered, if deemed important. The following aspects should be considered regarding wind loads on the RNA:

- Wind field perturbations due to the offshore wind turbine itself (wake-induced velocities, tower shadow, tower upwind effect, etc.).
- The influence of three-dimensional flow on the blade aerodynamic characteristics (e.g., three-dimensional stall and aerodynamic tip loss).
- Dynamic stall effects of the airflow for the profiles used.
- Unsteady aerodynamic effects.
- Aeroelastic effects.
- Aerodynamic asymmetries that can arise through production or assembly tolerances of the rotor blades.

3.6.4 Actuation Loads (Operational Loads)

Actuation loads (or operational loads) are generated by the operation and control of the RNA. The main source of actuation loads is the rotor speed control and/or the torque control through pitching the blades or adjusting other aerodynamic devices. Actuation loads also include the mechanical braking loads as well as the transient loads arising during the startup and shutdown of the rotor, engagement and disengagement of the generator, and nacelle yaw movements. As a minimum, the following should be considered:

- Static and load-dependent bearing friction moments (especially those at the blade pitch bearing and the yaw bearing).
- Behaviour of the control and safety systems of the RNA.

3.7 *Modelling of Wind Farm Wake Effects*

Within an offshore wind farm, the turbulence intensity associated with wake flow may be higher than the ambient turbulence intensity. In addition, wake flow is characterised by a reduced mean wind speed and an increased shear profile. In the absence of detailed analysis of the wind characteristics within an offshore wind farm, the design calculation may be performed using increased turbulence intensity based on experience and recognised calculation methods. The mutual influence of offshore wind turbines through the wake interaction behind the rotor should be considered in a wind farm configuration up to a distance of at least 10 rotor diameters from another wind turbine. Reference can be made to IEC 61400-1 for guidance on the wake effects from neighbouring wind turbines.

3.8 *Power Cable Considerations*

For Bottom-founded Offshore Wind Turbines, the power cable can be considered in the integrated load analysis as an appurtenance. The wave and current loads may be included in the analysis by the Morison element model.

3.9 *Modal Analysis*

3.9.1 General

Modal analysis is used to calculate the natural frequencies, damping, and mode shapes of the wind turbine blades, tower, and substructure. The finite element model is commonly used to perform the modal analysis. The purposes of the modal analysis are to:

- Obtain the mode shapes for the blades and the tower as input for the structural models.
- Obtain the natural frequencies for the tower and the support structure (substructure, tower, and foundation) design.
- Evaluate the damping level for the global model.

The modal superposition method is implemented in some ILA software tools to model the blade and tower flexibility. The mode shapes obtained from the modal analysis are used as input to the ILA software tools for blade and tower modelling. The overall damping level of the global model is a key element in the prediction of both fatigue and extreme loads for load-carrying components of a Bottom-founded Offshore Wind Turbine and should carefully be selected.

3.9.2 Bottom-Founded Offshore Wind Turbines

The global model for the ILA or a separate structural finite element model can be used for the modal analysis. Nonlinearity in the system should be linearised at a given load level for the modal analysis. Added mass on the submerged structural member should be included. Due to the nonlinearity of the foundation stiffness and foundation damping, it is important that the load level used for linearisation is comparable to the loads from the ILA of the design load cases. It should be verified that the fundamental frequencies of the support structure (integrated foundation, substructure, and tower) are away from the 1P rotor frequency and the 3P (for 3-blade turbine) blade passing frequency. The foundation model has significant effects on the natural frequencies of the support structure (tower, substructure, and foundation), while the effects on blade natural frequencies may be insignificant.

Modal Analysis of Tower

The global model for the ILA including an appropriate foundation model may be used for the modal analysis. Alternatively, modal analysis can be performed using the finite element method. The tower is modelled by the finite element model, while the hub and nacelle can be modelled as a lumped 6×6 mass matrix on top of the tower. The same foundation models as used in the ILA can be applied. A free decay test should be performed to verify the damping level. A force can be added to the top of the tower and then released to simulate the free decay. The displacement of the tower top can be recorded to estimate the damping level to be applied in the ILA.

Modal Analysis of Blades

The ILA model with the foundation model can be used for the modal analysis. Alternatively, modal analysis can be performed with a full structural model including blades, tower and substructure, and foundation, as discussed in "Modal Analysis of Tower".

4 Modelling of Environmental Loads

4.1 General

This chapter provides guidance on the global load calculation for the ILA for Bottom-founded Offshore Wind Turbines. Model or onsite test data may be employed to establish environmental loads. Alternatively, environmental loads may be determined using analytical methods compatible with the environmental condition models established in compliance with Chap. 21 or other relevant standards such as IEC 61400-3-1, API RP 2A, API RP 2T, API RP 2FPS, ISO 19902, ISO 19903, and ISO 19904-1. Any recognised load calculation method may be employed, provided it has proven sufficiently accurate in practice, and it is shown to be appropriate to the system's characteristics and site conditions. Environmental loads to be considered are listed below:

- Aerodynamic loads on the RNA.
- Wind loads on the tower and the substructure.
- Hydrodynamic loads on the substructures subject to waves and current.
- Dynamic amplification of wave and current loads on the flexible substructure.
- Ice and snow accumulation induced loads, if applicable.
- Earthquake loads, if applicable.
- Ice loads, if applicable.

Other loads, such as wave impact loads, wave run-up loads, boat impact loads, and loads due to vortex induced vibration (VIV) should be considered where appropriate.

4.2 Aerodynamic Loads on RNA

Aerodynamic loads induced by airflow passing through the rotor are determined by the mean wind speed and turbulence across the rotor plane, rotor rotational speed, air density, and aerodynamic shapes of wind turbine components, as well as interactive effects such as aero-elasticity and rotational sampling. Aerodynamic loads due to these effects should be calculated using recognised methods and computer programmes. Guidance on the calculation of aerodynamic loads and other loads as well as load effects on the RNA in a global model is given in Sect. 3.6. The 10-min mean wind speed at hub height (i.e., V_{hub} as defined in Chap. 3, Sect. 3.2, and the wind models defined in Chap. 3, Sect. 2 are used in the definition of design load conditions in Chap. 4, Sect. 3. For Bottom-founded Offshore Wind Turbines installed in a wind farm, the potential shadow effect and wake effect on the loads should be considered for both the strength and fatigue analyses. Guidelines for modelling wind farm wake effects can be found in Sect. 3.7. Engineering models of aerodynamic loads for the ILA (refer to Chap. 19, Sect. 4) are usually based on the blade element momentum (BEM) theory. Some software tools may apply medium fidelity methods such as the free-vortex wake (FVW) model which is more accurate but requires a much longer simulation time than the BEM theory. Advanced CFD analysis and wind tunnel tests can be used to obtain and calibrate the aerodynamic coefficients of air foils and semi-empirical aerodynamic correction models. In the BEM theory, the aerodynamic loads of the rotor are calculated based on predefined two-dimensional (2D) air foil aerodynamic coefficients with semi-empirical corrections. The BEM theory is a combination of the blade element method and momentum theory. With the blade element method, the blade is divided into small elements. Each element is independent of surrounding elements aerodynamically as a two-dimensional (2D) air foil. Aerodynamic forces are calculated based on the local flow conditions using the momentum theory with energy conservation of the upstream wake and energy extracted by the rotor from the wind. BEM theory is simple and efficient and predicts the global performance of wind turbines with acceptable accuracy for ILA. To account for the 3D effects and unsteady aerodynamics, the following semi-empirical corrections are applied to the BEM:

- Tip loss, hub loss.
- Dynamic inflow (also known as dynamic wake) models.
- Dynamic stall models.
- Skewed inflow (wake) correction.
- Tower influence models.

These corrections should be included in ILA for the aerodynamic loads on the rotor. In addition, the wind loads on the nacelle should also be included in the ILA. Wind loads on the nacelle can be calculated with the drag coefficients in the longitudinal and transverse directions, projected areas, and wind speed at the nacelle.

4.3 Wind Loads on Tower and Substructures

4.3.1 Wind Loads

Wind loads on the tower are calculated as distributed loads along the tower height. The tower is discretised into small elements, and wind loads are calculated with the drag coefficient and wind speed at the height of each small element. Wind loads on the substructure can be modelled as distributed loads like the wind loads on the tower. In this case, the substructure is modelled as a frame of tubular elements or a collection of discrete elements, each of which has a projected windage area. Wind loads are calculated at each discretised element with drag coefficients and wind speed at the height of the element.

4.3.2 Vortex-Induced Vibration

Vortex-induced vibration (VIV) due to wind on the tower and slender members of a substructure should be considered in the structural analysis. Both drag and lift components of the load due to vortex-induced vibrations should be accounted for. The effects of wind loading on structural members or components that are not normally exposed to wind loads after installation should be considered where applicable. This would especially apply to load-out or transportation phases. Guidance on the VIV effects of the tower can be found in IEC 61400-6 and IEC 61400-3-1.

4.4 Hydrodynamic Loads on Substructures

4.4.1 Hydrodynamic Loads

Hydrodynamic loads on substructures are modelled using the panel model, the Morison element model, or a combination of panel and Morison element models. For large bodies, the wave diffraction forces should be included in the analysis. Hydrodynamic analysis can be performed with a panel model based on the radiation/diffraction theory with a boundary element method. A second-order radiation/diffraction software tool is normally used for the hydrodynamic analysis with the panel model. Drag loads due to waves and current can be calculated using the Morison equation. The advantage of the Morison element model is to account for the current speed, wave kinematics, and substructure motions in a single model. For members with a diameter to wavelength ratio larger than 0.2, the Morison equation may not be appropriate. Diffraction effects should be considered for wave inertia loads. CFD analysis may be applied to calculate the hydrodynamic loads. However, CFD analysis is time consuming and often used as a verification and calibration tool. Wave impact loads and wave run-up loads are normally considered as local loads. Global effects of wave impact should be evaluated when there is a negative air gap. Global effects of wave impact can be evaluated with advanced numerical simulations, CFD analysis, and model tests. Guidance on shallow water wave theory, wave impact loads, and hydrodynamic loads on substructures of Bottom-founded Offshore Wind Turbines is available in Annex B and Annex C of IEC 61400-3-1. Further guidance on modelling methods of hydrodynamic loads for offshore structures can be found in API RP 2A, ISO 19902, ISO 19903, API RP 2T, API RP 2FPS and ISO 19904-1.

Gravity Based Substructure

Gravity based substructures are normally not slender structures, so the Morison method is not appropriate for calculating wave loads. Gravity based substructures are typically located in shallow water where nonlinear wave loads should be considered in the hydrodynamic analysis. The Froude-Krylov wave excitation forces induced by incident waves can be calculated to the instantaneous wave surface by integrating the dynamic pressure on the panel model. Nonlinear wave theories can be used for calculating the dynamic pressure. The diffraction forces can be calculated using the second-order diffraction/radiation theory.

Monopile Substructure

Monopiles are used in shallow and intermediate water depths. Nonlinear wave loads should be included in the ILA. The diameter of the monopile can be large. For large monopiles, the drag loads due to waves and current on the monopiles can be calculated using the Morison equation, and the inertia wave loads are calculated through the MacCamy-Fuchs formula for a cylinder. The MacCamy-Fuchs formula is based on diffraction theory to calculate wave inertia loads for large cylinders with a diameter to wavelength ratio larger than 0.2. Different wave theories and wave stretching methods can be used to account for the nonlinear wave loads.

Tripod Substructure

The wave and current loads on the structural members of tripod substructures are normally calculated using the Morison equation. Different wave theories and wave stretching methods can be used to account for the nonlinear wave loads. For members with a diameter (D) to wavelength (L) ratio larger than 0.2, the Morison equation may not be appropriate. Diffraction effects should be considered for wave inertia loads.

Jacket Substructure

The wave and current loads on the structural members of jacket substructures are normally calculated using the Morison equation. Different wave theories and wave stretching methods can be used to account for the nonlinear wave loads.

Other Substructures

In general, hydrodynamic loads on substructures can be modelled using the Morison element model. For general modelling methods, guidance can be found in IEC 61400-3-1, API RP 2A, ISO 19902 and ISO 19903.

4.4.2 Wave Kinematics

There are several approaches to model wave conditions in the global model:

- Irregular waves.
- Regular waves.
- User defined wave spectrum.
- User defined wave elevation time histories.

For irregular waves, a time history of the wave surface elevation can be generated based on a given wave spectrum. Wave kinematics are calculated using the linear superposition of several wave frequency bands. To account for the instantaneous wave surface, kinematic stretching methods including linear stretching, vertical stretching, and Wheeler stretching methods are normally used. The second order wave theory may be considered to improve accuracy in shallow water. For a user-defined wave spectrum or wave surface elevation time history, the same methodology for a given wave spectrum can be used. For regular waves, different wave theories including the Airy, Stokes, and Stream functions and Cnoidal and Solitary wave theories can be applied to account for different nonlinearity of waves. Guidance on the selection of suitable wave theories and calculation of wave kinematics is provided in Annex B of IEC 61400-3-1, API RP 2A, ISO 19902, API RP 2MET and ISO 19901-1.7.7.3 Marine Growth Marine growth could affect the hydrodynamic loads through:

- Increased hydrodynamic diameter.
- Increased surface roughness used in the determination of hydrodynamic coefficients (e.g., lift, drag, and inertia coefficients).
- Increased permanent load and inertial mass.

The amount of accumulation assumed for design should reflect the extent of and interval between cleaning of submerged structural parts. Marine growth on the power cables and members of substructures can be modelled using the Morison equation. Increase of diameter, mass, weight, drag, and added mass coefficients can be included in the Morison element model. If the marine growth thickness is such that certain assemblies of components are completely blocked, the effect should be properly incorporated in the modelling of the hydrodynamic loads on the substructure. Guidance to account for marine growth can be found in API RP 2A and Annex C of IEC 61400-3-1. General information on marine growth on offshore structures in different geographical offshore areas can be found in ISO 19901-1 and API RP 2MET.

4.4.3 Vortex-Induced Vibration

Vortex-induced vibration (VIV) due to current acting on structural members of a substructure should be considered in the structural analysis. Guidance on VIV analysis for cylindrical members can be found in IEC 61400-3-1 Annex C, API RP 2A, and ISO 19902.

4.5 Ice and Snow Accumulation

4.5.1 General

Where relevant, ice and snow accumulation should be assessed. Cold climate conditions can result in specific design conditions for a Bottom-founded Offshore Wind Turbine. The effects due to low temperature and icing conditions should be considered in the design load cases to be assessed in the ILA. Guidance on ice and snow accumulation for the RNA, tower and substructures can be found in the following standards:

- Design requirements for wind turbines in cold climate conditions, additional design parameters for cold climate, assessment and effects of icing climate and design load cases can be found in IEC 61400-1 Clause 14, Annex A and Annex L.
- For ice and snow accumulation on the structural parts, refer to ISO 19906, ISO 19901-1, and API RP 2MET for guidance.

4.5.2 Ice and Snow Accumulation Induced Loads

Icing on an offshore structure requires a combination of water or moisture and surfaces above sea level at sub-freezing temperatures. There are two major types of icing: atmospheric and sea spray. Atmospheric icing includes in-cloud icing and precipitation icing. In load calculations for the RNA in cold climate conditions, the turbine controller behaviour and loads are primarily affected by:

- Ice accretion on the rotor blades.
- Increased air density.

The effect of ice accretion on blade aerodynamic coefficients and inhomogeneous ice distribution on wind turbine blades should be considered for ice accretion induced loads on the blade. The air densities used for load calculation and in the determination of the power curve should be calculated by applying the ideal gas law as given in IEC 61400-1 Annex A.2. For the wind turbine blade, the ice accretion, and effects due to atmospheric icing should be assessed based on IEC 61400-1 Annex L. In the absence of other information, the blade aerodynamic coefficients may be modified for the effects of ice accretion based on IEC 61400-1 Annex L.1.5. The ice mass distribution (mass/unit length) for a wind turbine blade due to atmospheric icing may be calculated based on IEC 61400-1 Annex L.2. The sea spray icing on the blade may be assessed based on ISO 19906 and/or API RP 2N. Ice and snow accumulation on the nacelle, tower, and substructures could cause:

- Uneven distribution of ice and snow accumulation on the nacelle, tower, and substructure.
- Modification of the aerodynamic and hydrodynamic properties and dynamic response of the offshore wind turbine.

To model these effects, the following factors should be considered according to ISO 19906:

- Mass of the structure should be modified to account for uneven distribution of snow accumulation.
- Calculation of wind, wave, and current loads should be modified to account for ice and snow accumulation on the nacelle, tower, and substructure.
- Strouhal numbers for the assessment of vortex-induced vibration of slender members should be modified.
- Masses of structures and any added masses should be adjusted in accordance with the icing.

The ice and snow accumulation on the nacelle, tower, and substructures should be assessed based on guidance in ISO 19906 and/or API RP 2N. General information on ice and snow accumulation on offshore structures in different geographical offshore areas can be found in ISO 19901-1 and API RP 2MET.

4.6 Earthquake Loads

4.6.1 General

For a Bottom-founded Offshore Wind Turbine located in seismically active areas, the Strength Level and Ductility Level earthquake induced ground motions (refer to Chap. 3, Sect. 7.5) should be determined based on seismic data applicable to the installation site. Earthquake ground motions should be described by either applicable ground motion records or response spectra consistent with the return period appropriate to the design life of the structure. Available standardised spectra applicable to the region of the installation site are acceptable provided such spectra reflect site-specific conditions affecting frequency content, energy distribution, and duration. These conditions include the:

- Type of active faults in the region.
- Proximity of the site to the potential source faults.
- Attenuation or amplification of ground motion between the faults and the site.
- Soil conditions at the site.

As appropriate, effects of soil liquefaction, shear failure of soft mud, and loads due to acceleration of the hydrodynamic added mass by the earthquake, submarine slide, tsunamis, and earthquake generated acoustic shockwaves should be accounted for. Where relevant, earthquake loads (seismic loads) should be considered. A time domain simulation should be performed for the load calculation of the support structures of offshore wind turbines. The evaluation of seismic loading should consider the combination of seismic loading with other significant, frequently occurring operational loads. Further guidance on the assessment of earthquake loads and design conditions for the seismic analysis of Bottom-founded Offshore Wind Turbines can be found in IEC 61400-1 Subclause 11.6 and Annex D, IEC 61400-3-1 Subclause 7.3.7, API RP 2A, ISO 19902 and ISO 19903.

4.6.2 Seismic Load Model

Seismic loads can be provided in the form of a spectrum or a time history. The seismic analysis is performed in the time domain. The aerodynamic loads, hydrodynamic loads, and seismic loads can be applied simultaneously in time-domain simulations. The seismic loading can be modelled by predefined time histories of accelerations and/or displacements of the ground motion. The time histories can be measured from the earthquake or can be generated from a response spectrum of the ground motion. Seismic time histories can be applied at one point of the foundation, or at points of individual piles in the global model.

4.7 Ice Loads

4.7.1 General

Ice loads acting on Bottom-founded Offshore Wind Turbines may include both static and dynamic loads. Static loads can be generated by temperature fluctuations, ice features lodged against the structure, or changes in water level in a fast ice cover. Dynamic loads are normally caused by moving ice interactions with the substructure. The global forces exerted by ice on the structure and local concentrated loads on structural elements should be considered. The effects of rubble piles on the development of larger areas and their forces on the substructure should be considered. Ice jamming between legs should be accounted for where the substructure is designed to consist of multiple legs. Where relevant, sea/lake ice load should be assessed and included in the ILA. Guidance on the assessment of sea/lake ice conditions and calculation of ice loads for Bottom-founded Offshore Wind Turbines is provided in IEC 61400-3-1 Clause 6 and Annex D.

4.7.2 Ice Load Model

Sea ice loads acting on an offshore wind turbine are both static and dynamic loads. Static loads have their origin either in temperature fluctuations or changes in water level in a fast ice cover. Dynamic loads are caused by wind and current induced motion of ice floes and their failure in contact with the support structure. The following ice loads should be considered where relevant:

- Horizontal load due to temperature fluctuation in a fast ice cover (thermal ice pressure).
- Horizontal load from a fast ice cover subject to water level fluctuations and in terms of arch effect.
- Horizontal load from moving ice.
- Pressure from hummocked ice and ice ridges due to both subduction and ridging processes.
- Vertical force from fast ice covers subject to water level fluctuations.

Calculation of ice loads for the support structures of Bottom-founded Offshore Wind Turbines should be based on the criteria provided in IEC 61400-3-1 Annex D.

5 Integrated Load Analysis

The input data for each design load case (DLC) should be prepared prior to the start of the ILA. The number of simulations, simulation time duration, calculation of maximum values, and fatigue loads should be based on applicable design standards. After completing the ILA, the ILA results should be post-processed in a way that can facilitate the subsequent design evaluation. Output data including time histories of loads, statistics and maximum loads, and damage equivalent loads are typically derived from the ILA. These output data are further used for the design evaluation of the RNA, tower, substructure, and foundation. The simulation length, number of simulations, and calculation of maximum (or minimum) values should comply with the applicable design standards. When there are no specific requirements, guidance provided in Chap. 23 can be applied. Due to the considerable number of DLCs to be analysed through the ILA, automation of data preparation and analysis management helps improve efficiency and reduce the likelihood of errors. Data set-up templates and run scripts are normally used to assign parameters for each simulation of DLCs.

6 Results Post-processing

6.1 General

Time histories of the ILA results should be post-processed to obtain the maximum value for strength analysis and fatigue loads for fatigue analysis. The individual loads on the RNA, tower, substructure, and foundation should be determined. All DLCs for strength analysis are typically post-processed in a group to obtain the maximum and/ or minimum loads for each component of a Bottom-founded Offshore Wind Turbine. All DLCs for fatigue analysis together with the probability of occurrence of each DLC are grouped together to calculate the equivalent damage loads (DELs) or the total fatigue damage solely based on the time history of the loads for each component of a Bottom-founded Offshore Wind Turbine. Guidance on the assessment of the structural integrity of the RNA using the ILA results under site-specific conditions can be found in IEC 61400-1 clause 11.10. Both loads and deflections calculated from the ILA should be considered. For the fatigue loading conditions, the damage equivalent loads and moments on structural components should be calculated. The equivalent moment of the load duration distribution of the driving torque is normally considered sufficient for the verification of components. For extreme loading, a comparison of contemporaneous loads is not required. Design of the Bottom-founded Offshore Wind Turbine support structure, which includes the tower, substructure, and foundation, is site specific. Full sets of relevant loads including extreme loads, contemporaneous loads, and fatigue loads in terms of time histories are normally used for the subsequent structural analysis for the tower, substructure, and foundation. In general, the ILA results are provided in the following forms:

- Time histories.
- Statistics of time histories including mean, maximum, minimum, and standard deviations.
- Extreme loads for all DLCs for strength analysis including maximum and minimum values.
- Damage equivalent loads for the RNA and tower loads for all DLCs for fatigue analysis.
- Main shaft equivalent torque based on load duration distribution method.

6.2 ILA Results for Bottom-Founded Offshore Wind Turbines

6.2.1 General

The ILA results for Bottom-founded Offshore Wind Turbines include the RNA and tower loads, substructure loads, and foundation loads. The results for strength analysis are normally obtained in the form of maximum and minimum values calculated based on multiple simulations. The critical loads and the associated design load cases for different components should be identified and reported after incorporating applicable partial safety factors for loads. Fatigue loads for the RNA and tower can be provided as time histories and damage equivalent loads as discussed in Sect. 6.3. Fatigue loads for the substructure are provided in the form of time histories for the substructure fatigue analysis in accordance with applicable design standards.

6.2.2 RNA and Tower Loads

The RNA and tower loads include the following:

- Blade loads along the blade span.
- Generator torque and power.
- Main shaft loads.
- Nacelle acceleration.
- Blade-tower clearance.
- Tower load along the tower.
- Tower tilt angle.

6.2.3 Substructure and Foundation Loads

In addition to the RNA loads and the tower loads, the following substructure loads, and foundation loads should be included:

- Tower interface loads.
- Substructure mudline loads.
- Substructure member loads.

- Pile head loads, if applicable.
- Pile deflection and rotation at mudline, if applicable.
- Foundation horizontal and vertical forces and overturning moment.

6.2.4 Resultant Loads

Loads are normally given in the form of six (6) load components, including three forces, F_x, F_y, and F_z, and three moments, M_x, M_y, and M_z. When the structural stiffness and strength in response to loading in the plane are similar for the different loading directions, extreme loading can occur when both x and y components are large in magnitude but not at their largest values. Thus, the in-plane vector resultant values should also be calculated. These in-plane resultants loads are defined as:

$$F_R = \sqrt{F_x^2 + F_y^2}$$

$$M_R = \sqrt{M_x^2 + M_y^2}$$

where:

F_R = resultant force, N (lb).
F_X = force in x-axis, N (lb).
F_Y = force in y-axis, N (lb).
M_R = resultant moment, N-m (lb-ft).
M_X = moment about x-axis, N-m (lb-ft).
M_Y = moment about y-axis, N-m (lb-ft).

6.3 *Fatigue Loads for RNA and Tower*

6.3.1 General

Fatigue is the damage accumulation process in a component caused by cyclic loading. Fatigue loads for the RNA components and the tower are usually calculated as damage equivalent loads (DELs). The methods to calculate the DEL are provided in Sect. 6.3.2. The load range and the number of cycles of the load range are calculated by the rain flow counting method as described in Sect. 6.3.3. For the drive train components, gear and bearing system, the load duration distribution method is normally used for the main shaft fatigue loads calculation as presented in Sect. 6.3.4.

6.3.2 Damage Equivalent Load

Damage Equivalent Load (DEL) is the equivalent load range for a reference number of cycles that can generate the same amount of fatigue damage as the fatigue loads from all the fatigue DLCs during the design life. Damage equivalent loads are used to equate the fatigue damage caused by a single load range repeating at a single number of cycles to the total fatigue damage. The Damage Equivalent Load (DEL) is calculated by:

$$L_{eq} = \left(\frac{\sum n_i L_i^m}{N_{eq}} \right)^{\frac{1}{m}}$$

where:

L_{eq} = damage equivalent load (DEL), N (lb) for force, N-m (lb-ft) for moment.
$L_i = i^{th}$ i^{th} bin load range N (lb) for force, N-m (lb-ft) for moment.
n_i = number of cycles of load range L_i L_i.
m = slope of the S–N curve.
N_{eq} = reference number of cycles for the DEL.

The reference number of cycles, N_{eq}, is normally taken as 10^7. In addition, to compare the fatigue loads from different wind and wave conditions, the fatigue load is also calculated in the form of the 1 Hz damage equivalent load (DEL). The damage equivalent load is the cyclic load occurring N_{eq} times to cause the same fatigue damage as the fatigue damage for the wind turbine exposed to a wind and wave condition within a certain duration. For a typical fatigue load simulation that lasts for 600 s, 1 Hz DEL is calculated with N_{eq} equal 600 for the given duration of 600 s.

6.3.3 Rain Flow Counting Method

The rain flow counting method can be used to calculate the load range and load range cycle numbers for the fatigue loads. The time histories of the loads from the ILA are used for the calculations. For a load time history, the algorithm in the rain flow counting method decomposes the load time history into load cycles with different load ranges. The rain flow cycle distributions represent the probability of occurrence of each load range. The damage equivalent load (DEL) can then be calculated.

6.3.4 Load Duration Distribution Method

The load duration distribution method is used to derive the load range cycles for the drive train components from the time series of the loads from the ILA. The main components of the drive train include the hub, main shaft, main bearing, gearbox, coupling, generator, and bedplate. The load range and cycle counting for gears and bearing systems are different than for shafts or other components and depend on the magnitude of the driven load and the speed at which every geared shaft and bearing is turning. The load duration distribution method is a convenient way to describe fatigue loads for the design of bearings and gearboxes. The load duration distribution method is used to derive the load range cycles for the drive train components from the time series of the loads from the ILA. The shaft loads from the ILA are processed to obtain the load duration distribution (LDD) for the fatigue design of the gears and bearing system. As part of the results of the ILA, the main shaft equivalent torque is calculated using the load duration distribution method as below:

$$T_{eq} = \left(\frac{\sum n_i T_i^p}{\sum n_i} \right)^{\frac{1}{p}}$$

where:

T_{eq} = equivalent torque, N-m (lb-ft).
T_i = i^{th} bin torque range, N-m (lb-ft).
n_i = number of cycles of torque T_i.
p = slope of the S–N curve.

Chapter 21
Integrated Load Analysis for Floating Offshore Wind Turbines

1 General

The ILA is performed for the following purposes:

- Classification based on the Class requirements for the design, construction, and classification of Floating Offshore Wind Turbines when ILA is recommended.
- Certification and verification in compliance with local regulatory requirements, IEC certification scheme based on IECRE OD-502, and IEC 61400 series standards or other industry standards for floating offshore wind turbines.

The ILA involves aero-hydro-servo-elastic analyses that can account for the coupled dynamics of a Floating Offshore Wind Turbine including the RNA, control system, tower, floating substructure, station keeping system, and where applicable, power cables modelled and solved simultaneously in the time domain. In contrast to the uncoupled analysis methods, the ILA accounts for the coupling effects and damping effects including aerodynamic and hydrodynamic damping consistently. It can provide a more rational calculation of global motions and loads for the detailed design evaluation of the RNA components, tower, floating substructure, station keeping system, and power cables. Software tools for the ILA should be compatible with the modelling approaches applied to individual Floating Offshore Wind Turbine components and the load transfer strategy for detailed design evaluations for each component of the Floating Offshore Wind Turbine. The numerical modelling and simulation methods should comply with the applicable classification or certification requirements. Where there are no specific requirements, the guidance provided in this book may be applied.

F. Aderouane, *Design, Construction, and Global Performance Analysis of Offshore Wind Turbines*, Green Energy and Technology, https://doi.org/10.1007/978-3-031-86010-2_21

1.1 Design Data, Reports, and Drawings

The ILA requires the following design data, reports, and drawings:

- Site conditions and environmental data (water depth, seabed bathymetry, soil data, wind, wave, current, water level, wind/wave misalignment, fatigue sea states, etc.).
- RNA modelling data (geometry, dimensions, air foil aerodynamic data, drive train properties, blade material properties, stiffness, damping, mass distribution, natural frequencies, turbine power and thrust curves, etc.).
- Tower modelling data (geometry, dimensions, material properties, stiffness, damping, mass distribution, natural frequencies, etc.).
- Wind turbine control system modelling data (description of the wind turbine control software, the controller dynamic link library (DLL) to integrate to the load analysis software, and operation procedures, etc.).
- Sea state limit for shutdown in severe sea states, if applicable.
- Hull modelling data (dimensions, material properties, stiffness, mass properties, free decay, motion responses amplitude operators (RAOs), etc.).
- Hull damping from model testing, if available.
- Mooring system modelling data (pattern, fairlead and anchor locations, line components properties and pretensions).
- Modelling data of the hull and mooring active control system, if any.

In addition, the following design reports are needed for information for the ILA:

- Hull damaged (leakage) condition including the amount of flooded water and induced platform heel motions, if applicable.
- Other design considerations pertinent to the Floating Offshore Wind Turbines, if applicable.

2 Integrated Load Analysis Procedures

2.1 General

The integrated load analysis (ILA) consists of the following key steps:

(1) Build a global performance analysis model.
(2) Define environmental load models.
(3) Set up data files for each Design Load Case (DLC).
(4) Perform ILA simulations.
(5) Obtain time histories of loads and motions.
(6) Post-process and report results.

Modelling strategies for floating offshore wind turbines are outlined in Sect. 3. Modelling methods for environmental loads are provided in Sect. 4. Guidance on simulation and post-processing is provided in Sect. 5 and Sect. 6.3. Interface loads for detailed design evaluations are described in Sect. 6.

2.2 ILA Procedure for Floating Offshore Wind Turbines

The ILA procedure for Floating Offshore Wind Turbines is illustrated Fig. 1. The ILA model for a Floating Offshore Wind Turbine comprises the WTG (RNA and tower) model, floating substructure model and station keeping system model. The power cable model may also be included in the ILA model. The floating substructure model is normally simplified as a rigid body or multiple rigid bodies. The flexibility of the floating substructure may be modelled through a simple beam element model. The ILA results provide input for the WTG, floating substructure, station keeping system and power cable design evaluations. A separate structural analysis software tool is normally needed for the subsequent floating substructure design evaluation. Global motions and loads calculated through the ILA can be applied to the detailed structural analysis model of the floating substructure. The ILA for a Floating Offshore Wind Turbine typically includes the following steps:

(1) Build the global analysis model:

- Wind turbine generator (WTG) model including RNA and tower models.
- Wind turbine controller integration.

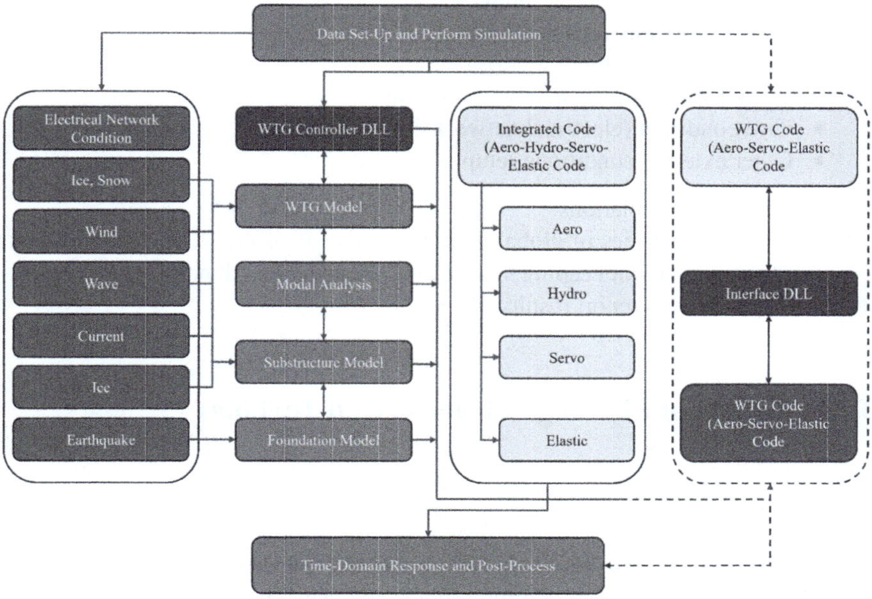

Fig. 1 ILA procedure for floating offshore wind turbines

- Floating substructure model.
- Station keeping system model.
- Power cable model, if applicable.
- Modal shapes calculated through the modal analysis.

(2) Define environmental load models:

- Wind load models.
- Wave load models.
- Current load models.
- Modelling of vortex induced motion (VIM), if applicable.
- Modelling of ice and snow accumulation induced loads, if applicable.
- Modelling of sea/lake ice loads, if applicable.
- Modelling of seismic loads, if applicable.

(3) Set up data files for each Design Load Case (DLC):

- Simulation time duration, time step, and number of random seeds.
- Random seeds for each simulation requiring applying random waves and winds.
- For a fatigue DLC, the probability of occurrence or number of occurrences of fatigue design conditions.
- Partial load safety factor.
- Wind, wave, current, and water level models and parameters.
- Wind farm wake models.
- Wind turbine controller setup.
- Wind turbine condition setup.
- Nacelle yaw misalignment setup.
- Hull condition setup.
- Station keeping system condition setup.
- Gird condition (electrical network condition) setup.
- Other external conditions setup.

(4) Perform ILA simulations.
(5) Obtain time histories of global motions and loads of the RNA, tower, floating substructure, station keeping system and, if applicable, dynamic power cables.
(6) Post-process and report results.

3 Modelling of Floating Offshore Wind Turbines

3.1 General

In general, dynamic analyses should be conducted to evaluate dynamic responses of the Floating Offshore Wind Turbine subjected to site-specific external conditions and operating conditions. While various simplified analysis methods may be used in the preliminary design, an integrated (coupled) dynamic analysis approach is recommended for the final detailed design. Prototype tests and model tests may be used to supplement the load calculation. To motion and load calculations in the global performance analysis and integrated load analysis, the analysis model should include the following aspects:

- Modelling of the hull.
- Modelling of flexibility of the tower.
- Modelling of station keeping systems.
- Modelling of rotor blades and control and safety systems.
- Modelling of dynamics of drive trains.
- Modelling of wind farm wake effects.
- Modelling of power cables, if needed.
- Modelling of mode shapes through the modal analysis.

The external conditions to be considered in global performance analysis and integrated load analysis include:

- Environmental conditions.
- Electrical network conditions.

In addition, various fault conditions relevant to the electrical system, control and safety systems and mechanical components should be properly modelled in the analysis.

3.2 Modelling of the Hull

3.2.1 General

The hull in general can be modelled as a rigid body or multiple rigid bodies with 6 degrees of freedom (6 DOFs) motions. Where deemed necessary, flexibility of the hull may be modelled using beam elements. The structural flexibility of the connection of the floating substructure to the tower should be included in the analysis. Modelling of the hull should consider the following loads and load effects:

- Hydrostatic loads.
- Gravitational and inertial loads.
- Wind, wave and current loads.
- Hull vortex induced motion (VIM), and vortex induced vibration (VIV), if applicable.
- Other loads (such as ice loads, ice and snow accumulation induced loads, earthquake loads, impact loads, etc.), if applicable.

Guidance on modelling of environmental loads on the hull is provided in Sect. 4.

3.2.2 Hydrostatic Model

Balancing mass with buoyancy in the vertical direction is usually the starting point for hydrodynamic analyses. The vertical component of mooring pretensions is part of this load balancing. Buoyancy of a large-volume floating hull can be calculated directly using the wetted surface of the panel element model created for diffraction-radiation analyses. When an analysis model includes both the panel elements and Morison elements, the buoyancy could still be calculated by most commercial software if actual locations and dimensions of the Morison elements are provided. Correct modelling of metacentric heights (GML, GMT) is as important as modelling the location of the centre of buoyancy. Free surface effects in partially filled internal tanks should be considered in determination of metacentric heights. Stiffness contributions from moorings and cables should be appropriately accounted for. The mass distribution of the hull can be represented by either a global mass matrix or a detailed mass distribution (e.g., FE model). The input coordinate system normally depends on the software employed and its origin may be placed at the vertical centre of gravity or on the still water plane. Proper input of roll and pitch radii of gyration is critical and requires a correct definition of reference axis systems.

3.2.3 Hydrodynamic Model

The hydrodynamic model of the floating substructure consists of the panel model, the Morison element model, or a combination of the panel and the Morison element models. Marine growth could affect the hydrodynamic loads and should be considered in the Morison element model. A diffraction-radiation analysis is usually performed using the panel model based on the boundary element method to calculate the hydrodynamic loads. The main output of a diffraction-radiation analysis includes the first-order excitation forces and moments in the form of load response amplitude operators (RAOs), hydrostatics, potential damping, added mass, first-order motions in 6 DOFs, and second-order forces and moments in the form of quadratic transfer functions (QTFs). The load RAOs and QTFs are used as input to calculate the hydrodynamic loads on the large volume hull in ILA. Further guidance on the hydrodynamic load calculation can be found in Sect. 4.4.

3.2.4 Instability

Mathieu's instability may occur for dynamic systems with time dependent stiffness. Several effects may cause such time dependent stiffness. Specifically, for the Floating Offshore Wind Turbine, there are two scenarios that may trigger instability:

- Non-constant heave stiffness caused by the geometric shape of the hull.
- Non-constant pitch stiffness caused by a nonlinear heave coupling term.

The heave/pitch-coupled instability in the second scenario could be critical for a single column hull with low heave damping. Instability can be identified through numerical simulations and/or model and field testing.

3.3 Modelling of Flexibility of Tower

The tower can be modelled based on linear elastic theory. However, nonlinear relationships between loads and load effects should be properly accounted for, when deemed important. Structural damping of the tower should be properly determined. The wind loads on the tower should be included. When wind loads on the Floating Offshore Wind Turbine are analysed, the influence of the tower shadow or tower upwind effects on the wind field perturbation should be properly modelled. The tower structures can be modelled using the finite element method. The model should account for the distributed stiffness and mass properties along the axial direction of the tower. The moment of inertia about the axial direction should be accounted for. The dynamic structural response of the tower may also be calculated by the modal superposition method. Sufficient modes should be included in the analysis. As a minimum, the first few orders of bending modes should be included. If the tower does not have sufficient torsional stiffness, torsional modes should also be included in the modelling. Modal analyses of the tower should consider the influences of the floating substructure and the station keeping system. For some types of Floating Offshore Wind Turbines, such as a catenary-moored Spar-type or Semi-submersible-type Floating Offshore Wind Turbine, the tower may be modelled as part of the rigid body of the floating support structure in preliminary mooring system designs. However, the tower flexibility should be adequately modelled for the final detailed design. Guidance on the tower modal analysis is provided in Sect. 3.9.

3.4 Modelling of Station Keeping System

3.4.1 General

There are two approaches for mooring system analysis (i.e., quasi-static analysis and dynamic analysis). Both methods can be pursued in either the frequency domain or the time domain. The quasi-static analysis may be used in the preliminary design stage. For the ILA, the time-domain dynamic method is usually required to account for nonlinearity and dynamic effects of the mooring (or tendon) system. The formulation for modelling the mooring (or tendon) system is based on the finite element method (FEM) or the lumped-mass method. Mooring lines are typically modelled as slender cylindrical members. Modelling of the mooring lines should consider the following loads and load effects:

- Hydrodynamic loads induced by current and wave.
- Ice-induced loads, if applicable.
- Vortex-induced vibrations of the mooring lines, if applicable.
- Marine growth.

Guidance on modelling environment loads on the mooring line can be found in Sect. 4.

3.4.2 Mooring Line Nonlinearity

There are four primary nonlinear effects that could affect mooring line behaviours:

- ***Nonlinear Stretching Behaviour of the Mooring Line***. The strain or tangential stretch of the mooring line is a function of the tension magnitude. Nonlinearity occurs mostly in synthetic materials such as polyester, while chain and wire rope can be regarded as linear. In many cases, this nonlinearity is simplified by a linearised behaviour using a representative tangent or secant modulus.
- ***Changes in Geometry***. The geometric nonlinearity is associated with large variations of the mooring line shape.
- ***Fluid Loading***. The Morison equation is most frequently used to represent fluid loading effects on mooring lines. The drag force on the line is nonlinear because it is proportional to the square of the relative velocity between the fluid and the line.
- ***Bottom Effects***. In many mooring designs, a considerable portion of the mooring line is in contact with the seafloor. The interaction between the line and the seafloor is usually considered to be a nonlinear frictional process.

In addition, the length of grounded line segment constantly changes, causing an interaction between this nonlinearity and the geometric nonlinearity. In the time-domain method, it is possible to model the non-linear effects described above—the elastic stretch can be mathematically modelled, the full Morison equation can be implemented, the position of the mooring line can be updated at each time step, and the bottom interaction can be simulated using a frictional model. Such a time-domain analysis requires recalculating mass, damping, and stiffness matrices and loading at each time step. Hence, the computation can become complex and time consuming. The frequency-domain method, on the other hand, is always linear because of the principle of linear superposition. Hence, all sources of non-linearity should be simplified by either a direct linearisation approach or an iterative linearisation approach.

3.4.3 Finite Element Analysis Approach

The Finite Element (FE) method can be an effective approach for modelling mooring (or tendon) systems in the global performance analysis. The key features that are desirable for adequate modelling and analysis of mooring (or tendon) systems normally include:

- 3D formulation.
- Conventional small strain slender beam and bar elements capable of considering material and geometric stiffness and nonlinear material properties.
- Hull/mooring (or tendon) connection formulation.
- Seafloor/mooring line contact formulation.
- Seabed/tendon connection formulation.
- Structural damping formulation.
- Hydrodynamic loading according to the Morison equation expressed by the relative water/structure velocity and acceleration.
- Regular and irregular loading due to waves and hull structure motions.
- Current modelling.
- Capability of modelling mooring components such as swivels, hinges, buoyancy modules, clump weights, flex-joints, etc.
- Capability of modelling constant (or variable) line tension devices.
- Nonlinear static analysis.
- Eigenvalue analysis.
- Nonlinear time-domain dynamic analysis.

The computational efforts of nonlinear time-domain dynamic analysis can be substantial. This is particularly the case for irregular wave analyses where long simulations are typically required to estimate extreme responses with sufficient statistical confidence. It is therefore beneficial to apply simplified analysis approaches as a supplement to achieve more efficient computer analyses (e.g., linearised time-domain analysis, frequency-domain analysis, etc.).

3.5 *Modelling of Dynamics of Drive Trains*

Drive train dynamics should be carefully considered in the ILA. As a minimum, the torsional mode of the drive train should be included in the analysis. The drive train includes all torque-transmitting components from the rotor to the generator as well as the elastic mounting of the drive train. The parameterisation assumptions of the drive train model used in the global load calculation should be verified by the calculation using more detailed drive train models. In most cases, the controlling parameters are "resulting drive train stiffness" and "moment of inertia of generator rotor." The verification can also be conducted through the comparison of the first eigen frequency obtained from the detailed drive train model to the corresponding value derived from the global load simulation model. Other verification techniques can also be used, if appropriate.

3.6 Modelling of Rotor Blade and Control and Safety Systems

3.6.1 General

Guidance on the modelling of rotor blades and control and safety systems can be found in Sect. 3.6. In addition, the following should be considered for Floating Offshore Wind Turbines:

- Inertial loads, including gyroscopic loads, are of special importance to Floating Offshore Wind Turbines due to their potentially additional compliance and increased dynamic response from aerodynamic and hydrodynamic loading.
- The interaction of the turbine control system with the low frequency motions of the floating substructure should be incorporated into the control system design and load analysis.

3.7 Modelling of Wind Farm Wake Effects

Within an offshore wind farm, the turbulence intensity associated with wake flow may be higher than the ambient turbulence intensity. In addition, wake flow is characterised by a reduced mean wind speed and an increased shear profile. In the absence of detailed analysis of the wind characteristics within an offshore wind farm, the design calculation may be performed using increased turbulence intensity based on the experience and recognised calculation methods. The mutual influence of offshore wind turbines through the wake interaction behind the rotor should be considered in a wind farm configuration up to a distance of at least 10 rotor diameters from another Floating Offshore Wind Turbine. Reference can be made to IEC 61400-1 for guidance on the wake effects from neighbouring wind turbines. As appropriate, the global yawing moment exerted on the floating support structure due to unbalanced rotor aerodynamic loads caused by the shade effect or the wake effect of neighbouring Floating Offshore Wind Turbines should also be considered. Such global yawing moment may be assessed separately and added to the total RNA aerodynamic loads.

3.8 Power Cable Considerations

The power cable system for exporting generated electricity could have a long-suspended segment extending from its connecting point on the hull to the seafloor. The power cable interacts with the floating substructure and the mooring in several aspects. Wave and current actions on the cable could increase the environmental actions to be resisted by the mooring, while the cable system stiffness aids the mooring. Furthermore, damping from the cable system decreases the low frequency motions and in turn reduces the mooring tensions. The net result of these effects depends on several factors such as the type of cable and the water depth. Mooring design should consider the cable loads, stiffness, and damping, unless it can be demonstrated that neglecting the cable in global performance analyses results in a more conservative mooring design.

3.9 Modal Analysis

3.9.1 General

Modal analysis is used to calculate the natural frequencies, damping, and mode shapes of the wind turbine blades, tower, and substructure. The finite element model is commonly used to perform the modal analysis. The purposes of the modal analysis are to:

- Obtain the mode shapes for the blades and the tower as input for the structural models.
- Obtain the natural frequencies for the tower on the floating foundation.
- Evaluate the damping level for the global model.

The modal superposition method is used in some ILA software tools to model the blade and tower flexibility. The mode shapes obtained from the modal analysis are used as input to the ILA software tools for blade and tower modelling. The overall damping level of the global model is a key element in the prediction of both fatigue and extreme loads for load carrying components of a Floating Offshore Wind Turbine and should carefully be selected.

3.9.2 Modal Analysis for Floating Offshore Wind Turbines

The global model for the ILA or a full structural model can be used for the modal analysis of the tower and the blade. Added mass on the floating substructure, hydrostatic stiffness, mooring stiffness on the floating substructure, and added mass due to mooring lines should be included in the modal analysis. Flexibility of the tower-to-hull connection should be included and can be determined using the structural analysis model for the detailed hull design. A free decay of the tower top displacement should be performed to verify the damping level used for the tower on a Floating Offshore Wind Turbine. The free decay test may be performed with the floating substructure fixed. Numerical free decay tests of a Floating Offshore Wind Turbine should be performed using the ILA model to verify the damping level for the 6-DOF motions. Available model testing, field testing, or computational fluid dynamics (CFD) analysis can be used to verify the damping level. It should be verified that the fundamental frequencies of the tower and the natural frequencies of 6-DOF motions of the Floating Offshore Wind Turbine are away from the 1P rotor frequency and 3P (for 3-blade turbines) blade passing frequency.

4 Modelling of Environmental Loads

4.1 General

This chapter provides guidance on the global load calculation for the global performance analysis and the ILA for Floating Offshore Wind Turbines. Model or onsite test data may be employed to establish environmental loads. Alternatively, environmental loads may be determined using analytical methods compatible with the environmental condition models established in accordance with Chap. 11 or other relevant standards such as IEC TS 61400-3-2, API RP 2A, API RP 2T, API RP 2FPS, ISO 19902, ISO 19903, and ISO 19904-1. Any recognised load calculation method may be employed provided it has proven sufficiently accurate in practice, and it is shown to be appropriate to the system's characteristics and site conditions. Environmental loads to be considered include:

- Aerodynamic loads on the RNA.
- Wind loads on the tower and the floating substructure.
- Hydrodynamic loads on the mooring lines (or tendons) and cables.
- Hydrodynamic loads on the floating substructures subject to waves and current.
- Dynamic amplification of wave and current loads on the flexible substructure members.
- Vortex-induced motion (VIM) of floating substructures, if applicable.
- Ice and snow accumulation induced loads, if applicable.
- Earthquake loads, if applicable.
- Ice loads, if applicable.
- Other loads, such as wave impact loads, wave run-up loads, boat impact loads, and loads due to vortex induced vibration (VIV) should be considered where appropriate.

4.2 Aerodynamic Loads on RNA

Aerodynamic loads induced by airflow passing through the rotor are determined by the mean wind speed and turbulence across the rotor plane, rotor rotational speed, air density and aerodynamic shapes of wind turbine components as well as interactive effects such as aero-elasticity and rotational sampling. Aerodynamic loads due to these effects should be calculated using recognised methods and computer programmes. Guidelines for calculation of aerodynamic loads and other loads and load effects on the RNA are given in Sect. 3.6. The 10-min mean wind speed at hub height (i.e., V_{hub} as defined in Chap. 11, Sect. 3.2) and the wind models defined in Chap. 11, Sect. 3.2 are used in the definition of design load conditions in Chap. 12, Sect. 3. For Floating Offshore Wind Turbines installed in a wind farm, the potential shadow effect and wake effect on the loads should be considered for both the strength and fatigue analyses. Guidance for modelling of wind farm wake effects can be found in Sect. 3.7. Further information on the aerodynamic loads on the RNA can be found in Sect. 4.2.

4.3 Wind Loads on Tower and Floating Substructures

4.3.1 Wind Loads

Wind loads on the tower are calculated as distributed loads along the tower height. The tower is discretised into small elements, and wind loads are calculated with the drag coefficient and wind speed at the height of each small element. Wind loads on the floating substructure can be modelled as distributed loads like the wind loads on the tower. The floating substructure is modelled as a frame of tubular elements or a collection of discrete elements each of which has a projected windage area. Wind loads are calculated at each discretised element with drag coefficients and wind speed at the height of the element. Wind loads can also be calculated using 6-DOF wind force and moment coefficients and windspeed at a given height. The wind and moment coefficient can be calculated with a given wind profile or from wind tunnel tests.

4.3.2 Vortex-Induced Vibration

Vortex-induced vibration (VIV) due to wind on the tower and slender members of a floating substructure should be considered in the structural analysis. Both drag and lift components of the load due to vortex-induced vibrations should be accounted for. The effects of wind loading on structural members or components that are not normally exposed to wind loads after installation should be considered, especially during the load-out or transportation phases. Guidance on the VIV effects of the tower can be found in IEC 61400-6 and IEC 61400-3-1.

4.4 Hydrodynamic Loads

4.4.1 Hydrodynamic Loads on Mooring Lines and Power Cables

Hydrodynamic loads on the mooring lines and power cables due to waves and current are modelled using the Morison equation for the drag and inertia load calculation. The current speed, the water particle velocity in the waves, and the velocity of the mooring lines or cables are added vectorially to compute the drag forces. The relative acceleration between the water particle in wave and the mooring lines and cables are used for computing the inertia forces. The type and accumulation rate of marine growth at the design site can affect mass, weight, hydrodynamic diameters, and drag coefficients of mooring lines. This should be considered for mooring systems not subject to regular marine growth removal.

4.4.2 Hydrodynamic Loads on Floating Substructures

General

For structures consisting of slender members that do not significantly alter the incident wave field, semi-empirical formulations, such as Morison's equation, may be used. For calculation of wave forces on structural configurations that significantly alter the incident wave field, appropriate methods which account for both the incident wave force (e.g., Froude-Krylov force) and the forces resulting from wave diffraction should be used. In general, Morison's equation is applicable for structures comprising slender members with diameters (or equivalent diameters giving the same cross-sectional areas parallel to the flow) less than 20% of the wavelengths. For installation sites where the ratio of water depth to wavelength is less than 0.25, nonlinear effects of wave action should be accounted for. This may be fulfilled by modifying linear diffraction theory to account for nonlinear effects or by performing model tests. Wave force calculations should account for shallow water effects which increase current due to blockage effects, change the system natural frequency due to nonlinear behaviour of moorings, and alter wave kinematics.

Current induced loads on immersed structural members should be determined based on analytical methods, model test data, or full-scale measurements. When currents and waves are superimposed, the current velocity should be added vectorially to the wave induced particle velocity prior to computation of the total force. Current profiles used in the design should be representative of the expected conditions at the installation site. Where appropriate, flutter and dynamic amplification due to vortex shedding should be considered. Drag loads due to waves and current can be calculated using the Morison equation. The advantage of the Morison element model is to account for the current speed, wave kinematics, and floating substructure motions in a single model. Drag loads due to waves and current can be modelled separately. Current forces and moments can be calculated through 6 DOFs of current force and moment coefficients with the current speed at a given depth from the water surface. The wind and moment coefficient can be calculated with a given wind profile or from wind tunnel tests. Additional 6×6 linear and nonlinear viscous damping matrices can be applied to account for viscous damping effects.

CFD analysis may be applied to calculate the hydrodynamic loads. However, CFD analysis is time consuming and often used as a verification and calibration tool. Wave impact loads and wave run-up loads are normally considered as local loads. Global effects of wave impact should be evaluated when there is a negative air gap. Global effects of wave impact can be evaluated using advanced numerical simulations, CFD analysis, and model tests. Marine growth could affect the hydrodynamic loads and should be accounted for. Further guidance on modelling methods of hydrodynamic loads for offshore structures can be found in API RP 2A, ISO 19902, ISO 19903, API RP 2T, API RP 2FPS and ISO 19904-1.

Hydrodynamic Loads by Diffraction-Radiation Analysis

The typical hulls are large-volume floating structures that are inertia-dominated with respect to global motions. Radiation/diffraction analyses are commonly used to determine the wave loads on such hulls. Some types of hulls such as semi-submersibles and truss spars may also have slender members and braces for which the Morison model is more appropriate. A linear radiation/diffraction analysis is usually sufficient. The term "linear" means that the average wetted area of a floating structure (up to the waterline) is used in the analysis. The main output of a radiation/diffraction analysis gives the first-order excitation forces, hydrostatics, potential damping, added mass, first-order motions in 6 DOFs and second-order drift forces/moments. Such an analysis can also provide information relevant to the slowly varying responses in roll and pitch that is important for Floating Offshore Wind Turbines based on Spars or other deep-draft floating structures, large semi-submersibles, and TLPs. Low frequency vessel motions are caused in part by nonlinear second order drift forces. If the natural period of Floating Offshore Wind Turbine motions is long (for instance, larger than 25 s), a linear frequency domain solution may be achieved by Newman's approximation, which eliminates the off-diagonal terms in the QTF (Quadratic Transfer Function) matrix. Newman's approximation gives satisfactory results for low frequency motions in the horizontal plane where the natural periods of Floating Offshore Wind Turbine motions are much longer than the wave periods. For low frequency motions in the vertical plane (e.g., the pitch motion of a Spar), Newman's approximation may underestimate the second order drift forces. If such responses are deemed important for the design, time-domain analyses using a full QTF matrix may be required. When the full QTF matrix approach is used, special attention should be paid to establishing a consistent damping level.

Second-order wave forces at the sum-frequencies in a random sea-state could excite resonant responses in heave, roll, and pitch of TLP-type Floating Offshore Wind Turbines. Such a resonant response, also known as springing, is a stationary time-harmonic oscillation of the TLP-type Floating Offshore Wind Turbine at a resonance period of one of the vertical modes (i.e., heave, roll, pitch). In addition, TLP-type Floating Offshore Wind Turbines in deep water may experience excessive resonant high frequency transient ringing response. Time-domain analysis is typically performed to evaluate these high frequency responses. There are methods and computer tools for calculating the sum-frequency QTF. The important aspects to be considered for springing analyses include:

- Discretisation (mesh) of wetted surface geometry.
- Discretisation of free surface and its extension.
- Number of frequency pairs in the QTF matrix.
- Damping level for the tendon axial response.

Wave periods and wave headings should be selected such that motions and forces/moments can be described as correctly as possible. Cancellation, amplification, and resonance effects should be properly captured. Modelling principles related to the panel mesh (size) should in general be followed, e.g.:

- Diagonal length in panel elements should not be larger 20% of the smallest wavelength analysed.
- Fine panel mesh should be applied in areas with abrupt changes in geometry.
- Finer panel mesh should be applied towards the waterline to capture correct wave drift excitations.

Hydrodynamic interactions between multiple (n) floating substructures in proximity may also be solved using radiation/diffraction analyses, where the floating substructures are normally solved together for motions in n $\times$ 6 DOFs.

4.4.3 Wave Kinematics

There are several approaches to model wave conditions:

- Irregular waves.
- Regular waves.
- User defined wave spectrum.
- User defined wave elevation time histories.

For irregular waves, a time history of the wave surface elevation is generated based on a given wave spectrum. Wave kinematics are calculated using the linear superposition of several wave frequency bands. To account for the instantaneous wave surface, kinematic stretching methods including linear stretching, vertical stretching, and Wheeler stretching methods are normally used. The second order wave theory may be considered to improve accuracy in shallow water. For a user defined wave spectrum or wave surface elevation time history, the same methodology for a given wave spectrum can be used. For regular waves, different wave theories including the Airy, Stokes, and Stream functions, and Cnoidal and Solitary wave theories can be applied to account for different nonlinearity of waves. Guidance on the selection of suitable wave theories and calculation of wave kinematics is provided in Annex B of IEC 61400-3-1, API RP 2A, ISO 19902, API RP 2MET and ISO 19901-1.

4.4.4 Marine Growth

Marine growth could affect the hydrodynamic loads through:

- Increased hydrodynamic diameter.
- Increased surface roughness used in the determination of hydrodynamic coefficients (e.g., lift, drag, and inertia coefficients).
- Increased permanent load and inertial mass.

The amount of accumulation assumed for design should reflect the extent of and interval between cleaning of submerged structural parts. Marine growth on the mooring lines, cables, and members of floating substructures can be modelled using the Morison equation. Increase in diameter, mass, weight, drag, and added mass coefficients can be included in the Morison element model. If the marine growth thickness is such that certain assemblies of components are completely blocked, the effect should be properly incorporated in the modelling of the hydrodynamic loads on the substructure. Guidance to account for marine growth can be found in API RP 2A and Annex C of IEC 61400-3-1. General information on marine growth on offshore structures in different geographical offshore areas can be found in ISO 19901-1 and API RP 2MET.

4.4.5 Vortex-Induced Motions (VIM) of the Floating Substructure

Floating substructures consisting of large diameter cylindrical components such as Spars, semisubmersibles, and TLPs can experience low frequency motions due to vortex shedding in the presence of currents. These vortex-induced motions (VIM) are most prominent for Spars where most of the industry experience has been acquired. Nevertheless, multi-column floating structures such as semi-submersibles and TLPs can also experience VIM, and this effect should be accounted for in the design. r_{VIM} could have three primary effects on the mooring design:

- Increase in the average in-line drag coefficient.
- Large low frequency VIM motion amplitudes relative to the total floating substructure responses.
- Additional low frequency oscillating mooring line tensions.

These effects should be considered for strength and fatigue design of Floating Offshore Wind Turbine station keeping systems as well as the power cables. The occurrence of the Loop Current and associated eddies in the Gulf of Mexico makes consideration of VIM particularly important for this geographic area. For example, unlike other extreme events (e.g., winter storms and tropical revolving storms), the Loop Current and associated eddies could affect a particular site for an extended period and thus cause a significant fatigue damage accumulation in mooring components. For Floating Offshore Wind Turbines, vortex-induced motion (VIM) in currents and waves should be considered where relevant. Since the hull VIM is a strongly nonlinear phenomenon, model testing has often been used to determine the hull VIM responses and calibrate the numerical simulations. The amplitudes of in-line and transverse VIM responses to the direction of current and wave can be calibrated against model tests and/or CFD analysis. VIM can be represented by sinusoidal forces in both inline and transverse directions of the current and waves. Drag augmentation of the floating substructure due to VIM should be considered. Wind, wave, and current forces can be applied together with the sinusoidal forces for VIM.

4.4.6 Vortex-Induced Vibration

Vortex-induced vibration (VIV) due to current acting on structural members of a floating substructure should be considered in the structural analysis. Guidance on VIV analysis for cylindrical members can be found in IEC 61400-3-1 Annex C, API RP 2A, and ISO 19902. For slender members, such as mooring lines, tendons, and cables, VIV should be assessed. Increased drag loads on individual components and fatigue damage of individual components due to VIV should be considered. The VIV analysis is normally conducted to evaluate the effect of VIV on the fatigue damage of slender members. Guidance on the VIV analysis for slender members can be found in ISO 19904-1 Subclause 7.4.6 and Annex A.7.4.6, API RP 2FPS Subclause 7.5.6 and Annex A.7.5.6, and API RP 2T Subclause 6.4 and Annex A.

4.5 Ice and Snow Accumulation

4.5.1 General

Where relevant, ice and snow accumulation should be assessed and included in the analysis model. Cold climate conditions can result in specific design conditions for an offshore wind turbine. The effects due to low temperature and icing conditions should be considered in the design load cases to be assessed in the ILA. Guidance on ice and snow accumulation for the RNA, tower and substructures can be found in the following standards:

- Design requirements for wind turbines in cold climate conditions, additional design parameters for cold climate, assessment and effects of icing climate and design load cases can be found in IEC 61400-1 Clause 14, Annex A and Annex L.
- For ice and snow accumulation on the structural parts, refer to ISO 19906, ISO 19901-1, and API RP 2MET for guidance.

4.5.2 Ice and Snow Accumulation Induced Loads

Icing on a Floating Offshore Wind Turbine requires a combination of water or moisture and surfaces above sea level at sub-freezing temperatures. There are two major types of icing: atmospheric and sea spray. Atmospheric icing includes, in general, in-cloud icing and precipitation icing. In load calculations for the RNA in cold climate conditions, the turbine controller behaviour and loads are primarily affected by:

- Ice accretion on the rotor blades.
- Increased air density.

The effect of ice accretion on blade aerodynamic coefficients and inhomogeneous ice distribution on wind turbine blades should be considered for ice accretion induced loads on the blade. The air densities used for load calculation should be calculated by applying the ideal gas law as given in IEC 61400-1 Annex A.2. For the wind turbine blade, the ice accretion, and effects due to atmospheric icing should be assessed based on IEC 61400-1 Annex L. In the absence of other information, the blade aerodynamic coefficients may be modified for the effects of ice accretion based on IEC 61400-1 Annex L.1.5. The ice mass distribution (mass/unit length) for a wind turbine blade due to atmospheric icing may be calculated based on IEC 61400-1 Annex L.2. The sea spray icing on the blade may be assessed based on ISO 19906 and/or API RP 2N. Ice and snow accumulation on the nacelle, tower, and substructures could cause:

- Uneven distribution of ice and snow accumulation on the nacelle, tower, and substructure.
- Modification of the aerodynamic and hydrodynamic properties and dynamic response of the offshore wind turbine.

To model these effects, the following factors should be considered according to ISO 19906:

- Mass, buoyancy, and stability of the floating substructure should be modified to account for uneven distribution of snow accumulation.
- Calculation of wind, wave and current loads should be modified to account for ice and snow accumulation on the nacelle, tower, and substructure.
- Strouhal numbers for the assessment of vortex-induced vibration of slender members should be modified.
- Masses of structures and any added masses should be adjusted in accordance with the icing.

Consequences for additional vertical actions, stability, static position, reductions of structure freeboard and righting moments of the floating substructure should also be considered. The ice and snow accumulation on the nacelle, tower, and substructures should be assessed based on ISO 19906 and/or API RP 2N. General information on ice and snow accumulation on offshore structures in different geographical offshore areas can be found in ISO 19901-1 and API RP 2MET.

4.6 Earthquake Loads

4.6.1 General

For a Floating Offshore Wind Turbine supported by a tendon system and located in a seismically active area, the Strength Level and Ductility Level earthquake induced ground motions (refer to Chap. 11, Sect. 7.5) should be determined based on seismic data applicable to the installation site. Reference should be made to API RP 2T for designing the tendon systems against earthquake loading. Earthquake ground motions should be described by either applicable ground motion records or response spectra consistent with the return period appropriate to the design life of the structure.

Available standardised spectra applicable to the region of the installation site are acceptable provided such spectra reflect site-specific conditions affecting frequency content, energy distribution, and duration. These conditions include:

- The type of active faults in the region.
- The proximity of the site to the potential source faults.
- The attenuation or amplification of ground motion between the faults and the site.
- The soil conditions at the site.

As appropriate, effects of soil liquefaction, shear failure of soft mud and loads due to acceleration of the hydrodynamic added mass by the earthquake, submarine slide, tsunamis and earthquake generated acoustic shock waves should be considered. For earthquake loads (seismic loads) for Floating Offshore Wind Turbines, refer to IEC 61400-3-2 Annex J, ISO 19901-2, API RP 2T and API RP 2EQ for guidance.

4.7 Ice Loads

4.7.1 General

Ice loads acting on a Floating Offshore Wind Turbine are both static and dynamic loads. Static loads are normally generated by temperature fluctuations or changes in water level in a fast ice cover. Dynamic loads are caused by moving ice interactions with the floating substructure. The global forces exerted by ice on the global floating substructure and local concentrated loads on structural elements should be considered. The effects of rubble piles on the development of larger areas and their forces on the floating substructure should be considered. Where relevant, sea/lake ice load should be assessed and included in the ILA. Guidance on the assessment of ice conditions and calculation of ice loads for Floating Offshore Wind Turbines can be found in ISO 19904-1 Subclause 7.4.7, API RP 2FPS Subclause 7.5.9, ISO 19906, and API RP 2N.

4.7.2 Ice Load Model for Floating Offshore Wind Turbines

Ice loads occur when the Floating Offshore Wind Turbine encounters sea ice or impacts icebergs. The collision load can be determined through theoretical models, model laboratory tests, or full-scale measurements. Ice loads should be considered in combination with motions of the Floating Offshore Wind Turbine due to aggregated loads from ice, wind, waves, and current. If the mooring/tendon system and power cable are exposed to ice loads, such loading should be considered. When determining the magnitude and direction of ice loads, the following factors should be considered:

- Geometry and nature of the ice.
- Mechanical properties of the ice.

- Velocity and direction of the ice.
- Geometry and size of the ice/structure contact area.
- Ice failure mode as a function of the structure geometry.
- Inertia effects for both ice and structure.

For Floating Offshore Wind Turbines, the ice conditions should be assessed based on ISO 19906 and/or API RP 2N. Guidance on the calculation of ice loads can be found in ISO 19906 Annex A.8.2 and API RP 2N Annex A.8.2.

5 Integrated Load Analysis

The input data for each design load case (DLC) should be prepared prior to the start of the ILA. The number of simulations, simulation time duration, calculation of maximum values, and fatigue loads should be based on applicable design standards. After completing the ILA, the ILA results should be post-processed in a way that can facilitate the subsequent design evaluation. Output data including time histories of loads, statistics, and maximum loads, and damage equivalent loads are typically derived from the ILA. These output data are further used for the design evaluation of the RNA, tower, floating substructure, station keeping system and, if applicable, dynamic power cables. The simulation length, number of simulations, and calculation of maximum (or minimum) values should comply with the applicable design standards. When there are no specific requirements, the guidance provided in this book may be applied. Due to the considerable number of DLCs to be analysed through the ILA, automation of data preparation and analysis management helps improve efficiency and reduce the likelihood of errors. Data set-up templates and run scripts are normally used to assign parameters for each simulation of DLCs.

6 Results Post-processing

6.1 General

Time histories of the ILA results should be post-processed to obtain the maximum value for strength analysis and fatigue loads for fatigue analysis. The individual loads on the RNA, tower, floating substructure, and station keeping system should be determined. All DLCs for strength analysis are typically post-processed in a group to obtain the maximum and/or minimum loads for each component of a Floating Offshore Wind Turbine. All DLCs for fatigue analysis together with the probability of occurrence of each DLC are grouped together to calculate the equivalent damage loads (DELs) or the total fatigue damage solely based on the time history of the loads for each component of a Floating Offshore Wind Turbine. Guidance on the assessment of the structural integrity of the RNA using the ILA results under site-specific conditions can be found in IEC 61400-1 clause 11.10. Both loads and deflections calculated from the ILA should be considered. For the fatigue loading conditions, the damage equivalent loads and moments on structural components should be calculated. The equivalent moment of the load duration distribution of the driving torque is normally considered sufficient for the verification of components. For extreme loading, a comparison of contemporaneous loads is not required. Design of the Floating Offshore Wind Turbine support structure, which includes the tower, floating substructure, and station keeping system, is site specific. Full sets of relevant loads including extreme loads, contemporaneous loads, and fatigue loads in term of time histories are normally used for subsequent structural analysis for the tower, floating substructure, and station keeping system. In general, the ILA results are provided in the following forms:

- Time histories.
- Statistics of time histories including mean, maximum, minimum, and standard deviations.
- Extreme loads for all DLCs for strength analysis including maximum, and minimum values.
- Damage equivalent loads for the RNA and tower loads for all DLCs for fatigue analysis.
- Main shaft equivalent torque based on load duration distribution method.

6.2 ILA Results for Floating Offshore Wind Turbines

6.2.1 General

The ILA results for Floating Offshore Wind Turbines include the RNA and tower loads, the floating substructure and mooring line or tendon loads. The results for strength analysis are normally obtained in the form of maximum and minimum values calculated based on multiple simulations. The critical loads and the associated design load case for different components should be identified and reported after incorporating applicable partial safety factors for loads. Fatigue loads for the RNA and tower can be provided as time histories and damage equivalent loads as discussed in Sect. 6.3. Fatigue loads for the station keeping system are provided as time histories for the fatigue analysis of the station keeping system in accordance with applicable design standards.

6.2.2 RNA and Tower Loads

The RNA and tower loads include the following:

- Blade loads along the blade span.
- Generator torque and power.

- Main shaft loads.
- Nacelle acceleration.
- Blade-tower clearance.
- Tower load along the tower.
- Tower tilt angle.

6.2.3 Floating Substructure and Station Keeping System

In addition to RNA loads and the tower loads, the following floating substructure and mooring line or tendon loads should be included:

- Tower interface loads.
- Floating substructure 6-DOF motions.
- Floating substructure offset and heel motions.
- Accelerations of the points of interest.
- Mooring line or tendon tension at the fairlead.
- Mooring line or tendon tension for different line components, if applicable.
- Mooring line tension at the anchor point or tendon tension at the foundation interface.

Loads are normally given in the form of six (6) load components and in-plane loads as discussed in Sect. 6.2. For the floating substructure, the 6-DOF motions are given as x, y, z and R_x, R_y and R_z. In addition, the offset and heel motions for the floating substructure should also be provided. The offset and heel motions (where heel angle $\leq 20°$) are defined as below:

$$Offset = \sqrt{x^2 + y^2} \quad Heel = \sqrt{R_x^2 + R_y^2}$$

where:

$Offset$ = floating substructure offset, m (ft).
x = floating substructure surge, m (ft).
y = floating substructure sway, m (ft).
$Heel$ = floating substructure heel motion (°).
Rx = floating substructures roll about x-axis (°).
Ry = floating substructure pitch about y-axis (°).

6.2.4 Resultant Loads

Loads are normally given in the form of six (6) load components, including three forces, F_x, F_y, and F_Z, and three moments, M_x, M_y, and M_z. When the structural stiffness and strength in response to loading in the plane are similar for the different loading directions, extreme loading can occur when both x and y components are large in magnitude but not at their largest values. Thus, the in-plane vector resultant values should also be calculated. These in-plane resultants loads are defined as:

$$F_R = \sqrt{F_x^2 + F_y^2} \quad M_R = \sqrt{M_x^2 + M_y^2}$$

where:

F_R = resultant force, N (lb).
F_x = force in x-axis, N (lb).
F_y = force in y-axis, N (lb).
M_R = resultant moment, N-m (lb-ft).
M_x = moment about x-axis, N-m (lb-ft).
M_y = moment about y-axis, N-m (lb-ft).

6.3 Fatigue Loads for RNA and Tower

Fatigue is the damage accumulation process in a component caused by cyclic loading. Fatigue loads for the RNA components and the tower are usually calculated as damage equivalent loads (DELs). The calculation method can be found in Sect. 6.

Chapter 22
Global Performance and Mooring Analysis for Floating Offshore Wind Turbines

1 General

Global performance analyses determine the global effects of environmental conditions and other loads on the Floating Offshore Wind Turbine (FOWT) and its components including the tower, floating substructure, mooring lines or tendons, anchors, power cable, etc. Global performance analyses should be conducted for all critical conditions in the pre-service and in-service phases, represented by the design load conditions specified in Part 2. Because significant interactions could occur among the Rotor-Nacelle Assembly (RNA) and control system, the tower, the floating substructure, and the station keeping system, an integrated ("coupled") model including all these components is recommended to be used for global performance analyses. An alternative method, where the dynamic analyses of the station keeping system are performed separately by using the responses of the floating substructure as boundary conditions, may also be acceptable, provided that the coupling effect of the station keeping system, and the floating substructure is adequately accounted for. The global performance analysis software should be capable of considering complex interactions among aerodynamic loads, hydrodynamic loads, actions of turbine safety and control systems, and structural dynamic responses of the Floating Offshore Wind Turbine. The analysis procedures should reflect the application limits of the selected software. Both publicly available, industry-recognised software and in-house software may be used for the analyses. However, in-house software needs to be adequately calibrated against model tests, field tests, or the industry-recognised software. General guidance on global performance analyses of the floating substructure and the station keeping systems can be found in:

- Floating Production Installations (FPI) Rules, API RP 2T, API RP 2FPS and ISO 19904-1 for the design of floating offshore structures.
- Class requirements for position mooring systems, API RP 2SK, API RP 2SM and ISO 19901-7 for catenary mooring and taut leg mooring systems.
- API RP 2T for TLP tendon systems.

F. Aderouane, *Design, Construction, and Global Performance Analysis of Offshore Wind Turbines*, Green Energy and Technology, https://doi.org/10.1007/978-3-031-86010-2_22

References are also made to other appropriate Class Rules, guides, and guidance notes. Global performance and mooring analysis for Floating Offshore Wind Turbines can be performed with integrated load analysis as given in Chap. 21. Simplified analysis methodologies in distinctive design phases as provided in this chapter may be used. This chapter provides guidance on different methodologies used in the following types of analysis for the floating substructure and the station keeping system of a Floating Offshore Wind Turbine:

- Global motion analysis of the floating substructure.
- Air gap analysis for the floating substructure.
- Mooring strength analysis.
- Mooring fatigue analysis.

For floating substructures, the global motion and air gap analysis should calculate the below items:

- Six degree-of-freedom (6 DOF) motions and accelerations at the points of interest of the hull.
- Air gap (or wave clearance) of the deck box and topside of the hull.

Mooring analysis is performed to evaluate the station keeping system performance as listed below:

- Maximum offset of the floating substructure.
- Maximum yaw motion of the floating substructure, if applicable.
- Mooring line tension at the top of the mooring line (fairlead).
- Mooring line tension at the different segments along the lines, if applicable.
- Mooting line tension at the anchor.
- Mooring line lift angle at the anchor, if applicable.
- Mooring line clearance to the seabed, water surface, if applicable.
- Mooring line clearance between mooring lines and other structures, if applicable.
- Clearance between a drag anchor and other installations, if applicable.
- Maximum and minimum tendon tension at the top and bottom of the tendon for TLP-type Floating Offshore Wind Turbines.
- Mooring strength design safety factor.
- Mooring fatigue design safety factor.

2 Characteristics of Floating Offshore Wind Turbines

2.1 General

A Floating Offshore Wind Turbine consists of several main subsystems including the floating substructure, the tower, the turbine RNA, and safety and control system, as well as the station keeping system including mooring (or tendon) system and anchoring systems. This chapter describes the key issues and characteristics of the main Floating Offshore Wind Turbine subsystems relevant to global performance analyses. Representative coupling effects between the components of the Floating Offshore Wind Turbine are also identified.

2.2 *Floating Substructures*

The floating substructure consists of the hull and topside structure. A common feature of the differing types of floating substructures is that they utilise excess buoyancy to support the tower and the RNA and to provide mooring (or tendon) system tension. For this reason, the design of the floating substructure tends to be weight sensitive. Depending on specific site conditions, ocean waves typically contain first-order energy in the range of 3–30 s. For the floating substructure, the natural periods of different modes of motion are of primary interest and reflect the design philosophy. Typical motion natural periods of several types of floating substructure are summarised Table 1.

2.2.1 TLP-Type Floating Substructures

A TLP-type floating substructure is a vertically moored, buoyant structural system wherein the excess buoyancy of the hull maintains tension in the station keeping system. A TLP-type floating substructure consists of the structural components of the hull connecting to the tendon system. It may also include a column top frame and topside deck. The hull consists of buoyant pontoons and columns. The tops of the columns may be connected to the tower directly or to a column top frame or a topside deck forming the global strength of the hull. The tendon system consists of a vertical mooring system that forms the link between the hull and the foundation for the purpose of mooring the floating support structure. The foundation system is used to anchor the tendons to the seafloor. The TLP-type floating substructure differs fundamentally from the other floating structure concepts, mostly because of the following reasons:

Table 1 Representative natural periods of typical floating offshore wind turbine floating substructure

Motions	Natural periods (s)		
	Spar-type	Semi-submersible-type	TLP-type
Surge	> 40	> 40	> 40
Sway	> 40	> 40	> 40
Heave	20–50	17–40	< 5
Roll	25–60	25–50	< 5
Pitch	25–60	25–50	< 5
Yaw	> 3	> 3	> 3

- Tendon stiffness, rather than the "water-plane" stiffness, governs the vertical motions.
- It is normally weight sensitive.
- It has low restraint to horizontal motions (surge, sway, and yaw), but is highly restrained in the vertical direction (heave, roll, and pitch).
- Higher order wave forces at different sum-frequencies may introduce resonant (springing) or transient (ringing) responses in the vertical direction. These effects may significantly increase tendon loads.
- Restrained by the tendon system, the TLP-type floating substructure moves along a spherical surface. This gives rise to the set-down effect, which is a kinematic coupling between the horizontal surge/sway motions and the vertical heave motion. The magnitude of the set-down affects the wave air-gap, tendon forces, and power cable responses.

The TLP-type floating substructure experiences horizontal wave frequency motions of the same order of magnitude as those of a semi-submersible of comparable size. On the other hand, the TLP-type floating substructure behaves like a fixed structure with no wave frequency vertical motion responses, because the wave frequency forces are counteracted by the stiffness of the tendon system. The flexibility of the tower could have significant influences on the natural periods of roll and pitch motions of the TLP-type floating substructure. The high frequency loads due to the rotor rotations and aeroelastic responses of the RNA and the tower could introduce resonant and/or transient responses in the vertical motions, which could significantly increase the tendon loads.

2.2.2 Spar-Type Floating Substructures

A Spar-type floating substructure is a deep draft, vertical floating structure, usually of cylindrical shape, supporting the tower and a topside structure (if any) and moored to the seafloor. A Spar-type floating substructure typically consists of an upper hull, midsection, and lower hull. The upper hull serves to provide buoyancy to support the topside and provides spaces for variable ballast. The midsection connects the upper hull with the lower hull. The midsection can be a cylindrical column or a truss space frame with heave plates. The heave plates are a series of horizontal decks between each bay of the truss space frame and are designed to limit heave motions by providing added mass and hydrodynamic damping. The lower hull normally consists of a fixed ballast tank and, in the case of a truss Spar, a flotation tank. The fixed ballast tank provides temporary buoyancy during a horizontal wet tow and provides the needed ballast in upending by flooding the tank. After upending, the ballast water may be replaced by fixed ballast (a substance with a density higher than water) to lower the Spar's centre of gravity. The ballast in the fixed ballast tank results in a vertical centre of gravity well below the centre of buoyancy, which provides the Spar with sound stability, as well as desired motion characteristics. The flotation tank is located adjacent to the fixed ballast tank to provide additional buoyancy for wet tow and ballast in upending. With a deep draft, the Spar-type floating substructure has a large area exposed to current forces, which is usually the dominant mean force on a Spar. Low frequency vortex-induced oscillations may increase the effective drag leading to even higher mean current forces. By installing strakes on the Spar hull, the vortex-induced cross-flow oscillation can be mitigated. However, the strakes increase the added mass and the drag forces on the Spar. The Spar-type floating substructure is usually compliant to surge, sway, heave, roll, pitch, and yaw motions. The natural periods of motions in all six degrees of freedom are usually outside the range of wave periods. In addition, the Spar-type floating support structure has a low level of vertical wave excitation due to its large draft, which exploits the fact that the first order wave motions and dynamic pressures decay exponentially with depth. These result in exceedingly small heave motions.

Due to small wave frequency motions, the Spar-type floating substructure is not subjected to large dynamic mooring line forces, although their actual effect should be evaluated by considering the actual location of the fairlead and the increase in horizontal wave frequency motion towards the waterline. The mooring system for the Spar-type floating substructure can be in the form of catenary moorings, semi-taut-line moorings or taut-line moorings. The motions and mooring loads of the catenary moored Spar-type floating substructure are normally insensitive to the high frequency loads generated by the rotor rotations and aeroelastic responses of the RNA and the tower. Since the Spar-type floating substructure has a slender hull in the vertical direction and the mooring system tends to have low yaw stiffness, the Spar-type Floating Offshore Wind Turbine may experience yaw instability, which should be avoided.

2.2.3 Column-Stabilised (Semi-submersible) Floating Substructures

A column-stabilised floating substructure, also known as the semi-submersible floating substructure, consists of a topside structure connected to the underwater hull or footings by columns or caissons. The floating substructure depends upon the buoyancy of columns or caissons for flotation and stability. Lower hulls or footings are normally provided at the bottom of the columns for additional buoyancy. The topside structure can be of an enclosed hull type or an open space frame truss construction. The topside structure is interconnected with the stability columns of the hull to form the overall strength. The tower may be connected directly to a column or caisson or to the topside structure. Current forces could be significant on column-stabilised floating substructure due to the bluff shapes of their underwater columns and pontoons. The column-stabilised floating substructure is characterised by having free modes of motion, which means that all natural periods of motion are outside the range of energetic wave periods. However, the wave frequency motions of the column-stabilised floating substructure may be significant, especially in extreme environmental conditions. For the column-stabilised floating substructure, wave impact underneath the deck due to insufficient air gap may influence the global motions and local structural responses. The spread mooring system in the form of catenary, semi-taut lines or taut lines is normally adopted as the station keeping system for the column-stabilised floating substructure. The column stabilised floating substructure with the catenary mooring system may experience significant dynamic mooring loads due to the wave frequency responses. The motions and mooring loads of the catenary moored column-stabilised floating substructure are normally insensitive to the high frequency loads generated by the rotor rotations and aeroelastic responses of the RNA and the tower.

2.2.4 Other Types of Floating Substructures

For the floating substructure having a hull configuration or a hull-mooring combination that does not belong to the those described above, general guidance may be found in the FPI Rules, API RP 2T, API RP 2FPS, ISO 19904-1, API RP 2SK and ISO 19901-7, as well as other appropriate Class Rules, guides, and guidance notes for the design of floating offshore structures.

2.3 Station Keeping Systems

2.3.1 General

A Floating Offshore Wind Turbine station keeping system is primarily designed to keep the Floating Offshore Wind Turbine within its position tolerances. It is also required to assist in achieving motion limits imposed by operational requirements of the RNA. Because the turbine RNA could exert large yaw moments to the floating substructure that normally has small yaw stiffness, the station keeping system may also need to be designed to provide sufficient yaw stiffness to mitigate the Floating Offshore Wind Turbine yaw motions. A station keeping system includes a mooring (tendon) system and an anchoring system. The mooring system provides resistance to environmental loading by mobilising reaction forces as the result of the gross changes in mooring geometry. It works as a collection of spring mechanisms where displacements of the floating substructure from its neutral equilibrium position introduce restoring forces to react to the applied loading. Each mooring line acts as a tension spring and provides its intended functions through one of the following two mechanisms:

- Hanging catenary effect due to gravity acting vertically on the mooring line.
- Line elastic effect due to elastic stretch over the length of the mooring line.

When Floating Offshore Wind Turbines are expected to be deployed in shallow water, fibre ropes, clump weights, and buoys may be used in addition to chains and wire ropes to provide the required stiffness of the mooring systems.

2.3.2 Catenary Moorings

The geometry of catenary moorings is related to the following parameters:

- Submerged weight of the suspended lines.
- Horizontal mooring load.
- Line tension.
- Line slope at fairlead.

The compliance of the catenary mooring line allowing for wave-induced floating support structure motions is provided by a combination of geometrical change and axial elasticity of the lines. Large line geometrical changes could lead to significant dynamic responses in the catenary mooring system due to the large transverse drag loads. The mooring lines in a catenary mooring system are commonly composed of steel rope and chain segments. If necessary, clump weights and buoys may be used to achieve desired line configurations.

2.3.3 Taut Moorings

In a taut mooring system, the mooring lines are straight between the anchor and the fairlead, and an uplift force is exerted on the anchor. The compliance of the mooring line allowing for wave-induced floating substructure motions is provided by line elasticity. Transverse geometric changes in a taut mooring system are typically much smaller than those in a catenary mooring system. Therefore, dynamic effects due to line drag loads are moderate. Synthetic ropes have been used as mooring lines in taut mooring systems to provide required stiffness and low weight. Compared to steel ropes, synthetic ropes exhibit more complex stiffness characteristics (e.g., hysteresis), which may significantly alter dynamic effects.

2.3.4 Tendons

The TLP tendons are vertical mooring lines like those in taut mooring systems. The main difference is that tendons are traditionally made of tubes that are hardly compliant in the axial direction. The TLP tendon system acts as an inverted pendulum, where the station keeping capacity is governed by tendon length and pretension. By restraining the Floating Offshore Wind Turbine at a draft deeper than what is required to balance its weight, the tendons are typically under a continuous tensile load that provides a horizontal restoring force when the hull is displaced laterally from its equilibrium position. With extremely high stiffness in the axial direction, the tendon system limits heave, pitch, and roll responses of the hull to small amplitudes, while its compliant transverse restraints can be designed to achieve surge, sway, and yaw responses within operationally acceptable limits. The tendons may take one of several forms, for example:

- ***Tubular Members with Threaded or Other Mechanical Connectors***. These members may be designed to be completely void, partially void, or fully flooded. The tubular member and the connectors may be fabricated as one piece or assembled from separate tubular segments that are joined with threaded connectors. The tendon components may be made of metal or fibre reinforced composites, with either integral or metallic connectors.
- ***Tubular or Solid Rod Members with Welded Connections***. The tubular members are fabricated from seamless or rolled and welded steel and are designed to be welded together, prior to or during offshore installation, to form a continuous tendon element.
- ***Tendon Strands***. These tendons are fabricated from small diameter high tensile strength wire or fibre strands and are formed into bundles. These tendons are designed to be installed offshore using a continuous one-piece spooling operation to minimise the need for intermediate connectors.

2.4 RNA and Control and Safety Systems

The RNA consists of the rotor and nacelle assembly, which includes all the associated mechanical, electrical, and control equipment and systems. A rotor may consist of two, three, or more blades typically made of fibre reinforced composites. A minimum clearance between the rotor blades and other parts of the Floating Offshore Wind Turbine and the expected highest wave elevation should be maintained. A deformation analysis should be performed by dynamic and aeroelastic means. The motions of the Floating Offshore Wind Turbine should be accounted for. The RNA control and safety system can optimise operations and keep the Floating Offshore Wind Turbine in a safe condition in the event of malfunction. The control system keeps the Floating Offshore Wind Turbine within the normal operating limits. The safety function of the system logically subordinates to the control function and is brought into action after safety relevant limiting values have been exceeded or if the control function is incapable of keeping the Floating Offshore Wind Turbine within the normal operating limits. For detailed descriptions of control and safety systems, refer to IEC 61400-1, IEC 61400-3-1, and IEC TS 61400-3-2. Due to its influence on mechanical loads, the performance of the control and safety system is critical not only for the safety of service personnel and normal operations but the integrity of structures and station keeping of the Floating Offshore Wind Turbine. The load-relevant functions of the control and safety system, which lead to various RNA operational conditions, should be considered in the load analysis.

2.5 Coupling Effects

2.5.1 General

Coupling effects that should be considered in Floating Offshore Wind Turbine global performance analyses include:

- Aero-elastic coupling effects.
- Aero-control coupling effects.
- Tower-hull and mooring (or tendon) coupling effects.
- Hull-mooring (or tendon) coupling effects.
- Other coupling effects.

The way of simulating these coupling effects is highly dependent on the actual software and analysis approaches employed in global performance analyses and needs to be evaluated on a case-by-case basis.

2.5.2 Aero-elastic Coupling

Aero-elastic coupling is the interaction between the aerodynamic loads on the rotor blades and the tower and the structural deformation due to the elasticity of the rotor blades, tower, and drive train, etc., of the RNA. In the calculation of aerodynamic loads on rotor blades, the aero-elastic coupling effects should be considered.

2.5.3 Aero-control Coupling

Aero-control coupling is the interaction between the behaviour of the control system and aerodynamic loads on the rotor blade.

2.5.4 Tower-Hull and Mooring Coupling

The coupling between the tower-hull and the mooring system includes:

- Influences of the hull and the mooring system on the tower modal shapes and natural frequencies.
- Influences of the tower flexibility on the hull pitch and roll motion, particularly for TLP-type Floating Offshore Wind Turbines.

2.5.5 Hull-Mooring Coupling

Hull-Mooring coupling is the interaction between the mooring line restoring, damping, and inertia forces and the hull mean position and dynamic responses.

- Restoring forces include static restoring force from the mooring and electrical cable system as a function of hull offset; current loading and its effects on the restoring force of the mooring and electrical cable system; and seafloor frictions if mooring lines and/or electrical cables have contact with the sea floor.
- Damping from the mooring lines and electrical cables due to their dynamics, current drag, etc.
- Additional inertia forces due to the mooring and electrical cable system.

More discussions on this coupling effect can be found in API RP 2SK for applications in floating offshore structures.

2.5.6 Other Coupling

Other coupling phenomena relevant to a Floating Offshore Wind Turbine may also exist, such as:

- Interaction between hull motions and behaviour of the control system.
- Interaction between hull motions and aerodynamic loads on the rotor blades.
- Interaction between hull motions and wind, wave, and current loads on the hull.
- Change in hydrodynamic loads, added mass, and damping on the hull as the hull is offset and set-down.
- Ringing and springing responses (TLP tendons).
- Loop Current/VIV effects/responses.

3 Global Motion Analysis

3.1 General

Dynamic global motions of Floating Offshore Wind Turbines can be simulated using frequency-domain analyses, time-domain analyses or combined frequency and time-domain analyses, as described in 19/3.2. All three approaches include various techniques of approximation and therefore may not yield consistent results. It is recommended that the approach selected for the mooring design be verified by model testing, full-scale testing, or the use of a different analytical approach. Environmental loads and load effects on the floating substructures can be categorised into the following types according to their frequency range.

- Steady loads such as wind, current, and wave drift forces that are constant in magnitude and direction for the duration of interest.
- Low frequency cyclic loads (often referred to as slow drift) with characteristic periods between 1 and 10 min. Low frequency cyclic loads typically induce dynamic excitations of the floating substructures at their natural periods of surge,

sway, yaw, and in the case of Spar-type Floating Offshore Wind Turbines, pitch and roll.
- Wave frequency cyclic loads with typical periods ranging from 3 to 30 s. Wave frequency cyclic loads result in wave frequency motions, which are normally independent of mooring stiffness. The approach of neglecting mooring stiffness in wave frequency motion calculations is considered conservative. However, for small floating substructures where the first order wave loads are not large, accounting for mooring stiffness can yield more realistic wave frequency motions. In addition, if the natural period of the Floating Offshore Wind Turbine is close to the wave periods, the wave frequency motions could be dependent on the mooring stiffness. In this case, the effect of stiffness should be properly accounted for.
- High-frequency cyclic loads with characteristic periods from 1 to 5 s. High-frequency cyclic loads typically induce dynamic excitation of TLP-type floating substructures at their natural periods of heave, roll, and pitch.
- High frequency vibration due to the RNA and the tower dynamic loads with frequency higher than 0.5 Hz (periods lower than 2 s). Such high frequency loads may be transmitted to the TLP-type floating substructure and its tendon system.

Floating Offshore Wind Turbine global motions contain the following components:

- Static and mean responses due to wind, wave, and current.
- Low frequency motions induced by wave/current effects and dynamic wind loads.
- Wave frequency motions induced by waves.
- High frequency motions induced by waves for TLP-type floating substructures.
- Vibrations induced by the loads exerted by the tower and the RNA, if applicable.
- Hull VIM, if applicable.

The steady or mean forces result in the mean offset of the Floating Offshore Wind Turbine. The low frequency forces excite the motions which are at frequencies lower than wave frequencies but close to the surge, sway, or yaw natural frequencies of the floating substructure. The wave frequency forces excite wave frequency surge and sway motions. Because of its weathervaning nature, a Floating Offshore Wind Turbine with a single point mooring may experience large low frequency yaw motions. These yaw motions may significantly affect Floating Offshore Wind Turbine and mooring system responses, and therefore should be considered in global motion analyses and model testing. Global motion analyses are normally performed in conjunction with mooring (or tendon) analyses. However, they may also be conducted separately when only global motions are of interest. In such cases, either the quasi-static or the dynamic analysis approach can be used to analyse the mooring system in global motion analyses. If the quasi-static analysis approach is used, special attention should be given to the appropriate simulation of the coupling effects.

3.2 Static and Mean Responses

Static and mean response analyses are conducted to determine the static equilibrium position with no wind, wave, or current present, as well as the mean positions with steady environmental loads acting on the Floating Offshore Wind Turbine. The determination of the equilibrium and mean positions is essential before any dynamic analyses are undertaken. The following steady forces should be considered in the analysis:

- ***Mean Wind Forces***. Mean wind forces should be calculated using the average wind speed measured for an extended period, normally about one hour, under a given design environmental condition.
- ***Current Forces***. The drag forces on the mooring lines (or tendons) and cables subject to the current under a given design environmental condition should be included in the calculation of total steady forces.
- ***Steady Wave Drift Forces***. Steady wave drift forces acting on the hull due to wave diffractions and second-order viscous effects. The interaction between waves and currents can also result in steady forces which should be considered.

3.3 Low Frequency Motions

Low frequency motions of Floating Offshore Wind Turbines can be obtained using either frequency-domain or time-domain analyses. In frequency-domain analyses, assuming that low frequency wind and wave forces are independent, the total low frequency force spectrum is given by the sum of the low frequency wave force spectrum and the low frequency wind force spectrum. This total force spectrum can then be multiplied by appropriate motion transfer functions to obtain desired motions. In time-domain analyses, because low frequency motions are dominated by the surge, sway, and yaw resonances with natural periods typically ranging from 60 to 150 s, a long simulation duration is necessary to obtain enough cycles for developing an accurate estimate of the statistical extreme. Time histories of low frequency forces could be generated either from wind and wave force spectra or directly from wind velocity and wave profile time histories. The estimation of damping is important because low frequency motions are dominated by resonant responses. Further discussion of damping is provided in 22/3.7.

3.4 Wave Frequency Motions

Wave frequency motions are induced by first-order wave forces. Wave frequency motions can be obtained using either the frequency or time-domain analyses.

3.5 High Frequency Motions

High frequency motions are induced by second and higher order wave forces. High frequency motions can be obtained using either the frequency or time-domain methods.

3.6 Tower and Turbine RNA Load Induced Vibrations

Due to the flexibility of the tower and the RNA, there are high frequency load components at the natural frequencies of the tower and the RNA. These loads may be transmitted to the hull and result in hull vibrations.

3.7 Damping

3.7.1 Damping of Low Frequency Motions

Low frequency motions of a moored Floating Offshore Wind Turbine are narrow banded in the frequency domain because they are dominated by the resonant responses at the natural frequency of the moored Floating Offshore Wind Turbine. Amplitudes of low frequency motions are highly dependent on the stiffness of the mooring system and damping. There is a substantial degree of uncertainty in the estimation of low frequency motions, particularly in damping that could in general come from five sources:

- Aerodynamic damping.
- Viscous damping of the floating support structure (tower and floating substructure), including wind, wave, and current drag.
- Wave drift damping of the floating substructure.
- Mooring system damping.
- Cable system damping.

The technology for estimating viscous damping has been well established, and viscous damping is normally included in the low frequency motion calculations. Wave drift damping, mooring system damping, and cable system damping, however, are more complex and are sometimes neglected because of a lack of understanding. Nevertheless, these damping components could be important and neglecting them may lead to significant over-estimation of low frequency motions. Where low frequency motions are considered important, it may be warranted to evaluate the damping from all these sources either by appropriate analytical approaches or model testing.

3.7.2 Damping of High Frequency Motions

The TLP-type Floating Offshore Wind Turbine could experience high frequency motions and vibrations in the vertical motion modes including heave, roll, and pitch motions. Damping in the vertical motion modes is affected by structural and soil damping, as well as hydrodynamic and aerodynamic damping, and should be estimated either by appropriate analytical approaches or model testing. This is especially important in the calculation of the springing and ringing responses and the tower and the RNA induced vibrations that could contribute significantly to the tendon fatigue damage.

3.8 Analysis Methods

For global motion analysis the following approaches may be used:

- Frequency-domain analyses.
- Time-domain analyses.
- Combined time-domain and frequency-domain analyses.

3.8.1 Frequency-Domain Approach

Mean, low frequency, and wave frequency responses of Floating Offshore Wind Turbines are calculated separately in frequency-domain analyses. The mean responses are calculated by the static equilibrium or mean position. The low frequency motions and set-down associated with offset are calculated in the frequency domain with appropriate consideration of damping effects from wind and current and mooring/cable. The wave frequency responses due to wave radiation/diffraction and wave asymmetry effects are calculated with consideration of wave loads only. Damping effects from wind and current and mooring/cable should be properly modelled. Total motions are obtained by combining the mean, wave, and low frequency motions. When the frequency-domain approach is used to compute the hull motion responses, the extreme values of the motion responses are defined as the mean plus or minus the maximum estimated value of the time-varying excursion due to combined wave frequency and low frequency hull motions. The extreme values of the hull motion, $S_{\max}$ and $S_{\min}$ can be determined by:

$$S_{\max} = S_{mean} + MAX\left(S_{dyn1}, S_{dyn2}\right)$$

$$S_{\min} = S_{mean} + MAX\left(S_{dyn1}, S_{dyn2}\right)$$

$$S_{dyn1} = S_{lfmax} + S_{wfsig}$$

$$S_{dyn2} = S_{wfmax} + S_{lfsig}$$

where:

MAX = the larger of the absolute values of the terms in parenthesis.
$S_{\max}$ = maximum hull motion.
$S_{\min}$ = minimum hull motion.
S_{mean} = mean hull motion.
S_{lfmax} = maximum expected value of low frequency hull motion.
S_{lfsig} = significant value of low frequency hull motion.
$S_{wfmasax}$ = maximum expected value of wave frequency hull motion.
$S_{wfsigax}$ = significant value of wave frequency hull motion.

The total offset should be defined as the vector sum of the individual offset components of different degrees of freedom (e.g., surge and sway). The total heel should be defined as the vector sum of the individual component angles of different degrees of freedom (e.g., roll and pitch).

3.8.2 Time-Domain Approach

The fully coupled, semi-coupled, and uncoupled analysis methods can be used when Floating Offshore Wind Turbine global motion analyses are conducted in the time domain. In the case of semi-coupled and uncoupled analyses, the applied damping level should be calibrated by fully coupled analyses and/or model testing. When the tower accelerations are of interest, a fully coupled analysis should be used, and flexibility of the tower should be carefully considered. Guidance on the integrated load analysis in the time domain can be found in Chap. 21.

3.8.3 Combined Time-Domain and Frequency-Domain Approach

Floating Offshore Wind Turbine mean, low frequency, and wave frequency responses are calculated separately when combined time-domain and frequency-domain analyses are employed in global motion analyses. The mean responses are calculated by the static equilibrium or mean position. The wave frequency responses due to wave radiation/diffraction and wave asymmetry effects are calculated with consideration of wave loads only in the frequency domain. Damping effects from wind and current, and mooring/cable should be carefully considered. The low frequency motions and set down associated with offset are calculated in the time domain. The total motions are obtained by combining the mean, wave, and low frequency motions.

4 Air Gap Analysis

4.1 General

Air gap analysis should be performed for the design load cases associated with the extreme environmental conditions (i.e., DLC 1.6, 6.1, 6.2, 10.1, 10.2, 10.3 as defined in Table 1 of Chap. 12) and the survival load cases (refer to Table 2 of Chap. 12). The most important parameters to be evaluated in the air gap analysis are the wave crest height and the hull roll, pitch, and heave motions. In the air gap analysis, the following global hull motions should be considered:

- Static and mean responses due to wind, wave, and current.
- Low frequency motions induced by wave/current effects and dynamic wind loads.
- Wave frequency motions induced by waves.

High frequency motions and vibrations, as well as the tower flexibility, may be neglected for the air gap analysis. For the wave crest height, the nonlinearity of wave profile, wave diffraction and run-up, tide and water level effects, and variations of draft of the floating support structure should be considered. General guidance for air gap and wave crest effects assessment can be found in in API RP 2FPS for semisubmersibles and spars and API RP 2T for TLPs.

4.2 Air Gap Analysis Methods

For the air gap analysis, the following methods may be used:

- Frequency-domain analyses.
- Time-domain analyses.
- Combined time-domain and frequency-domain analyses.

4.2.1 Frequency-Domain Analysis

Floating Offshore Wind Turbine mean, low frequency, and wave frequency responses are calculated separately in frequency-domain analyses. The mean responses are calculated by the static equilibrium or mean position. The low frequency motions and set-down associated with offset are calculated in the frequency domain with appropriate consideration of damping effects from wind, current and mooring/cable. The wave frequency responses due to wave radiation/diffraction and wave asymmetry effects are calculated with consideration of wave loads only. Damping effects from wind, current, and mooring/cable should be properly modelled. Total motions are obtained by combining the mean, wave frequency, and low frequency motions. The air gap is determined based on relative wave elevation and the pitch, roll, and set-down of the hull obtained from frequency-domain analyses.

4.2.2 Time-Domain Analysis

Fully coupled, semi-coupled, and uncoupled time-domain analyses may be used. In the case of semi-coupled and uncoupled analyses, the applied damping level used in the analyses should be calibrated with fully coupled analysis and/or model testing. Wave surface elevations should be calculated with consideration of the wave radiation/diffraction effects and wave asymmetry effects. Guidance on the integrated load analysis in the time domain can be found in Chap. 21.

4.2.3 Combined Time-Domain and Frequency-Domain Analysis

In this method, the wave frequency motions, and wave elevation are calculated in the frequency domain with consideration of radiation/diffraction effects and wave asymmetry effects. The high frequency motions of roll, heave, and pitch for TLP-type Floating Offshore Wind Turbines may be neglected. The low frequency surge, sway, and yaw motions, for Spar-type and semi-submersible-type Floating Offshore Wind Turbines the low frequency pitch, roll and heave motions, and for the TLP-type Floating Offshore Wind Turbines the set-down associated with offset, are calculated in the time domain. The air gap is determined based on relative wave elevation from frequency-domain analyses and the pitch, roll, and set-down of the hull obtained from time-domain analyses.

4.2.4 Other Considerations

The dynamic effects and coupling behaviours may not be important for the air gap analysis. As a result, any of the above three methods may be used in both preliminary and final detail designs provided the platform roll, pitch, and heave motions are properly calculated. For TLP-type Floating Offshore Wind Turbines, the set-down effect should be considered. For TLP-type Floating Offshore Wind Turbines, the minimum air gap, also known as the deck clearance, is governed by a combination of an increasing water level and a decreasing deck height. The increasing water level is caused by incident wave elevation, tide, storm surge, and radiation/diffraction effects from the hull. In very steep sea conditions, the diffraction effects could be significant. The decreasing deck elevation is caused by the hull set-down with offset. The hull set-down increases nonlinearly at large offsets, leading to a rapid decrease of the deck clearance with increasing environmental severity. In this case, a correction to the predicted offset should be made. Radiated or diffracted waves cause local wave impacts on the deck. In some cases, local impacts may need to be accounted for in the design and, as a result, may lead to local structural stiffening, increased weight of the deck, and tendon ringing, which affects the peak loads in the tendons and the porch structure. Increased tendon load responses could also affect the minimum tension, which in turn affects the design pretension. The deck height could significantly affect the vertical position of the centre of gravity and, in turn, the maximum and minimum tendon tensions. The deck elevation also affects the wind load and wind overturning moments. In general, a higher deck contributes adversely to the tendon tensions. On the other hand, if the deck does not have a sufficient air gap, large tendon tension variations could occur when waves strike the deck.

5 Mooring Strength Analysis

5.1 General

Mooring strength analyses are performed to predict extreme responses such as line tensions, anchor loads, and hull offsets under the design environment and other external conditions. These extreme global responses are then used to check against allowable values to verify the strength of the system against overloading and the sufficiency of clearance to avoid interference with other structures. The design load conditions and design criteria for station keeping systems are given in Chap. 8. The mooring line diameter used in the mooring analysis should be the nominal value (i.e., inclusive of corrosion and wear allowance) when calculating mass, weight, and hydrodynamic loads of the mooring line, unless otherwise noted. Installation tolerances of anchor placement and line length should be accounted for in the mooring system design. Dynamic excitations originating from the floating support structure, including the tower and the floating substructure and the RNA, could result in mooring system responses in the following distinct frequency bands:

- Mean responses.
- Low frequency responses.
- Wave frequency responses.
- High frequency responses for the tendons or taut lines.
- Vibrations induced by the loads exerted by the tower and the RNA, if applicable.

Responses of the mooring system to mean forces can be predicted by a static model of the mooring system. In general, the responses to low frequency motions can also be predicted by a static model because of the extended periods of these motions. The responses to wave and high frequency hull motions and hull vibrations may be predicted by either quasi-static or dynamic analyses. However, the quasi-static analysis method should only be used in preliminary designs. In final detailed designs, dynamic analyses should be employed.

5.2 Mooring Strength Analysis Methods

For the mooring strength analysis, the following methods may be used:

- Frequency-domain analyses.

- Time-domain analyses.
- Combined time-domain and frequency-domain analyses.

These methods incorporate different degrees of approximation and are affected by various limitations, and therefore do not necessarily yield consistent results. If verification of the approach selected for the mooring design is necessary, model test data or an alternative approach of similar reliability should be used. Because there is no established analytical method for determining Floating Offshore Wind Turbine motion responses undergoing VIM, model testing should, in general, be performed if VIM is deemed important in the design. In the case that current loads and VIM are determined to be important, it is usual practice to perform well planned model tests to determine motion amplitudes and drag coefficients for use in mooring designs. Full-scale infield measurement data may be employed to confirm the scalability of model testing. In general, time-domain analyses are recommended for the mooring strength analysis. When the effects of high frequency motions and vibrations are expected to be insignificant, frequency-domain analyses may yield satisfactory results, and the analysis procedures in this case are like those described in API RP 2SK.

5.2.1 Frequency-Domain Analyses for Spread Mooring Systems

When using frequency-domain analyses for the spread mooring system design, the mean position of the hull is first determined from static equilibrium calculations for surge, sway, and yaw. The surge, sway, and yaw responses to wave and low frequency excitations are then calculated and superimposed to the mean position. The procedure outlined in API RP 2SK may be used. Where the floating substructure has a small water plane area, alternative procedures incorporating all six DOFs of motion should be used. To obtain extreme (maximum) values, the combination of different components of the motions and mooring line tensions may follow the approach recommended in API RP 2SK.

5.2.2 Frequency-Domain Analyses for Single Point Mooring Systems

When using frequency-domain analyses for a single point mooring system design, assumptions on the hull's design heading should first be made. The design headings at which the mooring system responses are to be calculated should be determined with consideration of the mean equilibrium heading and low frequency yaw motions. The procedure outlined in API RP 2SK may be used.

5.2.3 Time-Domain Analyses

The following three analysis approaches as defined in 19/3.4 may be used for the time-domain mooring strength analysis:

- Fully coupled analysis approach.
- Semi-coupled analysis approach.
- Uncoupled analysis approach.

In general, the fully coupled (integrated) analysis approach is preferred for the following two important advantages:

- Low frequency damping from the hull, mooring lines, and cables can be internally generated in the simulation.
- Coupling between the floating substructure, the tower, the turbine RNA and control system, the station keeping system, and the cable system can be fully accounted for.

The fully coupled analysis approach should be used with a time-domain mooring analysis tool that can solve the equations of motion for the combined responses of the floating substructure, the tower, the turbine RNA and control system, the station keeping system, and, if applicable, the cable system. Guidance on the integrated load analysis in the time domain can be found in Chap. 21.

5.2.4 Combined Time-Domain and Frequency-Domain Analyses

In this approach, the mean and low frequency responses are typically simulated in the time domain to model the nonlinearity of the stiffness of the mooring lines and cables and the nonlinearity of the hull forces due to quadratic terms and yaw angle variations. Constant or variable thruster forces may also be modelled in the time domain. Transient responses resulting from line breaking, or thruster failure may be evaluated by specifying the time of failure in the time-domain analysis. Unlike the full time-domain approach, evaluation of low frequency damping cannot be included as part of such simulations because of the absence of wave frequency components. Damping will need to be evaluated separately and treated as an input parameter. Wave frequency hull motions are calculated separately in the frequency domain using the hull's motion RAOs and a given wave spectrum. These wave frequency motions may be combined with the low frequency motions in two ways:

- **Method 1**: The frequency-domain solution of wave frequency hull motions is first transformed to a time history, which is then superimposed to the mean and low frequency hull displacement time histories to obtain the combined hull displacement. In this process, the seed values for generating the wave frequency and low frequency time histories should be the same.
- **Method 2**: The mean and low frequency motion time histories are statistically analysed to determine the peak values, which are then combined with the peak values of the wave frequency motions to calculate the maximum hull offset, as described in API RP 2SK.

5.2.5 Maximum and Minimum Tendon Tensions

For the TLP-type Floating Offshore Wind Turbines, the tendon tension can be calculated either in the frequency domain or the time domain. In addition to wind, wave, and current loads, tidal effects, and weight variations, the calculation of maximum and minimum tendon tensions should also account for the effect of foundation mispositioning, individual tendon load sharing differential, and tendon VIV. Reference is made to API RP 2T for further guidance. A time-domain fully coupled (integrated) analysis should be used to calculate maximum and minimum tendon tensions.

5.3 Suggested Time-Domain Analysis Procedure

A suggested procedure for Floating Offshore Wind Turbine global performance analyses in the time domain is outlined below:

(1) Determine environmental criteria relevant to the global performance analysis of a specific Floating Offshore Wind Turbine station keeping system (refer to 15/3).
(2) Determine the external electrical network conditions.
(3) Identify the mooring pattern, the characteristics of chain, wire, and fibre rope to be deployed, and the initial tension.
(4) Define corrosion and wear allowance of the chain and wire rope used in analysis.
(5) Determine the tower dimensions, properties, modal shapes, and wind force coefficients.
(6) Define the RNA and control system configuration and create the models to be used in the simulation.
(7) Identify the startup and shutdown procedures and the associated settings of turbine RNA and control system.
(8) Identify the fault conditions in the RNA and control system.
(9) Define the hull structure's wind and current force coefficients and create the hydrodynamic model of the Floating Offshore Wind Turbine system including the hull, station keeping system, and if necessary, power cables.
(10) Perform time-domain simulations for the expected storm duration using a time-domain mooring analysis programme and applying different seed values for generating the wave and wind time histories.
(11) Perform statistical analyses to establish expected extreme values of hull offset, line tension, anchor loads, and grounded line length.
(12) Check the extreme hull offset, line tension, anchor load, and grounded line length from step against the design criteria (refer to 22/5.4, 22/5.5, and 22/5.6).

5.4 Design Checks

Extreme line tension values should be checked against the design criteria in Tables 1 and 2 of Chap. 15. The minimum breaking strength of the chain should be determined based on the diameter excluding the corrosion and wear allowance. For the TLP-type Floating Offshore Wind Turbines, the minimum tendon tension should be checked against the design criteria in 15/4.2.

5.5 Line Length and Geometry Constraints

Depending on the type of mooring system, the type of anchors, and the mooring line material, several line length and geometry parameters should be evaluated and assessed for compliance with design criteria. For catenary moorings with drag anchors not specifically designed to withstand uplift forces, the minimum length of each grounded line (i.e., the mooring line segment always resting on the sea floor) should be computed and compared with a design criterion to be prescribed on a site-specific basis. For anchors designed to withstand uplift forces, compliance with appropriate design criteria applicable to that specific type of anchor should be verified. For some types of mooring line, the grounded line is highly undesirable and, therefore, the portion of the line closest to the anchor is typically replaced by a chain. In such cases, the minimum elevation of the line-to-chain connection should be computed and verified against a minimum elevation criterion prescribed on a site-specific basis. For some types of mooring lines, exposure to the splash zone or to friction against the fairleads is undesirable and, therefore, the upper portion of the line is typically replaced by a chain. In such cases, the position of the upper termination should be evaluated and compared with a minimum depth criterion prescribed on a site-specific basis. For mooring lines in proximity to other underwater and surface installations, additional clearance requirements and geometric constraints may apply. In such cases, displacements at points of concern should be verified against applicable design criteria or be prescribed on a site-specific basis. Guidance online length and geometry constraints for fibre rope mooring lines can be found in the Class guidance relating to the application of fibre rope for offshore mooring.

5.6 Anchor Forces

In general, the highest mooring line tension obtained from the mooring analysis should be used to predict the maximum anchor force. Additional considerations should also be given to the case where a smaller mooring line tension acts in a less favourable direction. The results should be verified against the design criteria in 15/6.7, as applicable.

6 Mooring Fatigue Analysis

6.1 General

Typical mooring fatigue analysis procedures are presented below. Alternative procedures may be used, provided they can be demonstrated to achieve at least the same level of reliability as those presented herein. Miner's Rule can be used to determine accumulated fatigue damage. For main components of the mooring lines (i.e., chain, wire rope, and connecting links), Miner's Rule may be implemented in terms of tension range (the T–N curve approach) as described below, or stress range (the S–N curve approach). For other structural components of the station keeping system, such as anchor pile details and attachments, the S–N approach is normally used. T–N curves for chain and wire rope can be found in API RP 2SK. Special considerations of fibre rope fatigue analysis can be found in the Class guidance relating to the application of fibre rope for offshore. In addition to the tension-tension induced fatigue, the tension-bending induced fatigue for chain and wire rope should be carefully considered according to recognised industry criteria such as API RP 2SK. The Class guidance on position mooring systems will provide the requirements for the bending and tension induced fatigue for chain and wire rope in 1/7.5.3 and guidance on the mooring chain fatigue due to bending in Appendix 6. As indicated in 15/4.2, corrosion and wear should be suitably considered in the fatigue analysis. The general guidance is given in 22/6.2. In the mooring fatigue analysis, consideration should be given to the following aspects:

- Dynamic wind effects.
- Dynamic wave effects.
- Hull VIM, if applicable.
- Mooring line/tendon VIV, if applicable.
- Transient effects due to the startup and shutdown of the RNA.
- Coupling effects among the subsystems of the Floating Offshore Wind Turbine.
- Number of RNA startups and shutdowns in the Floating Offshore Wind Turbine design life.
- Duration of each operation such as installation, startup, power production, shutdown, parked, etc., that contributes to the mooring fatigue damage.

Either time- or frequency-domain dynamic analysis approaches may be applied for mooring tension range predictions. It is suggested that time-domain coupled analyses along with the rain flow counting method be used in the mooring fatigue analysis. Alternatively, mooring tension ranges may be obtained from model testing. In general, quasi-static analysis should not be used to calculate tension ranges due to its deficiency in estimating wave frequency mooring tensions. Guidelines for the mooring fatigue analysis which can be found in the Class guidance on position mooring systems, API RP 2SK and API RP 2T. Class guidance on position mooring systems, Appendix H of API RP 2SK and API RP 2T provide guidance on the hull VIM induced mooring fatigue.

6.2 T–N Curve

The fatigue life of a mooring line component is calculated by comparing the long-term cyclic loading in that component with its resistance to fatigue damage. For mooring systems, the T–N curve approach is commonly used to determine the number of cycles to failure for a specific mooring component as a function of constant normalised tension range. The T–N curve for a specific type of mooring component is established based on experimental results. The equation for a representative T–N curve is:

$$N \cdot T^m = K$$

where:

N = number of permissible cycles of tension range ratio, T.
T = ratio of tension range (double amplitude) to the reference breaking strength of the component (refer to the guidance given below).
m = inverse slope of the T–N fatigue curve.
K = constant coefficient or mean load dependent coefficient.

When determining the reference breaking strength of the mooring chain or connecting links, the diameter for different periods of service life can be established if the corrosion and wear rate can be predicted. If the corrosion and wear rate is uncertain, a conservative approach using the nominal diameter minus the corrosion and wear allowance should be considered for the fatigue analysis. The reference breaking strength for a wire rope should be its minimum breaking strength (MBS).

6.3 Accumulated Fatigue Damage

The total annual fatigue damage is the sum of the fatigue damage arising in a set of design states chosen to discretise the long-term environmental and operational conditions that the mooring system is subjected to:

$$D = \sum_{i=1}^{i=k} D_i$$

where:

D_i = annual fatigue damage of the component in the design state i.

Miner's Rule should be used to calculate the annual cumulative fatigue damage ratio D_i:

$$D_i = \sum \frac{n_j}{N_j}$$

where:

n_j = number of tension cycles within the tension range interval j encountered in the design state i per year.
N_j = number of cycles to failure under the normalised tension range j per year as determined by the T-N curve.

The discretisation into $i = 1, \ldots, k$ design states should have a sufficient resolution to avoid any significant error due to the discretisation. Each design state is defined in terms of the wind, wave, and current parameters and directions, as well as the RNA operating condition required to compute the mooring system responses. The probability of occurrence, P_i, should be determined for each design state. The calculated fatigue life of the mooring system is:

$$L = \frac{1}{D}(years)$$

The annual fatigue damage accumulated in each individual design state can be computed as:

$$D_i = \frac{n_i}{K} E\left[R_i^m\right]$$

where:

D_i = number of tension cycles encountered in the design state i per year.
$E\left[R_i^m\right]$ = expected value of the normalised tension range R_i raised to the power of m in the design state i per year.
m and K are defined in 22/6.2.

The number of tension cycles per year in each design state is:

$$n_i = v_i T_i = v_i P_i \cdot 3.15576 \times 10^7$$

where:

v_i = zero up-crossing frequency of the tension spectrum in the design state i, in Hz.
T_i = time (portion of years) spent in the design state i per year.
P_i = probability of occurrence of the design state i.

The effects of pretension and environmental loads due to wind, wave, and current, as well as the RNA operating conditions, should be considered when determining the normalised tension ranges. Total fatigue damage should be calculated with consideration of the fatigue damage from various fatigue load cases as described in 12/3. When suitably qualified, previously used mooring components are considered for re-use, fatigue damage from previous operations should be accounted for.

6.4 Time-Domain Fatigue Analysis Method

The tension ranges are obtained from time series as the results of time-domain dynamic analyses. The number of cycles should be obtained based on the rain flow counting method. Guidance on the integrated load analysis in the time domain can be found in Chap. 21.

6.5 Frequency-Domain Fatigue Analysis Method

The frequency-domain dynamic analysis method may be used in the fatigue analysis provided the nonlinear effects are properly linearised. For further guidance on fatigue analyses in the frequency domain, refer to the Class guidance on position mooring systems, API RP 2SK, and API RP 2T.

6.6 Tendon Fatigue Analysis

Either the S–N-curve-based fatigue analysis approach or the fracture mechanics analysis method can be used for tendon fatigue analysis. Total fatigue damage should be calculated by considering the fatigue damage from various fatigue load cases as presented in 12/3. In addition to the conventional fatigue life calculation using a set of design states based on discretised long-term environmental conditions and the turbine's operational conditions, the tendon components should demonstrate robustness against low-cycle, high-stress fatigue due to more prolonged events that exceed the probabilistic predictions. To confirm a minimum level of robustness, tendon fatigue damage should be assessed for all components considering a single event based on the 50-year return extreme storm, including the ramp-up and ramp-down before and after the storm. The duration to consider should be determined by the designer based on the data available and the response characteristics of the components. The unfactored damage accumulated during this event should be no more than 0.02. Other 50-year return period events that may induce substantial tendon fatigue, such as the Loop Current in the Gulf of Mexico, should also be considered in the same manner. Damage calculated for single event fatigue should not be added to the damage predicted in the long-term scatter-diagram-based fatigue analysis, nor compared to the associated safety factors. The intent, as a robustness check, is to screen for components that may incur excessive fatigue damage in prolonged extreme environmental conditions. Fatigue analysis for tendon components should consider local load effects in addition to global loads. Reference is made to API RP 2T for more details for tendon fatigue analysis.

6.7 Fatigue Design Checks

The calculated fatigue life, which is 1/D, should be greater than the design life multiplied by a fatigue design factor (FDF) defined in Table 1 of Chap. 15 and 15/5.2.

Chapter 23
Suggestions for Numerical Simulations

1 General

In time-domain numerical simulations of global performance analysis and integrated load analyses, appropriate settings of the following parameters are crucial to obtain reliable results:

- Time step.
- Transient time or ramp time.
- Grid size and range for wind field data generation.
- Number of wave components for generating random waves.
- For the deterministic events, the onset time of the events.
- For the stochastic events, the number of simulations with different random seeds and the time duration of each simulation.
- For fatigue sea states, the discretisation of fatigue bins.
- Setting of the control system, generator, blade pitch, nacelle yaw, rotor azimuth angle, rotor speed and degrees-of-freedoms for drive train, and tower and blade flexibilities for different turbine operating or parked conditions.
- Parameters for the aerodynamic load calculation, especially the tower shadow model and the static or dynamic stall models.
- Wind average speed, turbulence intensity, and wind shear for different wind models used in the wind field data generation.

Convergence test should be performed to determine:

- Time steps in time-domain simulations.
- Grid size and range for wind field.
- Number of wind and wave components and seed numbers.

Sensitivity analysis may be performed to determine the discretisation of the fatigue bins. For load cases involving start-up and shut-down of the RNA, the analysis procedure should be consistent with the actual operational procedure of the RNA.

F. Aderouane, *Design, Construction, and Global Performance Analysis of Offshore Wind Turbines*, Green Energy and Technology, https://doi.org/10.1007/978-3-031-86010-2_23

2 Time Step

Choice of time step is crucial for the stability and accuracy of the time-domain solution and is often dependent on the periods of the responses, degree of nonlinearity, and analysis formulation. Time step should be determined by a sensitivity check.

3 Initial Transient Response

The length of a time-domain simulation must allow for transient responses during the initial part of the simulation. The time required for dissipating transient responses is normally a function of the period and damping of the response. Since initial conditions used for the dynamic analysis could affect the statistics of the response during the starting part of the simulation time duration, the first 5 s to 10 min (or longer, if necessary, depending on the period and damping of the response) of time history data should be discarded when stochastic wind and/or wave are applied.

4 Wind Generation and Grid Size

10-minute, 1-hour or 3-hour simulation time duration may be used in numerical simulations for the unique design load cases (DLCs) and survival load cases (SLCs). Since statistical values of wind conditions for wind turbine design are normally based on 10-minute averaging time duration, input wind conditions should be adjusted for 1-hour and 3-hour numerical simulations to achieve statistical equivalence. In the absence of site-specific data for the wind conditions with an averaging time duration other than 10 min, Table 1 may be used to calculate the mean wind speed and turbulence intensity at hub height for 1-hour and 3-hour simulations by adjusting the 10-minute wind conditions at hub height.

Table 1 Adjustment of wind conditions with different averaging time durations

Simulation time duration	10 min	1 h	3 h
Adjustment factor applied to the 10-minute mean wind speed	1.00	0.95	0.90
Adjustment applied to the 10-minute turbulence intensity ($\sigma_{10\text{ min}}$)	1.00	$\sigma_{10\text{ min}}$ + 0.2 m/s	Case-by-case

Notes (1) Adjustments are applied to the hub height wind conditions. (2) Adjustment for the turbulence intensity is valid for the wind spectra recommended in IEC 61400-3-1. For the NPD (Frøya) wind spectrum and tropical revolving storm wind spectrum, see the guidance provided in this chapter. (3) For 3-hour simulations, additional guidance is provided in this chapter

When using the IEC-recommended wind spectrum to perform a 3-hour simulation, the following methods may be used to generate wind field data:

- ***Method 1***: Use 1-hour mean wind speed and turbulence intensity to generate time series of a turbulent wind field for 3 h.
- ***Method 2***: Use 3-hour mean wind speed and adjust the turbulence intensity (or use site measurement data) to generate time series of turbulent wind field for 3 h such that the resultant maximum wind loads or load effects match those obtained using either 10-minute or 1-hour simulations.

The adjustment relative to the 10-minute turbulence intensity, as shown in Table 1, is applicable to the wind spectra recommended in IEC 61400-1. When the NPD (Frøya) wind spectrum is used (refer to 3/3.3.2 and 11/3.3.2), the adjustment factor of the 1-hour average wind speed relative to 10-minute average wind speed still applies, but the turbulence intensity is a function of the 1-hour wind speed and is embedded in the wind spectrum formulation. When the NPD (Frøya) wind spectrum is used in 3-hour simulations, the mean wind speed and turbulence intensity may be taken the same as those used in 1-hour simulations. For the tropical revolving storm wind, the adjustment factor of the 1-hour average wind speed relative to 10-minute average wind speed should be calculated based on the wind gust factor given in Appendix 2 and Appendix 5. For the tropical revolving storm wind in a 3-hour simulation, the mean wind speed and turbulence intensity may be taken the same as those used in 1-hour simulations. The wind generation should consider the 3D effects. Sensitivity analysis may be needed to determine an appropriate grid size of the 3D wind field.

5 Simulation of Wave Conditions

10-minute, 1-hour or 3-hour simulation time duration may be used in numerical simulations for the unique design load cases (DLCs) and survival load cases (SLCs) (refer to 23/7). Since statistical values of wave conditions are normally based on a 3-hour time window, input wave conditions should be adjusted for 10-minute and 1-hour numerical simulations to achieve statistical equivalence. In the absence of site-specific data, the adjustment factors of 10-minute and 1-hour significant wave height relative to 3-hour wave conditions may follow Table 2.

Table 2 Adjustment of the significant wave height

Simulation time duration	10 min	1 h	3 h
Adjustment factor applied to the significant wave height H_s	1.00 (for simulations using the constrained wave method; or otherwise, 18 simulations with different random seeds)	1.09	1.0

For 10-minute simulations, the wave height is not factored. To obtain the same level of extreme value as a 3-hour simulation, 18 simulations (a total of 3 h long) with different random seeds should be performed and the maximum wave height should be taken as the maximum of the 18 simulations. If the constrained wave method is used, the number of random seeds may be reduced. The applicability of the constrained wave method should be verified with 3-hour simulations. The advantage of using 10-minute simulations is that wind conditions do not need to be adjusted. The number and range of discrete frequencies representing the hull transfer functions should be carefully chosen to cover the peaks in the transfer functions and the area of significant wave excitations. It is also important to identify how the employed computer programme handles excitations outside the frequency range of the hull transfer functions since this can be a source of error. A small frequency spacing may be required to avoid repeating a time history within one simulation. Variable frequency spacing may also be used, but the repetition period of a time history is more difficult to assess. The number, range, and spacing of discrete frequencies should be determined and verified by sensitivity analyses.

6 Flexibilities of RNA and Tower

For the tower, rotor blades and drive train, the mass distribution, stiffness, natural frequencies, and damping used for the calculation should be specified. In general, the DOFs of the tower, rotor blades, and drive train should be modelled in accordance with Table 3, as a minimum. The stability analysis normally is performed in the design of the RNA. This stability analysis should be used to assist in determining the modes that are important to the load calculations in the global performance analysis and the integrated load analyses.

Table 3 Suggested minimum DOFs for modelling flexibility of the RNA and the tower components DOFs

Components	DOFs
Tower	1st and 2nd fore-aft and side-to-side bending modes If necessary, 1st torsional mode should be included
Each blade	1st edge-wise, 1st and 2nd flap-wise modes If necessary, torsional modes should be included
Drive train	1st torsional mode

7 Simulation Length and Number of Random Seeds

For the design of floating offshore oil and gas platforms, a 3-hour duration is often used for model testing and time-domain global performance analyses. This duration is sufficient for calculating standard deviations of wave frequency responses because it represents approximately 1000 cycles of those responses having a natural period of around 10 s. Low frequency responses of floating systems, however, typically have natural periods of several minutes. A 3-hour simulation may contain less than 50 cycles of a low frequency response and therefore is insufficient to provide good statistical confidence for the calculated standard deviation. The required simulation time duration may even be longer for predicting extreme values of responses. To obtain statistically meaningful wave frequency extreme responses, for instance, several 3-hour simulations may be needed. An acceptable length of global load simulation time duration depends on many factors, such as natural periods of wave frequency and low frequency responses, contribution of wave frequency and low frequency responses to the total response, degree of nonlinearity, and system damping, etc. The simulation time duration should be determined and verified by sensitivity check. Recent studies indicate that, for typical floating oil and gas platforms in deep water, five to ten 3-hour simulations with different random seed numbers (equivalent to a continuous 15–30 h simulation) may be needed to obtain standard deviations and extreme values of responses with good confidence. For the design load cases (DLCs) and survival load cases (SLCs) defined in 4/2 and 12/3, time histories of turbulent wind and irregular wave conditions should be long enough to provide statistical reliability of the estimate of the characteristic responses. In general, at least eighteen 10-minute stochastic realisations (or a continuous 3-hour realisation) should be generated for each turbulent wind condition and sea state considered in the simulations. Additional guidance is given below:

- For the DLC1.2, 1.3, 1.4, 1.5, 2.1, 2.2, 2.3, 2.4, 2.5, 3.x, 4.x, and 5.1, as defined in Table 1 of Chap. 4, and for the DLC 1.2, 1.3, 1.4, 1.5, 2.1, 2.2, 2.3, 2.4, 2.5, 3.x, 4.x, 5.1 and 9.3, as defined in Table 1 of Chap. 12, at least one set of eighteen 10-minute stochastic realisations should be generated for each wind condition and sea state considered in the simulations. Three to six sets of eighteen 10-minute simulations are recommended for the DLC 1.4, 1.5, 2.1, 2.2, 2.3, 2.5, 3.2, 3.3, 4.2, 4.3 and 5.1 when the RNA involves a transient event. Fewer realisations may be used if it can be demonstrated that the estimated extreme response is not less than that obtained using 3-hour realisations or a total of 3 h of realisations. For DLC 1.2, 2.4, 3.1, and 4.1 for fatigue analysis, the number of simulations may be reduced.
- For the DLC 1.6, 6.1, 6.2, 6.3, 7.1, and the SLCs, as defined in Tables 1 and 3 of Chap. 4, and for the DLC 1.6, 2.6, 6.1, 6.2, 6.3, 7.1, 9.1, 9.2, 10.x and the SLCs, as defined in Tables 1 and 2 of Chap. 12, at least six 1-hour stochastic realisations should be generated for each turbulent wind and sea state considered in the simulations. Six 3-hour simulations are recommended for the DLC 6.1, 6.2, 6.3, 7.1, 10.x and the SLCs when the RNA is in the parked conditions. Realisations

with a shorter simulation time duration may be assumed if it can be demonstrated that the estimated extreme response is not less than that obtained using 1-hour or 3-hour realisations.

- For the DLC D.1, D.2, D.3, and D.6 as defined in Table 2 of Chap. 4, at least six (6) 10-minute stochastic realisations should be generated for each wind condition considered in the simulations. The values of mean wind speed, turbulence standard deviation and significant wave height referenced in the DLCs and the SLCs requiring dynamic simulations should be appropriate to the chosen simulation time duration. More specifically.
- For the DLC 1.6, 6.1, 6.2, 6.3, 7.1, and the SLCs, as defined Tables 1 and 3 of Chap. 4, and for the DLC 1.6, 2.6, 6.1, 6.2, 6.3, 7.1, 9.1, 9.2, 10.x and the SLCs, as defined in Tables 1 and 2 of Chap. 12, the values of mean wind speed, turbulence standard deviation and significant wave height should be adjusted for different simulation time durations in accordance with 23/4 and 23/5.
- For the load cases using 10-minute simulations, at least 18 simulations (a total of 3 h) should in general be performed. No adjustment is needed provided the wind conditions are based on a 10-minute averaging time duration. The suggested simulation time duration and number of random seeds for each DLC are summarised in Table 4. The suggested values for the DLC 6.1 may also be applied to the SLCs. The extreme statistical values for different load cases may be obtained as follows:
- For the DLC 1.4, 1.5, 2.3, 3.2, 3.3, 4.2 and 4.3 with the occurrence of events such as the extreme wind gust, extreme wind shear, severe sea state, etc., the maximum value of a response should be determined based on the transient value in each set of 18 10-minute simulations computed randomly in the worst case. The occurrence of the event can be at any moment during the simulation. For multiple sets of simulations, the maximum value of a response should be determined based on the average of the maximum responses of each set of simulations.
- For the DLC 1.3 and 9.3 with the turbulent wind together with irregular sea states, among all combinations of wind and wave conditions, the largest mean value of the maximum responses of stochastic realisations should be taken.
- For the DLC 2.1, 2.2, 2.5 and 5.1 and with the turbulent wind together with irregular sea states and with the occurrence of events such as a fault or a shutdown, the maximum value of a response should be determined based on the mean value of the largest half of the maximum responses in each set of 18 10-minute simulations computed randomly in the worst case. The occurrence of the event can be at any moment during the simulation. For multiple sets of simulations, the maximum value of a response should be determined based on the average of the maximum responses of each set of simulations.
- For the DLC 1.6, 2.6, 6.1, 6.2, 6.3, 7.1, 9.1, 9.2 and 10.x as well as the SLCs with the turbulent wind together with irregular sea states, among all combinations of wind and wave conditions, the largest mean value of maximum responses of stochastic realisations should be taken.
- For the DLC D.1, D.2, D.3, and D.6 for Bottom-founded Offshore Wind Turbines with the turbulent wind conditions, the largest mean value of maximum responses of stochastic realisations should be taken.

Table 4 Suggested simulation time duration and number of random seeds for the DLCs

Turbine condition		DLC	Type of analysis*	Simulation time duration and number of random seeds
1	Power production	1.2	F	18 × 10-minute simulation with different random seeds
		1.3	S	18 × 10-minute simulation with different random seeds
		1.4	S	Six sets of 18 × 10-minute simulation with different random seeds
		1.5	S	Six sets of 18 × 10-minute simulation with different random seeds
		1.6	S	1-hour simulation with 6 seeds
2	Power production plus occurrence of fault	2.1	S	Three sets of 18 × 10-minute simulation with different random seeds
		2.2	S	Three sets of 18 × 10-minute simulation with different random seeds
		2.3	S	Six sets of 18 × 10-minute simulation with different random seeds
		2.4	F	18 × 10-minute simulation with different random seeds
		2.5	S	Three sets of 18 × 10-minute simulation with different random seeds
		2.6	S	1-hour simulation with 6 seeds
3	Start-up	3.1	F	18 × 10-minute simulation with different random seeds

(continued)

Table 4 (continued)

Turbine condition		DLC	Type of analysis*	Simulation time duration and number of random seeds
		3.2	S	Six sets of 18 × 10-minute simulation with different random seeds
		3.3	S	Six sets of 18 × 10-minute simulation with different random seeds
4	Normal shutdown	4.1	F	18 × 10-minute simulation with different random seeds
		4.2	S	Six sets of 18 × 10-minute simulation with different random seeds
		4.3	S	Six sets of 18 × 10-minute simulation with different random seeds
5	Emergency stop	5.1	S	Three sets of 18 × 10-minute simulation with different random seeds
6	Parked (standing still or idling)	6.1	S	1-hour or 3-hour simulation with 6 seeds
		6.2	S	1-hour or 3-hour simulation with 6 seeds
		6.3	S	1-hour or 3-hour simulation with 6 seeds
		6.4	F	–
7	Parked and fault conditions	7.1	S	1-hour or 3-hour simulation with 6 seeds
		7.2	F	–
8	Temporary (installation, maintenance, and repair)	8.1	S	–
		8.2	S	–
		8.3	F	–
		8.4	F	–
9	Power production	9.1	S	1-hour simulation with 6 seeds
		9.2	S	1-hour simulation with 6 seeds

(continued)

Table 4 (continued)

Turbine condition		DLC	Type of analysis*	Simulation time duration and number of random seeds
		9.3	S	18 × 10-minute simulation with different random seeds
10	Parked (standing still or idling)	10.1	S	1-hour or 3-hour simulation with 6 seeds
		10.2	S	1-hour or 3-hour simulation with 6 seeds
		10.3	S	1-hour or 3-hour simulation with 6 seeds

Note * For each DLC in the Table, the "Type of Analysis" is denoted "S" for the strength analysis or "F" for the fatigue analysis

8 Analysis Methods

The suggested analysis methods for the global performance analysis for different subsystems are summarised in Table 5 for the design load cases specified in the applicable design standards. Although the time-domain fully coupled dynamic analysis method is preferred, global performance analysis using less sophisticated methods may still be used, especially in preliminary design stages. For the integrated load analysis in the detailed design stage, the time-domain aero-hydro-servo-elastic coupled analysis approach is used. The simulation software is normally developed by combining aeroelastic codes originally developed for land-based turbines with hydrodynamic codes and mooring dynamic codes used for designing offshore oil and gas platforms. A typical aero-elastic code contains an aerodynamic part, which is mostly based on the Blade Element Momentum (BEM) Method, and a structural part with various levels of complexity in modelling the drive train and elasticity of the blade and the tower. Control system modelling is usually designed by users and linked to the aero-elastic codes through a dynamic link library.

Table 5 Suggested analysis methods

OWT type	Design stage	Subsystem	Analysis type	Analysis method
Bottom-founded offshore wind turbine	Preliminary design stage	Substructure	Fatigue	Frequency-domain analysis Combined time- and frequency-domain analysis Sequential time-domain analysis Coupled time-domain analysis
			Strength	Sequential time-domain analysis Coupled time-domain analysis
	Final detailed design stage	Integrated support structure and RNA	Fatigue	Integrated load analysis
			Strength	Integrated load analysis
Floating offshore wind turbine	Preliminary design stage	Floating substructure	Global motion	Frequency-domain analysis Combined time- and frequency-domain analysis Time-domain analysis (Coupled, Semi-coupled, Uncoupled)
			Air gap	Frequency-domain analysis Combined time- and frequency-domain analysis Time-domain analysis (Coupled, Semi-coupled, Uncoupled)

(continued)

Table 5 (continued)

OWT type	Design stage	Subsystem	Analysis type	Analysis method
		Station keeping system	Mooring analysis (Fatigue and Strength)	Frequency-domain analysis Combined time- and frequency-domain analysis Time-domain analysis (Coupled, Semi-coupled, Uncoupled)
	Final detailed design stage	Integrated support structure and RNA	Global load analysis	Integrated load analysis
			Global motion	Integrated load analysis
			Air gap	Combined time- and frequency domain analysis Integrated load analysis
			Mooring analysis (Fatigue and Strength)	Integrated load analysis

Glossary

Added Mass An effective addition to system mass, which is proportional to the mass of displaced water.

Air Gap The clearance between the highest water surface that occurs during design environmental conditions and the lowest exposed structures not designed to withstand wave impingement.

Bottom-founded Offshore Wind Turbine A structure containing a site-dependent substructure and its foundation supporting the wind turbine RNA and the tower. The design of the Bottom-founded Offshore Wind Turbine is based on environmental, electrical network, and foundation conditions at a particular offshore site where it is intended to be installed. The sea floor attachment of the substructure may be obtained by use of pilings, direct bearing, or other types of foundation.

Catenary Mooring A mooring system wherein the restoring action is provided by the distributed weight of mooring lines.

Consultant Any person who, through education and experience, has established credentials of professionalism and knowledge in the stated field.

Compliant Tower A structure that consists of a slender tower supported at the sea floor by an installed foundation or by a large spud can. A compliant tower may also be partially supported by buoyancy aids. Guy lines may be used for lateral restraints.

Cut-In Wind Speed (V_{in}) The lowest 10-min mean wind speed at hub height at which the wind turbine starts to produce power in the *case of steady wind without turbulence.*

Cut-Out Wind Speed (V_{out}) The highest 10-min mean wind speed at hub height at which the wind turbine is designed to produce power in the case of steady wind without turbulence.

Design Life An assumed period for which a structure, a structural component, a system, or equipment is expected to be used for its intended purpose with anticipated maintenance, but without requiring substantial repair.

Dip and Thrash Zone A chain or wire rope section above and close to the seafloor that may touch the seabed due to the motions of the floating substructure.

F. Aderouane, *Design, Construction, and Global Performance Analysis of Offshore Wind Turbines*, Green Energy and Technology, https://doi.org/10.1007/978-3-031-86010-2

Dynamic Positioning (DP) A station keeping technique primarily using a system of automatically controlled onboard thrusters to generate appropriate thrust vectors to counter environmental actions and maintain an intended position within prescribed tolerances.

Emergency Stop Rapid shutdown of the wind turbine triggered by manual intervention.

Fabricator A Fabricator is any person or organisation having the responsibility to perform any or all the following: fabrication, assembly, erection, inspection, testing, load-out, transportation, and installation.

Floating Offshore Wind Turbine A structure comprising three principal areas: (1) The floating substructure for carrying the wind turbine RNA and the tower. (2) The station keeping system. (3) Onboard machinery, equipment, and systems including applicable marine systems and associated equipment and machinery, safety systems and associated equipment, and lifesaving appliances.

Floating Substructure A site-dependent offshore structure supported by buoyancy and maintained on location by the station keeping system. The floating substructure consists of the Hull and topside structures.

Floating Support Structure The Floating Support Structure consists of the tower and the floating substructure.

Foundation System (for Tendons) The structural, mechanical, and geotechnical components which are located on and beneath the sea floor and transfer the loads acting on the TLP Tendons into the seabed.

Gravity Structure A structure supported by the foundation structure resting directly on the sea floor. The geometry and weight of foundation structure are to be selected to utilise available cohesive and frictional strength components of the sea floor soil to resist loadings.

Gust A brief rise and fall in wind speed lasting less than 1 min.

Hull Combination of connected buoyant structural components such as columns, pontoons, and intermediate structural braces; refer to also "Monohull."

Hub Height The height of the centre of the swept area of the wind turbine rotor above the Still Water Level.

Idling A condition of a wind turbine that is rotating slowly and not producing power.

Load An external load applied to the structure (direct load) or an imposed deformation or acceleration (indirect load).

Load Effect Effect of a single load or combination of loads on a structural component or system (e.g., internal force, stress, strain, motion etc.)

Material Application Categories The application of structural members in the floating substructure is to be in accordance with the categories listed in this paragraph.

Mean Sea Level or Mean Still Water Level (MSL) The average level of the sea over a period long enough to remove variations due to waves, tides, and storm surges (refer to Fig. 3.2 and Fig. 11.2).

Mean Wind Speed The statistical mean value of the instantaneous wind speed over a specified time interval.

Minimum Breaking Strength (MBS) The certified strength of a chain, wire rope, fibre rope, or accessories.

Monohull A floating structure consisting of a single, continuous, buoyant hull, and geometrically like an ocean-going ship or barge.

Mooring Components A general class of components used in the station keeping system.

Normal Shutdown A wind turbine shutdown operation in which all stages are under the control of the control system.

Offshore Wind Farm A group of wind turbines installed at an offshore site. An offshore wind farm may also include other installations such as transformer/converter platforms, meteorological measurement facilities, power cables, and accommodation units.

Offshore Wind Turbine An Offshore Wind Turbine consists of a Rotor-Nacelle Assembly and its Support Structure.

Omni-Directional (Wind, Waves, or Currents) Acting in all directions.

Owner An owner is any person or organisation who owns Offshore Wind Turbines.

Parked The Condition of a wind turbine that is either in the Standstill or Idling condition, depending on the design of the wind turbine.

Pile-Supported Substructure A structure supported by foundation elements or suction type piles driven into the sea floor.

Pretension Tension applied to a mooring line or tendon when the floating support structure at its static equilibrium position in mean still water and still air.

Primary Application Structure Primary application structure refers to primary load carrying members of a structure where the occurrence of a fracture could induce a major structural failure.

Rated Power The quantity of power assigned, by a manufacturer, for a specified operating condition of a component, device, or equipment. For wind turbines, it is the maximum continuous electrical power output which a wind turbine is designed to achieve under normal operating and external conditions.

Rated Wind Speed (V_r) The minimum 10-min mean wind speed at hub height at which a wind turbine's rated power is achieved in the case of steady wind without turbulence.

Return Period (Recurrence Period) The average time duration between occurrences of an event or of a particular value being exceeded. A return period in years is equal to the reciprocal of the annual probability of exceedance of an event or of a particular value of a random parameter such as wind speed, wave height, or sea elevation.

Rotor-Nacelle Assembly (RNA) The Rotor-Nacelle Assembly (RNA) of a horizontal axis wind turbine, carried by the floating support structure, consists of: (1) The rotor components, including blades, hub, shaft, and spinner. (2) The nacelle, a housing which contains the mainframe, generator frame, drive train components, electrical generator components, wind turbine control and protection components and other elements on top of the tower.

Ringing A high frequency vertical vibration of the TLP spring-mass system excited by impulsive loading.

Secondary Application Structure Secondary application structure refers to less critical members due to a combination of lower stress and favourable geometry or where an incidence of fracture is not likely to induce a major structural failure.

Semi-submersible A floating structure normally consisting of a deck structure connected to submerged pontoons through several widely spaced, large cross-section supporting columns.

Self-Elevating Unit to Be Used as the Substructure A unit with movable legs capable of raising its hull above the surface of the sea and intended to be used as the site-dependent substructure of an offshore wind turbine. The hull of a self-elevating unit has sufficient buoyancy to transport the unit to a desired location. Once on location, the hull is raised to a predetermined elevation above the sea surface on its legs, which are supported by the sea floor. The legs of such unit may be designed to penetrate the seabed, may be fitted with enlarged sections or footings, or may be attached to a bottom mat.

Set-Down An increase in the draft of a floating structure (for example, TLP) with the increase in the offset due to mooring or tendon system restraint.

Single Point Mooring A mooring system that allows the floating structure to which it is connected to vary its heading (weathervane).

Spar A deep-draft, small water-plane area floating structure.

Special Application Structure Special application structure refers to highly stressed members, located at intersections of main structural elements and other areas of high stress concentration where the occurrence of a fracture could induce a major structural failure. (1) Primary Application Structure. Primary application structure refers to primary load carrying members of a structure where the occurrence of a fracture could induce a major structural failure. (2) Secondary Application Structure. Secondary application structure refers to less critical members due to a combination of lower stress and favourable geometry or where an incidence of fracture is not likely to induce a major structural failure.

Splash Zone Part of the mooring lines or tendons of the station keeping system above and below the Mean Sea Level and regularly subjected to wetting due to wave actions, motions of the floating substructure and, if applicable, tide and draft variations. Areas which are only wetted during major storms are not included.

Spread Mooring A mooring system consisting of multiple mooring lines terminated at separate locations on a floating structure and extending outwards, providing an almost constant heading to the floating support structure.

Springing A high frequency vertical vibration of the TLP spring-mass system excited by cyclic loading at or near the TLP pitch or heave resonant periods.

Standstill The condition of a wind turbine that is not rotating.

Station Keeping System A system capable of limiting the excursions of the floating substructure within prescribed limits, maintaining the intended orientation, and helping to limit motions at tower top.

Still Water Levels (SWL) Abstract water levels used for the calculation of wave kinematics and wave crest elevation. Refer to Fig. 3.2 and Fig. 11.2. Still Water Levels, which can be either above or below the Mean Sea Level, are calculated by adding to and subtracting from the effect of tide and surge on the Mean Sea Level.

Structural Critical Inspection Point (SCIP) Structure Critical Inspection Point (SCIP) is a structural point defined in the ISIP plan as a critical inspection area determined by structural assessment using applicable calculations and analysis. In general, SCIPs are locations with higher stresses and estimated lower fatigue life. These locations are identified from calculation to require monitoring or from the service history of the subject floating substructure or from similar sister floating substructures to be sensitive to cracking, buckling or corrosion which would impair the structural integrity of the unit.

Support Structure A support structure for a Bottom-founded Offshore Wind Turbine consists of the tower, substructure, and foundation, which are defined as follows: (a) *Tower*A structure extending upward from a location above the Still Water Level and connecting the substructure to the Rotor-Nacelle Assembly (RNA). (b) *Substructure*A structural component extending upward from the sea floor and connecting the foundation to the tower. (c) *Foundation*A structural and/or geotechnical component located on and beneath the sea floor and which transfers the loads acting on the structure into the sea floor.

Surveyor A person employed by Class whose principal functions are the surveillance during construction and the survey of marine structures and their components for compliance with the Class Rules or other standards deemed suitable by Class.

Tendon A system of components forming a link between the Hull structure and the foundation system for the purpose of restraining motion of the TLP-type floating support structure within specified limits in response to environmental and other loading.

Tendon Connector A device used to connect a tendon to the TLP Hull (top connector) or to the foundation template (bottom connector).

Tension Leg A collective group of tendons associated with one column of the platform.

Tower The structural component or assembly that connects the substructure to the Rotor-Nacelle Assembly.

Tension Leg A collective group of tendons associated with one column of the platform.

Turbulence Intensity The ratio of the wind speed standard deviation to the mean wind speed, determined from the same set of measured data samples of wind speed, and taken over a specified period.

Taut-Line Mooring A mooring system wherein the restoring action is provided by elastic deformation of mooring lines.

Thruster-Assisted Mooring A station keeping system consisting of mooring lines and thrusters.

Tropical Revolving Storm (Hurricane, Cyclone, Typhoon) A tropical storm possessing sustained wind speeds more than 33 m/s (64 knots or 74 mph). Such a storm is called a hurricane, typhoon, or cyclone based on the storm location. For example, tropical revolving storms are typically referred to as hurricanes in the

Gulf of Mexico and North Atlantic, while in the South China Sea and Northwest Pacific they are called typhoons. In the South Pacific and South Indian Ocean, however, they are commonly referred to as cyclones.

Uni-directional (Wind, Waves or Currents) Acting in a single direction.

Vortex-Induced Motion (VIM) The motion of the Hull structure induced by vortex when the hull is exposed to a current.

Vortex-Induced Vibration (VIV) A phenomenon of the vibration of cylindrical structures exposed to a current. Cylindrical structures exposed to a current may create alternating eddies, or vortices, at a regular period. The eddies create alternating lift and drag forces on the cylinder. When a natural period of a structure falls close to the period of vortex shedding, oscillations of the structure can occur.

Water Depth The vertical distance between the sea floor and the Still Water Level.

Wind Profile (Wind Shear Law) Mathematical expression for assumed wind speed variation with height above the Still Water Level.

Weathervaning The process by which a floating support structure passively varies its heading in response to time varying environmental actions.

Yawing Rotation of the rotor axis about a vertical axis for horizontal axis wind turbines.

Yaw Misalignment Horizontal deviation of the wind turbine rotor axis from the wind direction.

Bibliography

1. ACI 211.1, Standard Practice for Selecting Proportions for Normal, Heavyweight and Mass Concrete
2. ACI 213R, Guide for Structural Lightweight Aggregate Concrete
3. ACI 301, Specifications for Structural Concrete
4. ACI 318, Building Code Requirements for Reinforced Concrete
5. ACI 357.2R, State-of-the-Art Report on Barge-Like Concrete Structures
6. ACI 357R, Guide for the Design and Construction of Fixed Offshore Concrete Structures
7. ACI 359, Code for Concrete Reactor Vessels and Containments
8. AISC Steel Construction Manual
9. API Bulletin 2N, Interim Planning, Designing and Constructing Fixed Offshore Structures in Ice Environments
10. API RP 2A-LRFD, Recommended Practice for Planning Designing and Constructing Fixed Offshore Platforms – Load and Resistance Factor Design, 2nd Edition, 2019
11. API RP 2A-WSD, Recommended Practice for Planning Designing and Constructing Fixed Offshore Platforms – Working Stress Design, 22nd Edition, 2014
12. API RP 2EQ, Recommended Practice for Seismic Design Procedures and Criteria for Offshore Structures
13. API RP 2FPS, Recommended Practice for Planning Designing and Constructing Floating Production Systems
14. API RP 2GEO, Recommended Practice for Geotechnical and Foundation Design Considerations
15. API RP 2MET, Recommended Practice for Derivation of Metocean Design and Operating Conditions, 2019
16. API RP 2N, Recommended Practice for Planning, Designing, and Constructing Structures and Pipelines for Arctic Conditions
17. API RP 2SIM, Structural Integrity Management of Fixed Offshore Structures, 1st Edition, 2014
18. API RP 2SK, Recommended Practice for Design and Analysis of Station keeping Systems for Floating Structures
19. API RP 2SM, Recommended Practice for Design, Manufacture, Installation, and Maintenance of Synthetic Fibre Ropes for Offshore Mooring
20. API RP 2T, Recommended Practice for Planning, Designing, and Constructing Tension Leg Platforms
21. API SPEC 17E, Specification for Subsea Umbilicals
22. ASTM C172, Standard Practice for Sampling Freshly Mixed Concrete
23. ASTM C31, Standard Practice for Making and Curing Concrete Test Specimens in the Field
24. ASTM C330, Specification for Lightweight Aggregates for Structural Concrete

F. Aderouane, *Design, Construction, and Global Performance Analysis of Offshore Wind Turbines*, Green Energy and Technology, https://doi.org/10.1007/978-3-031-86010-2

25. ASTM C39, Standard Test Method for Compressive Strength of Cylindrical Concrete Specimens
26. ASTM C94, Standard Specification for Ready-Mixed Concrete
27. AWS, Structural Welding Code – Steel, D1.1
28. IEC 61400-1, Wind Energy Generation Systems – Part 1: Design Requirements, 4th Edition, 2019
29. IEC 61400-24, Wind Turbines – Part 24: Lightning protection, 1st Edition, 2010
30. IEC 61400-3-1, Wind Energy Generation Systems – Part 3-1: Design Requirements for Fixed Offshore Wind Turbines, 1st Edition, 2019
31. IEC 61400-6, Wind Energy Generation Systems – Part 6: Tower and foundation design requirements
32. IEC TS 61400-3-2 (Technical Specification), Wind Energy Generation Systems – Part 3-2: Design Requirements for Floating Offshore Wind Turbines
33. IEC61400-1, Wind Energy Generation Systems – Part 1: Design Requirements, 4th Edition, 2019
34. IECRE OD-501 (Operational Document), "IEC System for Certification to Standards relating to Equipment for use in Renewable Energy applications (IECRE System): Type and Component Certification Scheme", 2nd Edition, 2018
35. IECRE OD-502 (Operational Document), IEC System for Certification to Standards relating to Equipment for use in Renewable Energy applications (IECRE System): Project Certification Scheme
36. ISO 19901-1, Petroleum, and natural gas industries – Specific requirements for offshore structures – Part 1: Metocean design and operating considerations
37. ISO 19901-2, Petroleum, and natural gas industries – Specific requirements for offshore structures – Part 2: Seismic design procedures and criteria
38. ISO 19901-4, Petroleum, and natural gas industries – Specific requirements for offshore structures – Part 4: Geotechnical and foundation design considerations
39. ISO 19901-7, Petroleum, and natural gas industries – Specific Requirements for Offshore Structures, Part 7: Station keeping Systems for Floating Offshore Structures and Mobile Offshore Units
40. ISO 19902, Petroleum, and natural gas industries – Fixed steel offshore structures
41. ISO 19903, Petroleum, and natural gas industries – Fixed concrete offshore structures
42. ISO 19903, Petroleum, and natural gas industries – Concrete offshore structures
43. ISO 19904-1, Petroleum, and natural gas industries – Floating Offshore Structures, Part 1: Monohulls, Semi-submersibles, and Spars
44. ISO 19906, Petroleum, and natural gas industries – Arctic Offshore Structures
45. ISO 2533, Standard Atmosphere
46. NACE SP0108, Standard Practice: Corrosion Control of Offshore Structures by Protective Coatings
47. NACE SP0176, Standard Practice: Control of Submerged Areas of Permanently Installed Steel Offshore Structures Associated with Petroleum Production

The manufacturer's authorised representative in the EU is Springer
Nature Customer Service Centre GmbH, Europaplatz 3, 69115 Heidelberg,
Germany. If you have any concerns regarding our products, please
contact ProductSafety@springernature.com

Printed and bound by CPI Group (UK) Ltd, Croydon, CR0 4YY
30/09/2026
02225558-0018